climate

RODALE

climate

The force that shapes our world – and the future of life on Earth

George Ochoa • Jennifer Hoffman, PhD • Tina Tin, PhD

RODALE

This edition first published in 2005 by
Rodale International Ltd
7–10 Chandos Street
London W1G 9AD
www.rodalebooks.co.uk

Printed and bound in Spain by Artes Gráficas Toledo, S.A.U.

Rodale makes every effort to use acid-free paper from sustainable sources

1 3 5 7 9 8 6 4 2

A CIP record for this book is available from the British Library

ISBN 1–4050–8782–X

This edition distributed to the book trade by Pan Macmillan Ltd

NOTICE
Mention of specific companies, organizations or authorities in this book does not imply endorsement by the publisher, nor does mention of specific companies, organizations or authorities in the book imply that they endorse it. Internet addresses and telephone numbers given in this book were accurate at the time it went to press.

Produced for Rodale Books International by Hydra Packaging, 129 Main Street, Irvington, N.Y., 10533, U.S.A.

Publisher: Sean Moore
Creative Director: Karen Prince
Editorial Director: Lori Baird
Art Director: Edwin Kuo
Senior Editor: Aaron R. Murray
Editors: Liz Mechem, Glenn Novak, Sarah Novak, Jennifer Freeman, Suzanne Lander, Karen Cure, Elizabeth Poyet, Myrsini Stephanides, Nick Simonds
Designers: Shamona Stokes, Rachel Maloney, Gus Yoo, Amy Henderson, Patricia Childers, Edwin Tse
Production Director: Wayne Ellis
Production Manager: Sarah Reilly
Production: Brian MacMullen
Picture Researcher: Chrissy McIntyre
Cartography: Andrew Heritage, Alex Reay
Indexer: Stephen Callahan

RODALE
LIVE YOUR WHOLE LIFE™

We inspire and enable people to improve their lives and the world around them

COVER AND RIGHT *A towering glacier in Alaska.* HALF TITLE *Mist hovers in the treetop canopy in the foothills of the Andes, Ecuador.* TITLE PAGE *A Tuareg tribesman and his camels in the Sahara Desert.*

CONTENTS

Foreword **7**

Introduction **Is Climate Changing?** **8**

PART I: CLIMATES OF THE PAST
Chapter 1 **The Birth of Climate** **16**
Earth's early, poisonous atmosphere and its dramatic extremes in temperature gave rise to climate and a planet that could support life

Chapter 2 **Prehistoric Climates** **36**
Climates changed throughout the Paleozoic era largely because of continental drift, meteor storms, and ice ages, leading to the greatest mass extinction in Earth's history

Chapter 3 **Climate and Humankind** **52**
Climate has been an evolutionary force from the beginning, a powerful agent in human evolution against the frozen backdrop of the ice ages

PART II: CLIMATE TODAY AND TOMORROW
Chapter 4 **Human Impact on Climate** **84**
The facts behind the emerging Anthropocene theory make the case for a new geological era in which humans rival nature in their impact on the global environment

Chapter 5 **Too Hot to Handle** **96**
Increased temperature and chronic warming have a negative impact on the world, influencing people and ecosystems

Chapter 6   Melting Snow and Ice 112
Dramatic evidence shows that climate in the polar and alpine regions is
changing at alarming speeds, melting glaciers and reducing ice coverage
in crucial places, and exerting a profound effect on global climate

Chapter 7   Water: Too Much and Not Enough 130
By changing the water cycle, climate change could bring both drought
and flooding, with dire global implications

Chapter 8   Rising Seas 156
In the future, rapidly rising sea levels are likely to submerge islands, while warming
seas destroy coral reefs. Islanders will probably be the first climate refugees

Chapter 9   Changing Ecosystems 172
Changes in ecosystems – shifts in species ranges, the timing of migration and
other annual events, and disrupted interactions between species – are signs
that climate change is already affecting biological communities

Chapter 10   Cities Under Siege 192
Local climate change could adversely affect the existence of the world's
modern cities as we know them today

Chapter 11   Extreme Weather 208
Intense weather, including back-to-back hurricanes in the Caribbean, could
occur more frequently around the globe – and at great cost to society

Chapter 12   Our Future: Biome by Biome .222
What does the future hold? One thing is certain: climate change, in which
humans play a part, will alter the face of the globe, biome by biome

Chapter 13   Limiting Climate Change 238
Humanity's impact on Earth's climate can be curtailed. Individuals,
communities and governments must work together to become energy-smart
and use all available resources to put the brakes on climate change

Chapter 14   Adapt We Must 256
The global climate is changing, whether we like it or not, and some
communities, cultures and countries are taking steps to adapt to it

APPENDICES

A Chronology of Climate 270
Major climate events from the beginning of geological time and early
human history to the present day

Charting Climate Change 274
Comparison maps, charts and graphs

Glossary 276
Definitions of key terms involving climate and weather

Further Reading & Organizations 278
A summary of notable books and articles, and contact information
and websites for organizations worldwide devoted to monitoring
climate change and advocating practices to limit and adapt to it

Index 280

Acknowledgments & Picture Credits 288

FOREWORD

This book explains how the climate works – with some of the most powerful forces of nature known to humanity. The fact that these are altered by our own actions may be one of the most important discoveries ever made. The truth is that we are facing disruption of our climate on a scale we can scarcely imagine.

Radical steps will be needed to minimize this. The burning of fossil fuels – coal, oil and gas – releases carbon dioxide into the atmosphere. The carbon dioxide blankets the Earth, trapping heat and causing global warming. The UK government has set a target of a sixty per cent reduction in carbon dioxide emissions by 2050 but that may not be enough.

To be truly effective we must each play our part, in many different ways. Changing the way we grow our food is an important step because food accounts for almost a quarter of the UK's greenhouse gas emissions. But it is getting easier to make choices that will not add to the problem – with the rise of local and organic food.

Most of our food these days is transported huge distances. Almost half the freight in the United Kingdom is related to food, and emissions from air-freighted food are still rising fast. This is completely unnecessary because it is possible to buy local instead. Buying organic is also important. Organic production does not use artificial fertilizers and sprays, which add to climate change through the energy used in making them and through emitting nitrous oxide – the most potent greenhouse gas – when nitrogen fertilizer is made and then applied.

Organic farming also cares for the soil (hence the choice of name for one of the main organic food and farming organizations, the Soil Association). There is more carbon in soils than in the atmosphere – in fact around twice as much. Organic farming builds natural fertility in soils – trapping carbon and using soils as a carbon store, not a source.

The Soil Association exists to promote sustainable ways of producing food by working with the public, scientists and governments to ensure healthy food supplies for future generations. By joining them you will be taking an active step in helping limit global warming.

As someone who has been growing food organically for over a decade, I know that it results in better-tasting food with no harmful residues, much more wildlife, and a healthier working environment. If it also helps to avoid the climate chaos that this book so vividly foresees then that could be the most important benefit of all.

Monty Don
Gardener and broadcaster

IS CLIMATE CHANGING?

In 2002 and 2003, Australia experienced its worst drought in a century. At western Victoria's Lake Corangamite, water levels sank so low that much of the lake's bottom was exposed, and residents discovered, buried in the mud, bombs that had been dropped into the lake during a World War II bombing practice – undisturbed until uncovered by the ravaging drought.

The lack of rain devastated Australian farmers. Farm income dropped 70 percent. The population of sheep, a mainstay of the Australian economy, fell to its lowest level since 1947. Grain production plummeted by more than half. With temperatures climbing and water scarce, bush fires ran rampant. In January 2003, bush fires in the nation's capital, Canberra, killed four people and destroyed more than 500 houses.

Drought continues in parts of Australia, particularly in the south and southwest. In many areas of the country, rainfall has long been diminishing; it has been declining in New South Wales for about 50 years. Many Australians worry that drought has become a permanent condition.

Why is it so dry?

Trial by Fire

In 2003, the most devastating forest fires in 25 years blazed through Portugal. The flames damaged 2,500 houses and other buildings and destroyed 400,000 hectares (one million acres). This disaster was just part of a pattern: throughout the Mediterranean region, from Spain to Greece, wildfires have been on the increase for several decades.

Trees and property owners are not the only victims of the conflagrations. In Portugal's Monchique and Caldeirao mountains, animal habitats are being destroyed, putting further stress on already threatened species such as the Iberian lynx and the Bonelli eagle.

What is causing the fires?

Four Hurricanes

In 2004, Florida suffered more hurricanes in one season than any U.S. state in more than a century. First, on 13 August, Hurricane Charley battered southwestern Florida with winds of 233 km/h (145 mph) and waves of up to 3 m (10 feet). Charley killed 31 people and caused an estimated $6.8 billion in damage. Floridians were still recovering from Charley when Hurricane Frances slammed into eastern Florida on September 5 with winds of 169 km/h (105 mph). The storm uprooted trees, tore off roofs, smashed boats, and knocked out power for millions of people. Torrents of rain flooded streets and reduced visibility to zero. Walt Disney World's theme parks closed. The Kennedy Space Center suffered more damage than during any previous storm.

Those two hurricanes were followed by a third. Ivan, with winds that blew at 209 km/h (130 mph), hit Pensacola in Florida's western panhandle after making landfall in Alabama on 16 September. Finally, on 25 September, Jeanne, whose winds were clocked at 185 km/h (115 mph), swept ashore at almost the same location as Frances, scattering the debris created by that hurricane. Altogether, the four hurricanes killed 116 people and caused $17.5 billion in damage in Florida alone. Millions of Floridians were forced into evacuation shelters; thousands were left homeless. The last state to suffer four hurricanes in one season was Texas in 1886.

Why all the hurricanes?

Islands in Trouble

In Fiji, a South Pacific nation made up of hundreds of islands and atolls, people notice that things have changed. In the province of Nadroga, coastal flooding is increasing and the coastlines are eroding faster. Tides come in farther than they used to. Freshwater is scarcer, and the soil is less stable.

RIGHT *A firefighting airplane battles a forest fire in Castelo Branco, Portugal. Drought conditions caused by climate change have led to an increase in wildfires in recent years.* PAGE 6 *Lava flows from the vent of an erupting volcano. Natural geological events like this can contribute to climate change.*

The villagers depend on the sea for food, but the local marine life has gone haywire. Coral reefs are bleached white, having lost the microscopic algae that live inside their cells. These microalgae nourish the reefs, creating good habitats for fish. As reefs bleach and die, fish disappear. Octopuses spawn in March now instead of in November.

Villagers across the South Pacific islands tell similar stories of change: wind and rainfall patterns are not what they once were; small islets have been inundated by the ocean; potato and taro tubers rot in perpetually waterlogged soil. With sea levels rising, low-lying Tuvalu is considering moving its entire population to New Zealand.

Why are the islands in trouble?

Polar Bears at Risk

In the Arctic, polar bears are going hungry. In Canada's Hudson and James bays, in the southern range of these cold-adapted animals, cubs are starving to death for lack of food or because their nursing mothers lack the body fat they need to nourish their offspring.

The scarcity of food can be traced to changes in the ice. Polar bears normally live by hunting seals on the pack ice that covers the Arctic Ocean. In summer,

ABOVE *Tourists snorkel on the Great Barrier Reef, off the northeast coast of Australia. Warmer ocean temperatures have endangered coral reefs, killing the tiny organisms that give coral their colour. Scientists warn that if warming trends continue at the current pace, the world's coral reefs could be wiped out within a century.*

when much of the pack ice has melted, many bears stay on land, living off the body fat built up during hunting seasons. But the ice has melted earlier in recent springs and has formed later in fall, shortening the period when polar bears hunt and build up fat reserves for summer. For every week of hunting time lost, bears weigh about 10 kg (22 pounds) less. This weight loss leaves them in poorer condition, less able to reproduce or to produce healthy offspring – and in greater danger of local extinction.

Why is the ice melting?

Weather vs. Climate

Stories of extreme weather are always with us – floods, droughts, tornadoes and blizzards. But around the world, from Florida to Fiji, from Australia to Portugal to the Arctic, extreme weather appears to be becoming more frequent. What is causing all these dramatic events? Are they examples of climate change? Or are they merely coincidence, an unlikely run of bad weather? Is human activity a factor?

Weather is the atmospheric state in a given place and time; climate is a locality's weather as observed over a long period – in other words, the long-term behaviour of weather. Climate varies from region to region, but its longterm stability is part of what gives every region its special

RIGHT Vegetable sellers paddle through a floating garden in Dal Lake, Srinagar, in the Kashmir valley of India. The lake is one of many worldwide that are slowly dying through a combination of pollution, lack of freshwater, and warmer temperatures, which affect the entire ecosystem.

character. Climate is hot and dry in deserts, warm and wet in tropical rainforests; climate changes seasonally from warm to cold in temperate zones. The climate of the African savannah has been stable enough to make it a durable habitat for lions and zebras, while the snow that covers the region around Moscow for nearly six months falls reliably enough to make it part of that city's identity. Around the world, human beings, like all other organisms, depend on climate remaining about the same. They rely on the steadiness of climate to make travel plans, commute to work, plant crops, build houses

and start businesses. A change in the world's climate would be a universal cause for concern. Climate has changed before in Earth's history. The question that needs an answer is, 'Is climate changing now.'

Because climate can be measured only over the long term, no single weather event can be reliably attributed to climate change. Weather varies naturally from year to year, and not every variation signals a change in climate. Startling as it was for four hurricanes to strike Florida in one season, that episode alone cannot lead us to conclude that from now on Florida will experience four hurricanes

annually. Longterm study of a region's weather is necessary to establish whether the climate is changing, and in what direction.

It is even harder to determine what is causing climate-related events. On the broadest scale, climate is influenced by five components: the atmosphere (air), the lithosphere (solid Earth), the hydrosphere (liquid water), the cryosphere (ice and snow) and the biosphere (the totality of life on Earth). These can be considered five parts of one climate system, with the Sun as the power source that drives most of the phenomenon. The interplay of these factors is extremely complex. Causation is difficult to trace.

Hard Evidence Everywhere

The news has been full of stories about climate change; many contend that it is driven by human activity. Global warming, they claim, is being caused by the burning of fossil fuels that emit heat-trapping greenhouse gases. Some scientists say global warming can cause fiercer hurricanes, but does that mean that hurricanes Charley, Frances, Ivan and Jeanne were caused by climate change? Are rising sea levels, which will make hurricanes more damaging, also caused by climate change?

For many years now, evidence from around the world has been mounting that the global climate is indeed changing. Conditions and events that have been documented by climate researchers indicate that temperatures are rising around the globe. Glaciers are melting faster, and the danger to low-lying coasts like those of the South Pacific islands is real. The patterns of winds around the Antarctic are changing, dragging rain away from Australia and bringing drought. The fires ravaging Europe's forests, though probably aggravated by poor land use practices, are being fed by growing dryness, growing heat. Global warming is melting the pack ice on which polar bears depend.

Is human activity contributing to this change? Or could it be part of a natural cycle? Is it different from past climate changes?

The answer to the first question has become increasingly clear as evidence continues to mount. That answer is yes. Human activity is indeed contributing to climate change; natural processes alone cannot account for the massive changes that scientists have observed worldwide.

Since 1988, The UN Intergovernmental Panel on Climate Change (IPCC) has involved over 3,000 scientists worldwide to examine our understanding of climate change. In 2001, it reported that, 'Earth's climate system has demonstrably changed on both global and regional scales since the pre-industrial era, with some of these changes attributable to human activity.' The level of carbon dioxide – a gas normally present in the atmosphere but also produced from burning fossil fuels – is higher now than it has been for millennia. Scientists have little doubt that this, and other gases released by human activity, is preventing heat from escaping from the planet, causing global temperatures to rise faster than at any time during the past thousand years.

The Long View

The climate is getting warmer, and human activity is contributing to that warming – that much is clear. What, if anything, can be done to slow or stop global warming? How bad will it get if nothing is done – or not enough? What consequences has climate change had in the past? And how can people help themselves and the world's ecosystems to adapt to changes that are already too late to prevent?

To try to answer these questions, it is helpful to take a long view. Earth's climate went through many radical changes, travelling between ice ages and hothouse periods, long before humans appeared on the scene. The present crisis makes some sense in the context of that history, which shows that the forces that shape climate are complex, and that climate's effects on people and other living things are far-ranging and profound. The aim of this book is to provide that long view, the necessary perspective for grappling with the changes and challenges now confronting humankind.

This book is the story of Earth's climate, from the formation of the atmosphere to the beginning of modern industrial society. It explores how climate influenced the evolution of life, the emergence of humans, and the history of civilization; and how living things, including people, have influenced climate. It surveys the world's present climates, from Amazon rainforests to coastal metropolitan areas to the Arctic, not only as we have known them but also as they are becoming under the influence of climate change. Finally, these pages examine how human actions during the past century have driven climate change, how these changes have affected local weather and ecosystems, and how people are adapting to these changes.

Looking to the future, *Climate* explores the effects human-driven climate change are likely to have on people's everyday lives – and on the survival of Earth's species – if the current trends continue. Finally, it considers what can be done to reduce the extent of this climate change, and how human society can adapt to those changes that are inevitable.

If nothing is done to arrest the current global warming, the pace of change in the natural world will probably quicken dramatically; the price may be heavier than our society is willing or able to pay. The cost of change will be measured not only economically but also in intangible ways – not only agricultural losses from droughts and insurance costs from floods, but also increased deaths from heatstroke and malaria epidemics and the extinction of species that are valuable to humans for what they provide or symbolize—including, perhaps, iconic ones like the polar bear. *Climate* attempts to provide the background to understand the impact our actions today can have on our common future, and to suggest what we still have the power to do, in order to avert the worst while there is still time.

BELOW *A herd of zebra in Amboseli National Park, Kenya. Many of the world's most beloved animals, and the plant and animal species that support and depend on them, are at risk from the effects of human-driven climate change.*

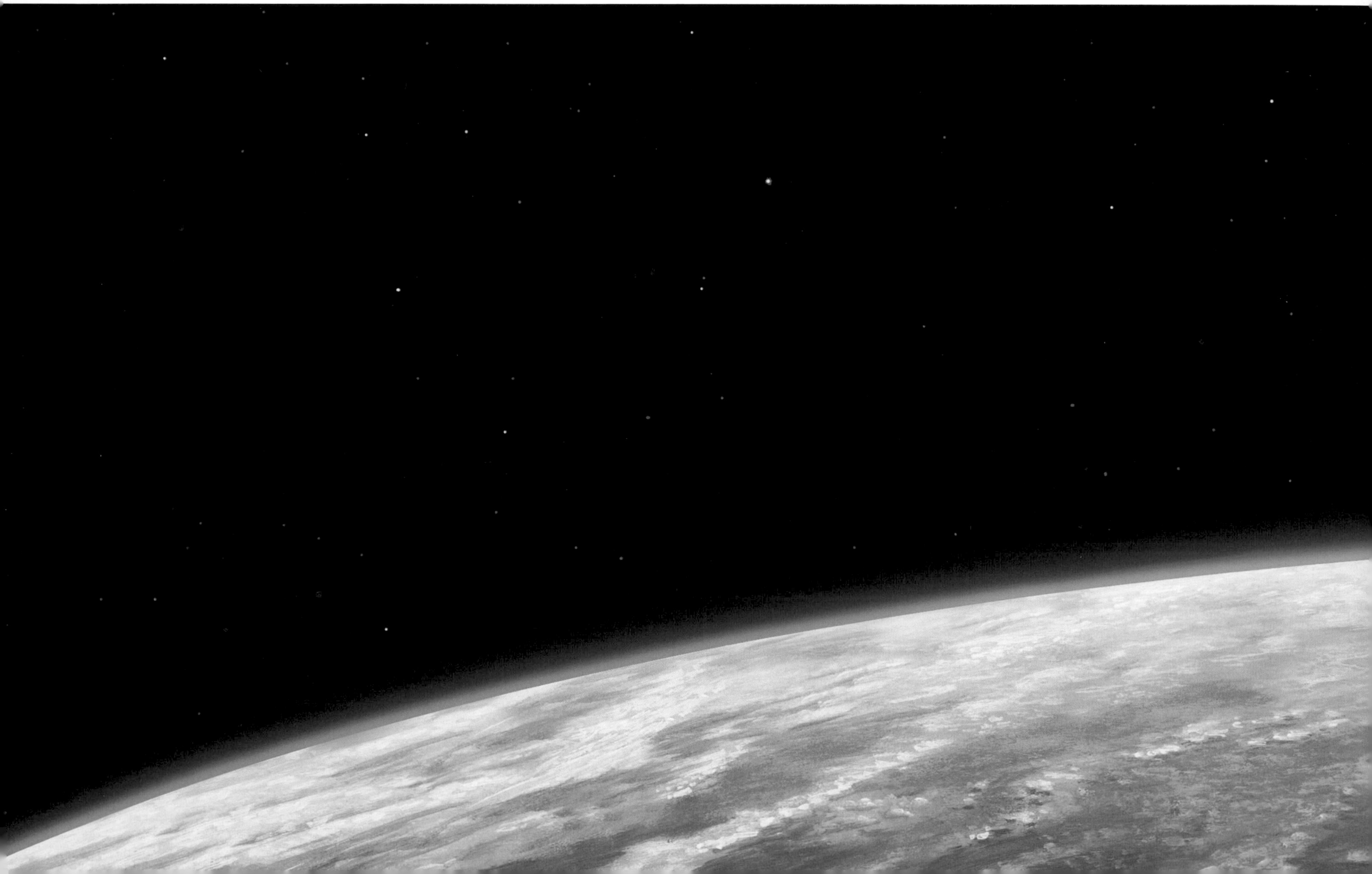

CLIMATES OF THE PAST

THE BIRTH OF CLIMATE

The Formation of Earth and Its First Climates

Climate has shaped the world ever since it was born in the emission of the atmosphere from the hot interior of primordial Earth. Rain produced the oceans on which life would depend. Atmospheric carbon dioxide maintained the planet within the range of temperature necessary for life to flourish. Climate developed as a complex system powered by the Sun and affected by winds, ocean currents, continental movements and the perpetual flow of carbon. Under the influence of climate, life emerged and the earliest animals evolved.

THE FORMATION OF EARTH

Climate is nearly as old as Earth, but not quite. During much of the period in which Earth formed, it had no atmosphere and so no climate as we understand it. The fundamentals of how climate came to be, how it works and how it undergoes long-term change are key to an understanding of the history of Earth, from before the first appearance of life to the evolution of the earliest animals and beyond.

Earth, along with the rest of the solar system, originated about 4,600 million years ago in a nebula, or vast cloud of interstellar dust and gas. The gases in this roughly spherical, slowly rotating nebula were mostly hydrogen and helium. Gravity caused the cloud to contract and rotate faster, eventually forming at its centre a proto-Sun, the precursor of our own Sun. A small percentage of the matter travelling around the centre formed the planets of the solar system.

The planets originated through accretion – the clumping together of pieces of matter. Small particles of dust and condensed gas, skidding in chaotic orbits and mutually attracted by gravity, collided and merged with one another to form larger bodies called planetesimals. These, in turn, crashed into one another, coalescing into yet larger bodies, massive enough to sweep up any chunks in their paths through gravitational force.

While the planets were forming, the proto-Sun turned into a full-fledged star. Its material continued to compact, and its heat increased until it was great enough to ignite thermonuclear fusion – in this case, the fusing of hydrogen atoms to form helium, a process that releases tremendous amounts of energy. That energy streamed through the solar system, becoming the central force in the evolution of Earth's climate.

Era of Huge Impacts

The early solar system was a chaotic place and, in the last stages of its formation, Earth was constantly being struck by meteorites. This period of time is known as the era of huge impacts, and it continued for 600 million years, until about 4,000 million years ago. It is also sometimes called the Hadean eon (4,600–3,800 million years ago), because of the hellish conditions that prevailed.

Soon after Earth had formed, a body at least as big as Mars struck the new planet with cataclysmic force. Great quantities of debris hurtled into space and formed the Moon. The same event knocked Earth into its present tilt. Before the impact, Earth's axis was vertical with respect to the plane of its orbit around the Sun. Afterwards, Earth's axis was inclined about 23 degrees from the vertical. That tilt is an important element in the story of climate: without it, there would be no seasons. Earth is slanted with respect to the Sun, so one half of the planet always receives more sunlight than the other half.

At this time, Earth was extremely hot. When the numerous bodies streaming in from space collided with Earth, most of the energy of their motion was converted to heat, causing widespread melting of Earth's surface. The impact that formed the Moon may alone have melted most of the planet. In addition, radioactive elements generated heat as their atoms decayed. As a result, Earth was in a largely molten state, its outermost layer a magma ocean hundreds of kilometres deep.

Below the magma ocean, the interior was in a soft state in which substances easily separated into layers. The lightest materials rose to the top and the heaviest sank toward the centre of the planet. During this process Earth developed three primary layers: the core, the innermost layer principally composed of the dense element iron; the mantle, made up of rock of intermediate density; and the

RIGHT *The nebula surrounding the star Eta Carinae, believed to be the most massive star in the Milky Way galaxy. Our solar system, also part of the Milky Way, was formed about 4,600 million years ago.* PREVIOUS PAGE *Clockwise from upper left: erupting volcano, Hawaii, USA; stratocumulus rain clouds; dry earth resulting from drought conditions, Kickapoo reservation near Horton, Kansas, USA; reef with tropical fish, Palau Islands.* PAGES 14–15 *The Moon rises above a limb of the Earth.*

crust, the outermost layer, composed mostly of light compounds of oxygen with elements such as aluminum, calcium, iron, magnesium and silicon.

While Earth's layers formed, gases rose to the surface. Expelled from the interior, just as gases today are spewed from erupting volcanoes, these gases became sources of Earth's first lasting atmosphere.

The First Climate

At this point, Earth's gravity was sufficiently strong to hold on to its new atmosphere, which was probably composed of the same materials that volcanoes eject today: principally water vapour; carbon dioxide; nitrogen; hydrogen, which is so light it tends to escape into space; and trace gases. The atmosphere would have been poisonous to humans, but in the beginning no life existed. And poisonous though that atmosphere was, it had the essential properties for a true climate to take shape.

The first climate, with Earth's first lasting atmosphere, originated around 4,000 million years ago. It was moist and hot, up to 110° C (230° F). The atmosphere was heavy with water vapour released from the interior. It exerted a pressure at the surface similar to that found today in the deep oceans, almost a hundred times greater than today's atmospheric pressure and certainly enough to crush a human.

Gradually, as the number of huge impacts declined and Earth began to cool, the atmospheric water vapour did what it does today: it condensed into clouds. These were vast, thick clouds, laden with volcanic ash and dust and they blackened the surface of the planet. Rain fell in torrents, but Earth was probably still too hot for the water to do more than evaporate or boil away on contact with the hot ground. However, the rain had the effect of cooling Earth further, eventually creating the right conditions for water to remain liquid on the newly formed crust. The rainwater pooled in low-lying areas, giving rise to oceans. Rain also dissolved chemicals in the Earth's crust. Over time, this caused the oceans to become salty.

In time, Earth was covered by one global ocean, almost 3.2 km (2 miles) deep, dotted with volcanic islands. The salty water was warm, heated by the Sun above and by active volcanoes below. The global ocean interacted with the atmosphere, absorbing much of its carbon dioxide. Through chemical reactions speeded by the intense heat, most of the carbon dioxide was trapped in the form of carbonate rocks such as limestone and marble. The atmosphere on Earth was now mostly composed of nitrogen, and because nitrogen molecules scatter blue light, the red sky turned blue.

This reduction in atmospheric carbon dioxide was vital to the future of Earth's climate because carbon dioxide is a greenhouse gas, which tends to trap the Sun's heat on the surface. If Earth had been a little warmer, water vapour would never have condensed into liquid and formed the ocean. If the ocean had not formed, it could not have absorbed most of the carbon dioxide from the atmosphere and locked it up in rock. Blanketed by carbon dioxide, Earth would have become hotter and hotter in a runaway greenhouse effect, and its water would have been lost by photolysis – the breaking up of water molecules by sunlight. The hydrogen in the water would have escaped Earth's gravity, eliminating the possibility of oceans ever forming. (This is probably what happened on Venus, the next-nearest planet to the Sun.) On the other hand, if Earth had had less atmospheric carbon dioxide, it might have been too cold for life, which is the case with Mars.

Soon after, continents began to form when, through the process of melting and remelting, lighter parts of the Earth's crust floated to the top of heavier layers. Over time, these lighter parts were eroded by water and wind, the debris spreading to low-lying areas, thus making the continents more extensive. Molten rock, streaming from the active interior, also added to the mix.

Thus, about 4,000 million years ago, Earth had oceans, atmosphere and the beginnings of continents. The story of climate had begun.

IN TIME, EARTH WAS COVERED BY ONE GLOBAL OCEAN, ALMOST 3.2 KM (2 MILES) DEEP, DOTTED WITH VOLCANIC ISLANDS. THE SALTY WATER WAS WARM, HEATED BY THE SUN ABOVE AND BY ACTIVE VOLCANOES BELOW

Solar System Climates: A Tour

Not every planet in our solar system has a climate. It takes a certain amount of mass for a planet to have enough gravity to keep an atmosphere from floating away.

THE INNER PLANETS: Mercury, the closest planet to the Sun, is too small to maintain a stable atmosphere; the thin layer of gases and particles around the planet is constantly changing. At 58 million km (36 million miles) from the Sun, it is a planet of fierce temperature extremes. During its day, which lasts about three Earth months, temperatures can hit 427°C (800°F) and can drop as low as -173°C (-280°F) at night.

Venus, at 108 million km (67 million miles), is second in distance from the Sun but the hottest of all the planets. Behind its extreme heat is a runaway greenhouse effect. Venus's atmosphere is 97 per cent carbon dioxide, which allows solar radiation to reach the surface but prevents heat from leaving. As a result, the temperature at the surface, is about 460°C (860°F). The atmospheric pressure is 90 times that on Earth's surface.

Mars is the fourth planet from the Sun. Like Venus, it has an atmosphere composed mostly of carbon dioxide.

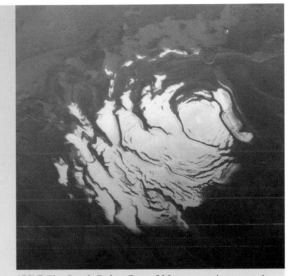

ABOVE *Jupiter's Great Red Spot, recorded by the Voyager 2 probe in 1979. Along its longest axis, this vast, hurricane-like storm is twice as wide as Earth.*

But Mars's atmosphere is thin, only one per cent of the density of Earth's. At 228 million km (142 million miles) from the Sun, Mars is generally cold, with temperatures falling as low as -125°C (-193°F) and rarely rising above the freezing point of water. However, variations exist at different latitudes and in different seasons. Like Earth, Mars is tilted on its axis (about 25°), so it too has summers and winters. Although the atmosphere is very thin, Mars has hurricane-force winds and massive, violent dust storms.

THE OUTER PLANETS: The next four planets in order of distance from the Sun — Jupiter 778 million km (483 million miles), Saturn 1,427 million km (886 million miles), Uranus 2,900 million km (1,800 million miles), and Neptune 4,500 million km (2,800 million miles) — are gas giants, enormous planets composed primarily of a thick atmosphere around a small, rocky core. Their atmospheres are mostly hydrogen and helium and are extremely dense. On Jupiter, the largest planet, the atmospheric pressure about 970 km (600 miles) below the top of the clouds is great enough to compress hydrogen into a liquid, forming an ocean. Temperatures on Jupiter increase with depth, from -140°C (-220°F) at the top of the clouds to about 24,000°C (43,000°F) at the core.

ABOVE *The South Polar Cap of Mars, seen in a mosaic of Viking Orbiter images. Temperatures on Mars rarely rise above the freezing point of water.*

With their extensive atmospheres, it is not surprising that the gas giants have weather — winds hundreds of kilometres per hour and vast storms. Jupiter's Great Red Spot is a massive, hurricane-like storm that is three times as wide as Earth. Wind systems are responsible for the colourful light and dark bands that ring Jupiter.

Astronomers disagree on whether Pluto should be considered a planet. If it is, it is the most remote one at 6,000 million km (3,700 million miles) from the Sun. Pluto is small and icy, with a thin methane atmosphere and very low surface temperatures — as low as -234°C (-390°F).

BELOW *The surface of Venus, seen in computer-generated artwork from NASA. A dense atmosphere of 97 per cent carbon dioxide results in a runaway greenhouse effect, making Venus the hottest planet in our solar system.*

THE SUN AND CLIMATE

The Sun drives weather and climate. It is a power plant, churning out energy in all directions in the form of electromagnetic radiation. This includes not only visible light, but also infrared rays, which we detect as heat and ultraviolet rays, which we cannot detect without special equipment. That radiation streams out to the planets of the solar system, including Earth. Without it, there would be no life on Earth.

The Sun's rays travel well in the vacuum of space, moving unobstructed at the speed of light – 298,000 km (186,000 miles) per second. Earth orbits the Sun at a distance of 149 million km (93 million miles), receiving the Sun's rays eight minutes after they leave the Sun. Partly because of Earth's distance from the Sun, only a tiny portion – less than one 2,000-millionth – of the Sun's radiation reaches Earth. That has been enough to sustain life throughout Earth's history.

When the Sun's rays reach Earth, three things may happen to them: they may be reflected back to space, scattered or absorbed. The amount of radiation reflected back to space depends on cloud cover and the condition of the Earth's surface where the Sun's rays meet it. Some of the radiation is scattered – redirected by molecules of gas or dust in the atmosphere. These rays are bounced around until they either make their way down to the surface or are sent back into space.

The rest of the solar radiation is absorbed by either the atmosphere or the Earth's surface. The latter absorbs more, because the atmosphere is fairly transparent, allowing most incoming solar radiation to pass through. However, the atmosphere does absorb ultraviolet (UV) radiation, preventing most but not all of it

from reaching the Earth's surface, where it would harm living things. Even the fraction of UV radiation that reaches the Earth can do considerable harm on its own, causing skin cancer. Earth's surface, in turn, reradiates back into the atmosphere much of the energy that it receives. From there the energy would be lost into space, if it were not for the greenhouse effect (see page 29).

The Earth's Albedo

Some of Earth's surfaces reflect a great deal of solar radiation. Such surfaces have a high albedo, or degree of reflectivity. Fresh snow, for example, has a high albedo; it reflects 80 to 90 per cent of the light it receives, meaning its albedo is 0.80 to 0.90. Clouds, on average, reflect about 50 to 55 per cent of the light they receive, but individual clouds vary tremendously, reflecting from 40 to 90 per cent of incoming radiation. By contrast, dark soil, green forests, and blacktop roads have albedos in the range of 0.05 to 0.15. Although albedo varies greatly by region, Earth's overall albedo is about 0.30. In other words, a little less than 30 per cent of the solar radiation that Earth receives is reflected straight back into space.

Earth's albedo also varies over time, and that can cause climate change. Ice ages, for example, have a built-in 'positive feedback' effect – when, for

RIGHT *Sunlight shining through clouds in the atmosphere. Clouds reflect about 50 per cent of the light that they receive. Variations in the overall albedo of the Earth's surfaces can cause climate change.* UPPER LEFT *Rendering of a sun surrounded by swirling, planet-building dust. This is how our solar system formed around the Sun.*

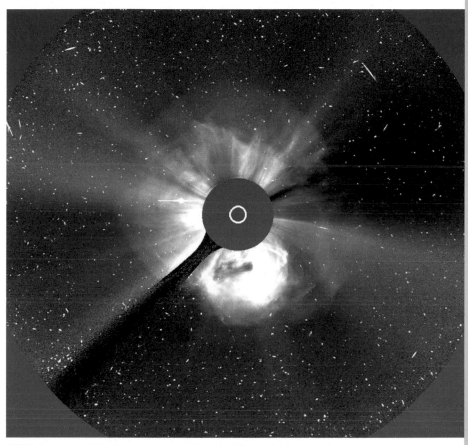

ABOVE *An image of the Sun made by an Extreme Ultraviolet Imaging Telescope. The white patches are the hottest areas of the Sun, while the handle-shaped prominence is an eruption of cool, dense plasma.*

whatever reason, the planet cools, snow and ice form. Snow and ice have a high albedo, so they reflect more sunlight, cooling the globe further. That, in turn, produces more snow and ice, which reflect more sunlight – and so on in a cycle that may produce runaway glaciation. The reverse of this process, when loss of snow and ice leads to greater absorption of heat by Earth, may be contributing to today's global warming.

Paradoxically, albedo can also work to stabilize climate. During a period of global warming, for example, more water evaporates and condenses into clouds, which tend to reflect away more sunlight. That may exert enough of a cooling effect to keep temperatures from rising too drastically.

Solar Power and Sunspots

Since the Sun is the energy source for climate, changes in the Sun's energy output can have important climatic consequences. In the early years of Earth's history, the Sun shone less brightly than at present and that probably influenced climate. Even since reaching full strength, the Sun's power output has fluctuated in subtle ways. Measurements made in the early 1980s showed that the total solar radiation reaching Earth decreased by 0.1 per cent in just an 18-month period. At this rate it would take only 15 years to bring a decrease of one per cent. If such a change became a long-lasting trend, the climatic effects might be significant. According to climate models, computer simulations of climatic interactions, a change in solar energy of one per cent per century would change Earth's average temperature by 0.5°C to 1.0°C (1°F to 1.8°F).

Sunspots may also be related to climate change. Appearing to us as dark areas on the surface of the Sun, sunspots are thousands of degrees cooler than their surroundings. Sunspots are associated with other kinds of activity, including explosive outbursts called solar flares and coronal mass ejections and clouds of flaming gas called prominences. The number of sunspots varies considerably over an 11-year cycle. Times of high sunspot activity are associated with a slight increase in energy from the Sun, but how much impact that increase has on Earth's complex climate system is not known.

One piece of evidence comes from a period of very low sunspot activity from the mid-1600s to the early 1700s, a time known as the Maunder Minimum after the British astronomer E. Walter Maunder, who discovered it. This period of low sunspot activity was a time of severe cold in Europe. The Maunder Minimum was the coldest part of what is known as the Little Ice Age (see page 77).

HOW WEATHER AND CLIMATE WORK

In Earth's large-scale system of winds, air circulates around the globe and drives currents of water on the surface of the ocean. This global circulation of air arises from the Sun's unequal heating of Earth. This leads to differences in air pressure and redistribution of heat through convection in a cycle in which warmer, lighter air rises and cooler, denser air sinks. When the air is warmed, it expands and rises, creating an area of low pressure. When it is cooled, it becomes denser and sinks, creating an area of high pressure. Air moves from an area of high pressure to one of low pressure. That movement is wind. The warming of the air near the surface of Earth is the main way that solar radiation drives the everyday atmospheric events we know as weather. As the Sun shines on the surface of the Earth and the air near the surface warms, surface water evaporates and warm, moist air rises, leaving an area of low pressure. Air from surrounding cooler regions moves in to fill the vacuum. When that cool air rolls in, we experience it as wind. When the warm, moist air rises high enough, it cools and the water vapour condenses to form clouds. Precipitation, such as rain or snow, occurs when the water droplets or ice crystals that compose the cloud become large and heavy enough to fall.

The Sun's effect on climate is not uniform across the globe but varies with latitude or distance from the equator. This is because the intensity of solar radiation reaching Earth depends on how it travels through the atmosphere. On average, solar radiation travels through less atmosphere at low latitudes, nearer the equator, because the angle at which it strikes the atmosphere is closer to a right angle.

Global Winds

Wind serves as a built-in mechanism for redistributing some of the unequal heat in the atmosphere. A system of winds blows in belts around Earth – the trade winds, westerlies and polar winds – and they influence climate. Since hot air is less dense than cold air, it rises at the equator to a high altitude, from where it flows north and south. Meanwhile, air flows from northern and southern regions along the surface towards the equator, towards the low-pressure area left by the rising air. The path of these winds is not directly north or directly south, but is influenced by the Coriolis effect, which deflects currents of air or water in the Northern Hemisphere to the right and currents in the Southern Hemisphere to the left. This effect results from Earth's rotation: by the time a bundle of air blowing south toward the equator reaches its destination, Earth will have rotated under it to the east, so that the wind will appear to be blowing southwest.

Nearly all weather takes place in the lowest layer of the atmosphere, the troposphere, which on average extends to a height of a little more

LEFT *The growth pattern of this tree was shaped by strong prevailing coastal winds. One function of global wind is to redistribute some of the unequal heat in the atmosphere.*

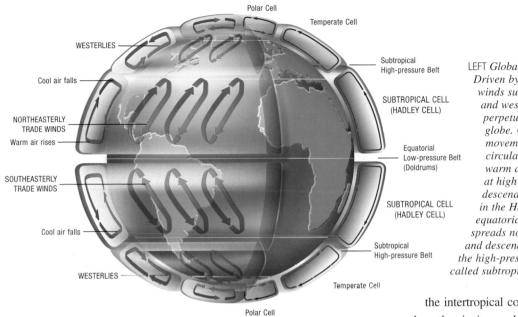

Polar Cell

Temperate Cell

WESTERLIES

Cool air falls

NORTHEASTERLY
TRADE WINDS

Warm air rises

SOUTHEASTERLY
TRADE WINDS

Cool air falls

WESTERLIES

Polar Cell

Subtropical
High-pressure Belt

SUBTROPICAL CELL
(HADLEY CELL)

Equatorial
Low-pressure Belt
(Doldrums)

SUBTROPICAL CELL
(HADLEY CELL)

Subtropical
High-pressure Belt

Temperate Cell

LEFT Global wind belts. Driven by solar radiation, winds such as the trades and westerlies circulate perpetually around the globe. Central to their movements are cells, circulations in which warm air rises, flows at high levels, then descends. For example, in the Hadley cell, hot equatorial air rises, spreads north and south, and descends to produce the high-pressure zones called subtropical highs.

than 11 km (7 miles). Near the equator, in a low-pressure zone called the equatorial low, hot air rises until it is stopped by the tropopause (the boundary between the troposphere and stratosphere) and then spreads north and south. In the phenomenon called the Hadley cell, it reaches a zone of about 30 degrees latitude north or south. Then part of it descends, producing high pressure and hot, dry conditions. That high-pressure zone is called the subtropical high and is associated with deserts in areas as far-flung as Australia, Arabia and North Africa.

Air flows outward from the subtropical highs, moving toward the poles or the equator, but, because of the Coriolis effect, is deflected either east or west. North and south of the equator, the deflected airflow becomes the trade winds, which blow from the north-east or south-east towards the equator. The trade winds meet in a narrow region near the equator called the intertropical convergence zone, where the air rises and begins again the cycle of recirculation north and south.

North and south of the subtropical highs, the westerlies are generated. These winds blow from the west to the east. (Winds are named for the direction from which they blow.) Cold air sinks at the poles and spreads toward the equator as polar easterlies, blowing from the east. At 60 degrees north, around the latitude of St Petersburg, Russia and 60 degrees south, at the latitude of the South Orkney Island, these cool winds interact with the warm westerlies, creating a stormy belt called the polar front.

In addition to this large-scale system of air circulation, there are local winds, small-scale movements of air based on local differences in air pressure. For example, a sea breeze is a local wind that blows from the sea toward the land. It arises because the land, and the air above it, heat more quickly than the sea and its accompanying air. The hot air above the land rises, creating a low-pressure area and the high-pressure, cool air over the sea rolls in to fill the relative void.

Ocean Currents

Heat is also distributed around the globe by ocean currents. Driven by wind, surface ocean currents moderate climates by transporting warm water to cold places and cold water to warm places. Their movement, like that of wind, is influenced by the Coriolis effect, so that the currents are deflected from flowing in the direction they would take if Earth were not rotating. Where the winds drive away warm surface water, cooler water rises to replace it – a phenomenon that is called upwelling. Another kind of ocean circulation, called thermohaline circulation, involves differences in the density of seawater (see page 26).

Surface ocean currents are driven primarily by wind. Two westward-moving currents, the equatorial currents, arise north and south of the equator, propelled by the trade winds. Because of the Coriolis effect, these currents are deflected away from the equator, toward the poles, forming a clockwise gyre, or large circular pattern, in each major ocean basin of the Northern Hemisphere and an anti-clockwise gyre in each basin of the Southern Hemisphere.

Oceans and large lakes heat and cool more slowly than land surfaces. They act as moderating forces on the climate of places downwind of them, causing milder summers and winters than would otherwise occur. More generally, the oceans moderate the climate by serving as heat sinks – they store heat during the summer and gradually release it into the atmosphere during the winter.

CLIMATE CHANGE

At any moment in time, the climate appears stable. Weather may vary, but climate in a given age seems fixed: the Sahara is dry and the South Pole is icy. Yet there have been times when the Sahara was lush and verdant and there were no glaciers at the South Pole.

Climate change rarely has a single cause; usually numerous factors interact in a complex way. A change in any variable in a climate system causes feedback, positive or negative. Positive feedback moves a climate system further in the direction it was already going. For example, when increased cold causes ice caps to grow, the ice or snow reflects more sunlight than ice-free terrain would, which makes the planet even colder, causing further freezing. Negative feedback reverses the direction of change, thus tending to stabilize the system. For example, a warming climate may promote photosynthesis (the process by which plants convert the energy from sunlight and carbon dioxide from the atmosphere into food), thereby reducing the amount of carbon dioxide in the atmosphere and helping to counteract the warming process.

Ocean Changes

The vast reservoirs of the oceans hold 97 per cent of the Earth's water and they are the main sources of water evaporated by the Sun's heat. In fact, the oceans generally lose more water by evaporation than they gain by precipitation, while the opposite is true for land. But this inequality balances out in what is called the hydrologic cycle. Water evaporates from the ocean, is transported as vapour through the atmosphere, condenses as clouds, falls as precipita-

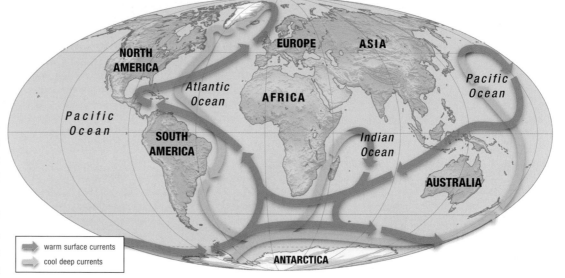

warm surface currents
cool deep currents

ABOVE *The great ocean conveyor belt moves warm water to higher latitudes, and redistributes cold water deep beneath the surface. Movement is caused by differences in density between cold, salty water and warmer, less salty water. Any change in balance between the two can powerfully influence climate.*

tion onto land, flows into rivers as runoff and from there returns to the sea. Some water flows underground or stays on the land. But the total amount of water on land, underground and in the oceans remains more or less constant.

Climate can be powerfully influenced by changes in the flow of ocean waters – including both surface currents, such as the Gulf Stream, and other deeper-reaching movements such as the thermohaline circulation, also known as the great ocean conveyor belt (see illustration, above).

This vast system of water movement depends on the fact that cold, salty water, such as that found in the North Atlantic and the Antarctic, is denser than warmer, less salty water and tends to sink beneath it. That deeper water then moves slowly toward the equator, while warmer water from lower latitudes moves toward the poles to replace it. The motion is like a conveyor belt – eventually the cold water finds its way to the surface as warm water that then becomes cold and dense again as the cycle continues.

By moving warm water from lower to higher latitudes, the great ocean conveyor belt keeps some mid- and high-latitude regions warmer than they would otherwise be. If it stops, the results are potentially catastrophic. The North Atlantic circulation stopped about 12,900 years ago, possibly because of an upset in the ocean's salt balance caused by an influx of too much fresh water from melting ice. The result was a period of glaciation – the Younger Dryas event. In a decade or less, this event plunged the world from temperatures similar to today's into an ice age.

RIGHT *The hydrologic cycle is a series of six processes: condensation, precipitation, infiltration, runoff evaporation, and transpiration. Water vapour condenses to form clouds (condensation); as the air gets more moist, the droplets that form the clouds get larger. Finally, the droplets fall to the surface (precipitation). Some of the precipitation is absorbed into the soil (infiltration) or flows into ponds, streams or oceans (runoff). While surface water returns moisture to the atmosphere (evaporation), plants return water to the air by releasing moisture from their pores (transpiration). The total amount of water in the hydrologic cycle remains more or less constant.*

Water vapour forms clouds through condensation

Rain falls from clouds as precipitation

Ground water is released to surface water bodies such as lakes, rivers and oceans

Surface water is returned to the air through evaporation

Water infiltrates the ground and is stored in underground aquifers

Water is stored in mountain snow and ice

Melting snow and ice release water as runoff

Runoff water flows into ponds, streams and oceans

Plants release moisture from their pores through transpiration

Freshwater is stored in lakes and rivers

Continents in Motion

Another agent of long-term climate change is the movement of continents. According to the theory of plate tectonics, the outer shell of Earth (the lithosphere) is composed of tectonic or structural plates. These vast sections of rock move around individually on a hot layer beneath Earth's crust, called the asthenosphere, with the continents embedded in their tops. Plates under the continents may spread apart, causing a rift in the continental crust that fills with ocean water. This is how the Red Sea was formed millions of years ago, when Africa and the Arabian Peninsula separated. Or plates may collide into each other to create mountain ranges. For instance, the Himalayas began to form 45 million years ago when the plate carrying India collided with the plate bearing Europe and Asia – the mountain range is still rising today. The movement of the plates is slow: the fastest plate moves at about the same rate as the growth of a fingernail. But over millions of years, this movement changes the map of the world.

Tectonic motion can affect climate in a number of ways. For example, at various times in Earth's history, plate movements have pushed continents into higher latitudes, causing glaciers to form and spread. Glaciers are thick, moving masses of ice that form on land when snow, accumulated from year to year, compacts and recrystallizes. An ice age is when glaciers spread over a large part of Earth's land area. For this to happen, there must be land in the colder parts of the planet – at or near the poles – because glaciers can form only on land.

Tectonic activity can also influence climate by raising mountain ranges that are high enough to affect the circulation of wind and rain. This is why the region of Asia north of the Himalayan mountains is a desert: the Himalayas block moisture-laden winds from reaching areas to the north.

And new land formations can block or free up ocean currents that circulate warm or cold water. For example, at present the Gulf Stream flows from the western Caribbean Sea, carrying warm water north and east and warming the westerly winds that cross its path. This keeps winters in western Europe milder than they would otherwise be. If, in the future, tectonic movements closed the channel between the Bahamas and Florida through which the Gulf Stream now flows, western Europe would likely become much colder.

Evolutionary biologists have established that plate tectonics can drive evolution by bringing together biological communities from different landmasses or splitting up previously unified populations. But plate tectonics also influences evolution through its effect on climate. If two or more continents collided to form one larger landmass, the climate in much of that new landmass would be more extreme than before, because it would be farther from the moderating influence of the sea. Summers would be hotter, winters colder. The living organisms on that landmass would have to adapt or

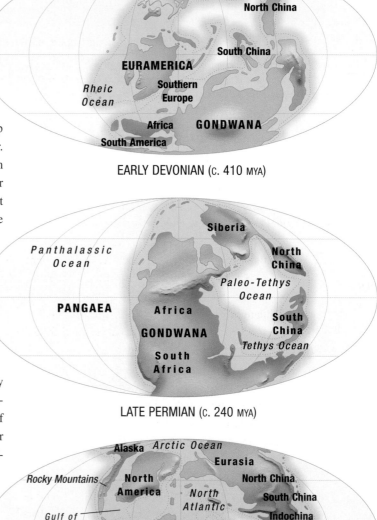

RIGHT *These maps show tectonic plates moving over time: Devonian (417–354 million years ago); Permian (290–248 million years ago); Cretaceous (144–65 million years ago). Tectonic motion can affect climate in a number of ways, including breaking up or unifying continents and raising mountain ranges, which change the circulation of wind and rain.*

EARLY DEVONIAN (C. 410 MYA)

LATE PERMIAN (C. 240 MYA)

LATE CRETACEOUS (C. 65 MYA)

become extinct. The formation of the supercontinent Pangaea in the Permian period (290–248 million years ago) spurred the proliferation of reptiles, which were better suited to the drier conditions than the amphibians that had formerly been dominant. When a large continent broke up, the smaller pieces would tend to have a milder, ocean-moderated climate. The breakup of Pangaea in the Jurassic period (206–144 million years ago) contributed to a wetter, greener environment that encouraged the evolution of new kinds of herbivorous dinosaurs. Moreover, if the orientation of a continent changed – for example, if it became longer from north to south – this might change the variations of climate throughout that continent. Climate is more likely to change with latitude than longitude because solar radiation varies with distance from the equator.

The Greenhouse Effect

Changes in the composition of the atmosphere can also bring on climate change. The most important gases in the atmosphere are the so-called greenhouse gases, especially water vapour and carbon dioxide. The greenhouse gases trap some of the heat that reaches Earth from the Sun, preventing that heat from escaping back into space. This process warms the planet.

In a greenhouse, the glass allows sunlight to enter but traps heat inside. Similarly, Earth receives the Sun's energy in the form of short wavelengths, which penetrate the atmosphere. But Earth reradiates the energy as long wavelengths, which do not escape the atmosphere. So Earth is warmed by direct heat from the Sun as well as recycled heat

Tides

The daily rise and fall of sea levels are called tides. They result from the gravitational pull of the Moon and, to a lesser extent, the Sun. The effect of the Moon's pull is to produce two bulges of ocean water, one on the side of Earth closest to the Moon, the other on the opposite side. Those bulges, as they pass over the rotating Earth, are called high tides. The Sun also causes tides, but its effects are most noticeable when they reinforce the Moon's tides. When the Sun and Moon are lined up with Earth, about every 14 days, the highest tides, spring tides, appear. The weakest tides, neap tides, also appear about every 14 days, when the Sun and Moon are at right angles to each other with respect to Earth.

Regular as they are, the tides change over long periods of time. Early in Earth's history, the Moon was closer to Earth, so tides would have been considerably stronger. They might have contributed, for example, to the spreading of inland seas across low-lying areas.

Surveying recent history – the last thousand years – geochemist Charles Keeling and geophysicist Timothy Whorf of the Scripps Institution of Oceanography, California, have linked climate change to an 1,800-year cycle that involves the alignment and closest approach of the Sun and Moon with Earth. As orbital characteristics shift, stronger ocean tides would increase a type of water movement known as vertical mixing, by which cold ocean water is transported from the depths to the surface. This vertical mixing cools the ocean surface, which in turn cools the air and land as well, making for cooler climates. Weak tides lead to less vertical mixing and warmer climates.

ABOVE *Two photos showing low (left) and high (right) tides on Campobello Island in New Brunswick, Canada. The highest tides on Earth occur here in the Minas Basin, the eastern extremity of the Bay of Fundy, where the tides can reach almost 16.4 m (18 yards).*

from the atmosphere. This effect has made the planet habitable. Today it makes the air near Earth's surface about 33ºC (60ºF) warmer than it would otherwise be.

The level of greenhouse gases can rise and fall as a result of natural processes. Human activities, such as the burning of fossil fuels also contribute to an increase in greenhouse gases.

Variations in Earth's Orbit

The Serbian mathematician Milutin Milankovitch pointed out three ways in which Earth's orbit around the Sun varies, all of which may affect climate.

First, Earth's orbit is not a circle; it is elliptical, or oval in shape. Variation in the orbit's eccentricity (how elliptical it is) alters how much change there is in the distance of the Earth from the Sun over the course of a year. Second, the tilt of Earth's axis changes about three degrees over a cycle of

41,000 years; since the tilt is responsible for the seasons, a change in the tilt affects the degree of contrast between the seasons. Third, in a phenomenon called precession, the Earth's axis wobbles like that of a spinning top. Over a period of about 20,000 years, this changes the time of Earth's closest approach to the Sun from summer to winter and back again. The net effect of all three orbital cycles is to change the degree of difference between winter and summer.

The Carbon Cycle

The carbon cycle – the perpetual global exchange of carbon in both short-term and long-term cycles – is important in maintaining the amount of atmospheric carbon dioxide necessary for a more or less stable climate. The short-term organic cycle involves living things, both terrestrial and marine. As plants on land photosynthesize, they remove carbon dioxide from the atmosphere and incorporate the carbon into their tissues. Animals eat them, incorporating the carbon into their own bodies. Through respiration and the decay of their dead bodies, plants and animals return carbon dioxide to the atmosphere. At sea, phytoplankton (tiny

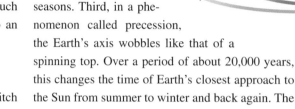

Variation in the eccentricity of the Earth's orbit (how elliptical it is) changes the distance between the Earth and Sun over the course of a year. The orbit becomes more and less elliptical every 100,000 years and every 433,000 years.

Earth's axis wobbles, and traces a complete circle in about 23,000 years

The tilt of Earth's axis changes from 22° to 24.5° every 41,000 years

ABOVE *The Milankovitch theory links changes in Earth's position relative to the Sun with climate. Earth's axis wobbles, tracing a circle once every 23,000 years. The tilt of the Earth's axis also fluctuates between 22° and 24.5° every 41,000 years. Finally, Earth's orbit becomes more and less elliptical every 100,000 and 433,000 years. The net effect of these cycles is to change the degree of difference between winter and summer.*

photosynthesizing organisms) near the surface take in carbon dioxide. When they, and the creatures that feed on them, die and decay, they settle into deeper waters, taking carbon with them. Deep ocean circulation slowly brings the carbon back toward the surface. There is also a long-term organic carbon cycle, in which organic matter is buried and converted over millions of years into fossil fuels or other geological deposits. Eventually these deposits weather away, returning the carbon to the atmosphere. Finally, there is a long-term inorganic carbon cycle, in which carbon dioxide emitted by volcanoes is transformed into rocks that eventually release carbon dioxide back into the atmosphere.

Incoming solar radiation

Infrared radiation emitted by Earth

Radiation reflected by atmosphere

Radiation reflected by surface

Absorbed radiation

Surface cools by radiating heat energy upward.

Infrared radiation reemitted back to Earth

LEFT *Although some incoming solar radiation is diffused in the atmosphere or reflected back to space, some is absorbed by Earth's surface, warming it. The surface cools by reradiating the energy as heat. Some of that heat is trapped by greenhouse gases, which prevent it from escaping into space. The effect is that Earth is warmed not only by the Sun but by recycled heat from the atmosphere.*

The Long-Term Carbon Cycle

CO₂ discharged by volcano into atmosphere

Rainwater removes CO₂ from atmosphere as carbonic acid

Chemical weathering occurs as carbonic acid weathers silicate rocks, releasing calcium and bicarbonate ions

CO₂ rises in the magma

Calcium carbonate sediment on the ocean floor is carried underground by tectonic processes

Runoff carries calcium and bicarbonate ions to the ocean

Intense pressure and high temperatures deep in the Earth's mantle convert calcium carbonate sediment to CO₂

Mid-ocean ridge, where partial melting of calcium carbonate sediment converts it into CO₂

Mantle   Oceanic crust   Coal deposits

ABOVE *In the inorganic cycle, rain removes carbon dioxide from the atmosphere; the carbon cycles under the ocean floor through tectonic processes; volcanism then releases the carbon back into the atmosphere. In the organic cycle, carbon in living things is buried as coal or other geological deposits that weather away.*

The long-term carbon cycle can be both organic (dealing with living organisms) and inorganic. In the long-term inorganic carbon cycle, carbon dioxide discharged by volcanoes is eventually converted into limestone, which weathers to return carbon dioxide to the atmosphere. This process is linked to the carbonate-silicate cycle, which helps to stabilize Earth's climate. In that cycle, rainwater removes carbon dioxide from the atmosphere and changes it to carbonic acid. In what is called chemical weathering, the acid reacts with silicate rocks on land to release calcium and bicarbonate ions. Water carries the chemicals to the ocean, where marine organisms use them to make shells of calcium carbonate or where they are converted to carbonate rocks. The shells settle as sediment on the sea floor. Tectonic processes carry the sediment and the rocks underground, where intense pressure and high temperatures reconvert the chemicals to carbon dioxide. After a cycle of more than 500,000 years, that carbon dioxide is returned to the atmosphere by volcanic eruptions.

This cycle is sensitive to temperature variations. If Earth warms up, the rate of evaporation, rainfall and chemical reactions will increase, leading to more chemical weathering and more burial of carbon in ocean sediments. The result is a cooling effect. If Earth cools, less water will evaporate, less rain will fall, less carbon dioxide will be removed from the atmosphere and the gas will build up again in the atmosphere – resulting in a warming effect.

LIFE BEGINS

The earliest period of Earth's history is called the Precambrian, and it extended from the formation of Earth until 543 million years ago. On the geologic timescale, it is usually divided into two eons, the Archean, which lasted from the beginning up to 2,500 million years ago, and the Proterozoic, which followed the Archean.

The Precambrian comprises nearly 90 per cent of all Earth's history, yet few fossils from this age have been discovered. Precambrian rocks that have not been destroyed by erosion or subduction (the forcing of one tectonic plate beneath the edge of another), have been heavily deformed and reworked by time. As a result, our knowledge of climate during the Precambrian is sparce.

One certainty is that the environment must have been hospitable enough for life to begin. Microscopic fossils of bacteria have been found in rocks dating to 3,500 million years ago, indicating that, by that time, life had appeared on Earth, probably in the oceans. Life may have begun even earlier, as long ago as 3,800 or 3,900 million years ago. The climate on early Earth was more extreme than that of today: according to some estimates, the strong greenhouse effect may have boosted Archean temperatures to nearly 60°C (140°F). But however severe the climate was, it was neither too hot nor too cold for the existence of liquid water. This was important because the relatively stable oceans became a haven for life, places where moisture was plentiful and temperature and ultraviolet radiation were moderated.

In what part of the sea did living things first develop from organic molecules? Some scientists think it happened in the surface waters, where sunlight would have provided an energy source. Yet the Sun's ultraviolet radiation would have been more intense than it is today, with little or no ozone in the atmosphere to block it. To avoid damage, early organisms would have needed a shield of water above them, perhaps 9 m (30 ft) deep. Other scientists think the first living things appeared much deeper than that, around hydrothermal vents. These superheated openings in the ocean floor are found at midocean ridges, places where continental plates are spreading apart and new crust is being produced. Today, these vents support communities of exotic organisms that draw their energy from hydrogen sulphide, carbon dioxide and oxygen in the hot water, rather than from sunlight. Life on Earth might have begun in these depths and then migrated upwards.

A few scientists believe that life arrived on Earth from outside, carried on comets. But most doubt that any living organism could have survived the trip through space.

In any case, once living things emerged, they evolved, changing gradually into new kinds of organisms as a result of natural selection and other processes. In natural selection, those organisms with variations better suited to their environment – including climate – survived and reproduced, thereby passing on the genes (hereditary instructions for replication) that distinguished them from less successful individuals. Over time, these processes brought about the diversity of life that exists on Earth today.

Modern Stromatolites

Like relics of Precambrian time, stromatolites still form today in places such as Shark Bay in Western Australia and the Sinai shore of the Red Sea. There, cyanobacteria form mounds similar to those raised by their predecessors 3,500 million years ago. The water where these mounds take shape is warm and salty – so salty that animals such as sea urchins and snails cannot live there. If such animals were present, they would graze away the cyanobacteria, making stromatolite growth impossible. The evolution of these grazing animals is probably the reason why stromatolites, once widespread, are now limited to only a few salty locations.

ABOVE *Stromatolites in Hamelin Pool, Shark Bay, Western Australia. Formed from blue-green algae, or cyanobacteria, these mineralized microbial communities are described as living fossils.*

The first organisms known from the fossil record are bacteria. Ubiquitous today, bacteria are single-celled organisms that possess a cell membrane and deoxyribonucleic acid, or DNA – the chemical carrier of genetic information – but lack a cell nucleus – that part of a cell in more complex organisms that functions as the cell's brain. The first bacteria lived off organic compounds in the ocean water, but in time some evolved that had the capacity for photosynthesis – the process by which food is produced from sunlight. The earliest fossils include stromatolites, mounds or domes left by blue-green bacteria, or cyanobacteria (also called blue-green algae), as they formed in mats in shallow water, trapping sediment in layers and living by photosynthesis. These organisms were active by 3,500 million years ago and became more common after 2,500 million years ago.

Stromatolites today grow in intertidal zones, between the high tide and low tide marks. If the same were true in those earliest times, stromatolites would have encountered much more powerful tides. Since the Moon was closer to Earth, tides

mats, sediment and slime in those shallow waters may have shielded the bacteria from those rays.

Oxygen Abounding

Bacteria were making changes in the atmosphere that would profoundly affect the history of life and climate. Before life emerged, the atmosphere had little or no free oxygen and the first bacteria were anaerobic – they did not require oxygen to live. But photosynthesis gives off oxygen as a waste product. The photosynthetic bacteria, thriving along the shores and possibly across the ocean surfaces, released large quantities of oxygen, though this did not go into the atmosphere right away. First it oxidized or combined with substances that had an affinity for oxygen, especially iron. Iron was then widely dissolved in the sea; as it became oxidized (or rusted) it settled to the ocean bottom. After this period of rusting, there was nowhere else for the oxygen to go, and it passed into the oceans and air. By about 2,000 million years ago, oxygen had begun to accumulate in the atmosphere.

The change was momentous. Today's proportion

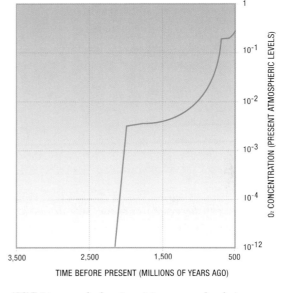

RISING OXYGEN LEVELS

TIME BEFORE PRESENT (MILLIONS OF YEARS AGO)

O_2 CONCENTRATION (PRESENT ATMOSPHERIC LEVELS)

ABOVE *Line graph showing rising oxygen levels in the atmosphere over the last 3,500 million years to 500 million years ago. O_2 concentrations are measured in comparison with present atmospheric levels.*

BACTERIA WERE MAKING CHANGES IN THE ATMOSPHERE THAT WOULD PROFOUNDLY AFFECT THE HISTORY OF LIFE AND CLIMATE

would have flooded coastal areas a long way inland. However, bacteria are adaptable, evolving rapidly in response to environmental pressures, and could have survived considerable buffeting. Stromatolites may have formed as they did in response to the Sun's heavy ultraviolet radiation. Their layers of

of oxygen in the atmosphere – 21 per cent – is essential to our survival and that of countless other aerobic (oxygen-breathing) organisms. This production of oxygen derives from the photosynthesizing work of bacteria that, ironically, did not need the oxygen and indeed found it poisonous. Some bac-

teria died off; others found shelter in anaerobic habitats such as deep mud or decaying material. But others adapted, evolving into organisms that could shield their cellular machinery from oxygen and used the gas to release energy from food. Such aerobic respiration is more efficient than the anaerobic variety; and it caught on, leading to a proliferation of new forms of life.

The accumulation of atmospheric oxygen was important for another reason – it led to the formation of an ozone layer. The Sun's ultraviolet radiation reacted with free oxygen (O_2) high in the atmosphere, producing ozone (O_3). Ozone is an effective shield against the most harmful ultraviolet radiation.

This protection permitted the evolution of new organisms more complex than bacteria.

Proterozoic Ice Ages

A radical climate swing occurred from about 2,700 to 2,300 million years ago, when there were one or more episodes of widespread glaciation. We know about this ice age from sediments that date to that period left behind by these glaciers in places as widespread as Canada, Africa, India and Europe. (Because of plate tectonics, the continents at the time would have been arranged differently.) How long this ice age lasted is unknown, but it was apparently followed by a long period of milder climates. Then, between about 850 and 600 million years ago, in the late Proterozoic, there was a series of global ice ages, leaving traces of glaciation right across the continents.

The Proterozoic ice ages are unusual because they seem to have affected low latitudes, near the equator.

at this time as 'snowball Earth'. Widespread extinctions may have resulted.

The causes of these ice ages are not well understood. One important factor may have been the grouping of continental landmasses in middle to low latitudes. Today, the tropical oceans play a vital role as heat-sinks, absorbing much of the Sun's heat and distributing it worldwide through ocean currents. But continents overall have a higher albedo than ocean and if the continents were in the tropics, they would have reflected away more of the Sun's heat, cooling the planet.

New Forms of Life

Whatever the causes of the Proterozoic glaciations, life persisted. Even under snowball Earth conditions, life could have survived around the superheated hydrothermal vents in the ocean depths, or in tropical puddles, places where the sea had not frozen over.

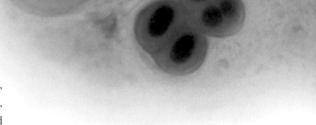

ABOVE *A photo of microscopic blue-green algae, or cyanobacteria. This colony of reproduces solely by fission, a form of asexual division, as seen in the cells on the left.*

DURING THESE EARLY ICE AGES, EARTH MAY HAVE BEEN ALMOST ENTIRELY COVERED BY ICE AND SNOW – SCIENTISTS REFER TO THE PLANET IN THIS TIME AS 'SNOWBALL EARTH'

Since that time – that is, for the past 600 million years – glaciers have been restricted to high latitudes or high mountains, while the tropics have remained warm even during ice ages. During these early ice ages, Earth may have been almost entirely covered by ice and snow – indeed, scientists refer to the planet

Some organisms may have evolved to survive in and on the ice and snow. Even today there are cold-loving bacteria and algae capable of living in snow and on rock particles in sea ice. Photosynthesis might have continued in places where the sea ice was not too thick, or in or on ice and snow.

In any case, evolution continued during this period. By 2,100 million years ago, eukaryotes, or organisms with a cell nucleus, emerged, quite unlike the prokaryotic bacteria that preceded them. The earliest eukaryotes were one-celled, like bacteria, but multicellular forms evolved by 1,300 million years ago in the form of algae. Multicelled animals, or metazoans, followed by about 600 million years ago. The metazoans included organisms that reproduced sexually.

What are called the Ediacaran fauna – named for the Ediacara Gorge in South Australia from which fossils of this age were first identified – lived 590 to 545 million years ago, a period after the end of the late Proterozoic glaciations, when the climate was probably mild. By that time, the supercontinent Rodinia, which had formed earlier in the Proterozoic, was in the process of breaking up, creating space for new seas. The glaciations themselves might have driven the evolution of the Ediacaran fauna by causing mass extinctions that left a biologically emptier world once the calamity was over. It is a characteristic of evolution that, in such periods, new species tend to appear explosively, filling the niches that the extinct species left behind. Time and again in Earth's history, climate change has driven evolution in just this way, by wiping out species and clearing the way for a diversification of the remaining life.

The Ediacaran fauna included jellyfish, sponges, plantlike animals attached to the sea floor (similar to modern sea pens) and some animals that cannot be easily categorized, such as Tribrachidium, a disc-shaped organism with three radiating ridges that curved as they approached the disc's edge. They also included worms that, like humans and most later animals, had a body cavity, the coelom, which could contain internal organs. One common feature of animals at this point is that they had no mineralized hard parts – no shells and no skeletons. Those would await the next era, the Paleozoic. The lion's share of evolutionary change was still to come, but the Precambrian had seen some important developments. By the end of this era, Earth had formed, life had emerged and, after great tumult, the planet's climate had settled down to a time of relative mildness and warmth.

Investigating Past Climates

Since it is not possible to jump in a time machine and go back to a prehistoric era to take the temperature of the air, scientists investigating climates of the remote past rely on indirect evidence. Plant and animal fossils provide clues. Plants, for example, need sunlight. Likewise, certain rocks form only in certain kinds of climates. Evaporites, the remains of dried-out seas and lakes, require arid conditions, while tillites are left behind by glaciers.

Another line of evidence comes from sea floor sediments. These sediments contain the shells of organisms that once lived near the ocean surface. Many of these organisms are sensitive to changes in water temperature. When climate changes, the numbers and types of organisms near the surface change too – and that evidence is recorded in the layers of sea floor sediments. Scientists obtain samples of these sediments and the rock below by drilling into the sea floor. Millions of years of climate history can be accessed this way.

Ice cores are also obtained by drilling, this time into the ice sheets that cover Antarctica and Greenland. The layers of ice there formed when snow fell each year and remained frozen, to be covered by more layers of snow until compacted into ice. By examining this ice, and bubbles of air trapped inside it, scientists can investigate atmospheric conditions going back hundreds of thousands of years. The ice cores preserve a record of such matters as temperatures when the snow fell and the chemical composition of the atmosphere at the time.

ABOVE *Drilling for ice cores from within the Antarctic ice sheet. This type of electromechanical drill can reach depths of nearly 0.4 km (0.25 miles), yielding ice from several thousand years ago.*

RIGHT *Scientists collecting an 3.4 m (11 ft) ice core. Chemical analysis of ice cores can reveal past changes in climate and current trends in the pollution of the atmosphere.*

PREHISTORIC CLIMATES

Continental Drift, Asteroids and Ice Ages

After the Precambrian age came what is sometimes called the Phanerozoic eon, spanning 543 million years of Earth's history. The Phanerozoic saw the evolution of the plants and animals familiar to us today, along with many that are now extinct, from trilobites to dinosaurs. This time also brought the shifting of continents, the building of mountains and the rearranging of oceans into the forms we know today. At many points, climatic change drove the evolution of new forms of life; often it led to mass extinctions. At every point, climate reacted to and shaped the changing world.

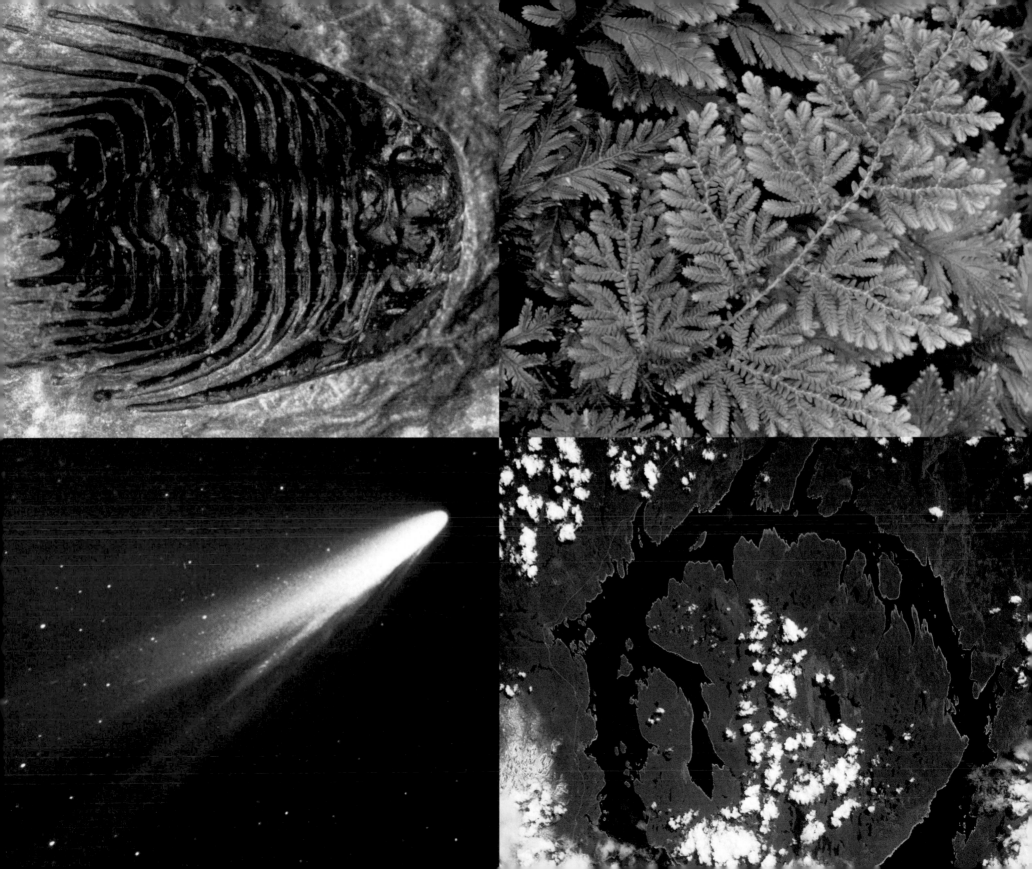

THE PALEOZOIC ERA

The Phanerozoic eon is divided into three eras, each separated from its predecessor by a mass extinction event: the Paleozoic era (543–248 million years ago), the Mesozoic era (to 65 million years ago) and the Cenozoic era (to the present). The Paleozoic era, which saw the expansion of life from the oceans to land and the later extinction of most species, is divided into six periods: the Cambrian, Ordovician, Silurian, Devonian, Carboniferous and Permian.

The Cambrian Period

The earliest period of the Paleozoic era, the Cambrian (543–490 million years ago), saw a great diversification of life known as the Cambrian explosion. This flourishing of life took place in a climate that was warmer than today's. The supercontinent Rodinia, formed during the Proterozoic eon, had broken up to leave behind several continents. Most of these continents were near the equator. That positioning allowed freer water circulation around the poles, which in turn helped to keep the oceans relatively ice-free and hospitable to life. The continents were flooded with inland seas that reached deep into their interiors, providing ample shorelines and warm, shallow habitats for new species to flourish. No living organisms had yet invaded the land.

The Cambrian animals were distinguishable from their predecessors by the hard parts of their bodies, including shells, exoskeletons (hard outer structures), and claws – all far more likely than soft parts to become fossilized. The most characteristic Cambrian animal was the trilobite, which makes up about 70 per cent of the period's known fossil record.

The land was still barren. It may have hosted bacteria, but not in the kind of soil we know, since modern soil is composed partly of organic particles from the decay of land plants and animals. The Cambrian continents were wastelands of boulders, stones, pebbles and sand, buffeted and eroded by wind and rain.

In the Southern Hemisphere, a portentous change was occurring: several landmasses were assembling into a large southern continent called Gondwana. This included present-day Africa, Antarctica, Australia, India, South America and possibly China.

The Ordovician Period

Sea levels were high in the Cambrian period, but early in the Ordovician period (490–443 million years ago), the seas reached farther inland than ever before or since. Laurentia (present-day North America), which then lay across the equator, was almost entirely covered by a shallow sea.

The climate remained warm for most of the Ordovician and invertebrates, or animals without backbones, continued to proliferate. Among them were the cephalopods, a group of mollusks that includes the modern squid, octopus and nautilus. The cephalopods produced the first large organisms, with one species reaching nearly 9 m (30 ft) in length. This period also saw the evolution of the first vertebrates, animals with spinal columns made of bone or cartilage. The first vertebrate fossils, dating from the early Ordovician, are those of jawless, armour-plated fish called agnathans.

About 450 million years ago, near the end of the Ordovician, plants first became established on land. The atmospheric ozone had continued to increase and it might have reached a level possible for organisms to brave the Sun's ultraviolet radiation without benefit of even a shield of water. The first land plants may have resembled modern liverworts, which lack true roots and grow close to the ground.

Near the close of the Ordovician, after ages of largely ice-free conditions, a major glaciation

RIGHT *Volcanic eruptions did much to shape the Earth during the Phanerozoic eon.* PREVIOUS PAGE *Clockwise from upper left: fossil of the trilobite* Leonaspis maura, *dating from 438–380 million years ago; spike moss,* Selaginella versicolor, *a plant which reproduces via asexual spores; an impact crater in Canada, the result of a meteorite impact some 212 million years ago; a comet in the western sky.*

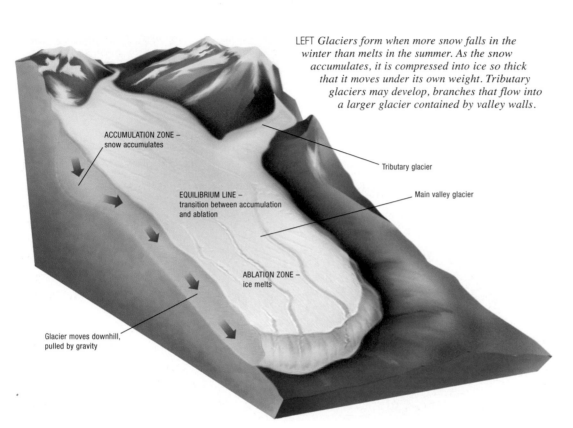

LEFT *Glaciers form when more snow falls in the winter than melts in the summer. As the snow accumulates, it is compressed into ice so thick that it moves under its own weight. Tributary glaciers may develop, branches that flow into a larger glacier contained by valley walls.*

ACCUMULATION ZONE –
snow accumulates

Tributary glacier

EQUILIBRIUM LINE –
transition between accumulation
and ablation

Main valley glacier

ABLATION ZONE –
ice melts

Glacier moves downhill,
pulled by gravity

occurred in the Southern Hemisphere. It may have been triggered by the movement of Gondwana to the south polar region; at the time what is now North Africa was at the pole and glaciers covered the Sahara. Glaciers form only on land and the presence of a large landmass in the Antarctic made it possible for snow to fall and remain frozen, building up in layers of ice to form glaciers. The high albedo of ice contributed to the ice age by reflecting more of the Sun's heat back into space. Sea levels fell as water that had evaporated from the oceans and fallen as snow became locked in

the glaciers rather than flowing back to sea.

The ice age brought about a mass extinction. As climates cooled, ocean temperatures fell and many species could not adapt to the change. More than half of Earth's species were eliminated.

The Silurian Period

After the glaciation at the end of the Ordovician, mild temperatures returned in the Silurian period (443–417 million years ago). As the glaciers melted, sea levels rose again, placing about two-thirds of present-day North America underwater. Life began

to thrive again in the shallow inland seas.

The world's continents, as always, were in motion. Gondwana moved away from the South Pole, only to return at the end of the period. Laurentia and Baltica (present-day Europe) lay along the equator, with Siberia just to their north. By the end of the Silurian, Laurentia and Baltica collided, forming a single continent, Laurasia.

The collision raised mountains at the point of impact in what is called the Caledonian orogeny, or mountain-building episode. In Europe, the mountains extended from southern Wales through Scotland to northern Norway; in North America, they included what are now the northern Appalachians and reached from Alabama to Newfoundland and as far west as Iowa and Wisconsin. Towering like the Alps, the Caledonian mountains were still growing in the Devonian period that followed.

The first jawed fish emerged in this period. On land, vascular plants appeared, with tissues specialized for transporting water and minerals from the roots upwards. As land plants expanded their domain, they created a food supply for any animals hardy enough to venture out of the water. Yet the first land animals, arthropods dating from the late Silurian, may have come searching not for plants but for dead and decaying organic matter washed up on the shores by storms and tides. Scavengers such as millipedes, mites and springtails fed on organic debris, while predators such as spiders and centipedes fed on the scavengers.

The move onto dry land was a momentous shift for both plants and animals and a milestone in the story of climate. By venturing for the first time to live in the open air, the land organisms of the Paleozoic

LEFT *The line of the Atlantic Fault can be clearly seen near Pingvellir, Iceland. Here the North American tectonic plate (on the left) meets the European tectonic plate, creating deep ravines and cliffs. The plates are slowly moving apart by the process of continental drift, a process which was responsible for the creation of mountain ranges during the Silurian Period.*

were exposed directly to the atmospheric conditions we know as weather and the long-term weather patterns called climate. The struggle to adapt to life in a world directly exposed to climate would drive new rounds of evolution. In turn, the evolution of new forms of life would propel changes in climate.

These changes in climate were already happening. By the end of the Silurian, sea levels dropped again, and large parts of the continents became dry land.

The Devonian Period

During much of the Devonian (417–354 million years ago), large areas of the globe experienced a warm, uniform climate. Primitive land plants were widely distributed and amphibians first appeared.

The Devonian is often known as the Age of Fishes because of the many new kinds of fish that evolved. The armour plates of many fish evolved into lighter-weight scales, allowing for greater speed in the water. Devonian fish included sharks, with their cartilage skeletons, and bony fishes such as ray-finned fish, which included the ancestors of most present-day fish. Another group of bony fish evolved lungs, which permitted them to survive in shallow, oxygen-poor water, such as lagoons and deltas. By the late Devonian, some of these fish evolved into the first amphibians, animals that live part of their lives in water and part on land.

On land, the earliest insects evolved. By the

Precambrian time

Cambrian period

Ordovician period

Silurian period

Devonian period

Carboniferous period

Permian period

Triassic period

Jurassic period

Cretaceous period

Tertiary period

Quaternary period

RIGHT *This timescale presents the history of Earth in units of time. At top is the Precambrian time, lasting from the origin of Earth 4,600 million years ago to 543 million years ago. In the Precambrian the first living things emerged. The Cambrian period brought a great diversification of life, including the hard-shelled trilobites. In the Silurian, vascular plants developed on land. In the Devonian, the first amphibians evolved. The Carboniferous saw a flourishing of forests. From the Triassic to the Cretaceous periods, dinosaurs reigned supreme. In the Quaternary, humans evolved.*

middle Devonian, the first trees were growing, reaching heights of up to 18.3 m (60 ft).

Toward the end of the Devonian, two or three mass extinctions occurred. South America was at the South Pole and thick with glaciers, which probably led to a lowering of sea levels and global cooling. However, the glaciation was not as substantial as the one that would follow in the Carboniferous period, and it is unlikely that this alone accounted for the global extinction.

Another possibility is that one or more comets or asteroids struck the Earth with catastrophic results. Evidence exists that there was a shower of comet or asteroid impacts at the end of the Devonian, including one that

created Woodleigh crater, near Shark Bay, Australia. Tectonic activity may also have played a role. The last pieces of Laurasia were moving into place and Gondwana and Laurasia were just beginning to collide. The movement of continents may have altered the flow of ocean currents, with devastating effects on temperature in many parts of the globe.

The Carboniferous Period

The Carboniferous period (354–290 million years ago) began with sea levels rising and shallow seas expanding across Laurasia (including present-day North America and Europe). The climate in the tropical areas of Laurasia was warm and moist, encouraging the growth of vegetation and the spread of insects and amphibians. Trees, which had first appeared in the Devonian, multiplied to become dense, swampy forests. Towering to nearly 46 m (150 ft) high, the vegetation, which included giant ferns, horsetails and clubmosses, produced habitats for smaller species. The first seed-bearing plants evolved alongside many others that reproduced by the spreading of spores.

Under the swampy conditions in the Carboniferous forests, dead plant material piled up as peat, was buried in sediment and in time became coal. The greatest period of coal formation in Earth's history was the Carboniferous.

Insects abounded in the forests, including beetles, termites, grasshoppers and cockroaches. The first winged insects evolved, including mayflies and dragonflies. Some were unusually large, including dragonflies with wingspans greater than 60 cm (2 ft).

Amphibians proliferated. In fact, the Carboniferous is sometimes called the Age of Amphibians. There were large ones – Eryops was nearly 1.8 m (6 ft) long and may have lived and fed like a crocodile. And there were small ones – microsaurs, similar to small lizards, fed on insects.

A new form of life was also emerging: reptiles. Unlike amphibians, reptiles could lay their eggs, which were more durable than those of amphibians, on land. By the late Carboniferous, many kinds of reptiles had developed, including the ancestors of mammals and dinosaurs.

The rampant growth of Carboniferous plants led to changes in the atmosphere, as oxygen levels rose and carbon dioxide levels fell. In the process of photosynthesis, plants absorb carbon dioxide and emit oxygen. Later, the organisms that eat the plants or cause their decay use up oxygen and return carbon dioxide to the atmosphere. This is part of the carbon cycle, by which a balance of carbon dioxide and oxygen is maintained (see pages 30–31). In the Carboniferous, this process was interrupted as extensive peat deposits buried much of the carbon, allowing oxygen levels to climb and carbon dioxide levels to fall. The result was a cooling effect, bringing on glaciation by the end of the period.

The Permian Period

As the Permian period (290–248 million years ago) began, the glaciers in the southern continent of Gondwana kept expanding and retreating. Gondwana continued to collide with the northern continent of Laurasia, a process that had begun in the Devonian. By the end of the Permian, all the world's major continents had joined together to form Pangaea, a single supercontinent that stretched from the South Pole almost to the North Pole.

As the landmasses fused, their climates changed. Large parts of the northern lands rose in elevation, as inland seas receded. Many of the forest-swamps of the Carboniferous dried up. Pools and lagoons evaporated, leaving behind deposits of gypsum and salt. Deserts spread. Because most of the interior was far from the moderating influence of the sea, the weather became more seasonal, with extreme fluctuations.

The climatic changes spurred evolutionary developments, especially the spread of reptiles, which were better adapted than amphibians to the deserts of Pangaea's interior.

Climate also drove the development of warm-blooded animals. Reptiles are cold-blooded, relying on the environment to supply the necessary body temperature for peak metabolic performance; thus they are at a disadvantage when the local weather is either too hot or too cold. Changing climates can be catastrophic. Warm-blooded animals, by contrast, can adjust more readily to different weather.

By the end of the Permian, the cynodonts appeared. They were warm-blooded animals, yet some of them may have had fur, which would have enabled them to penetrate colder climates in the southern regions. They were still reptiles, but they were probably ancestors of mammals. Their ability to spread to the south indicated another reason why the formation of Pangaea was important to evolution: it permitted life forms from different regions to mix. With minimal barriers to migration, the supercontinent was one large arena for the contests of evolution.

Paleomagnetism

The study of paleomagnetism, or changes in Earth's magnetic field over time, can be a great aid to the investigation of past climates. The permanent magnetization of certain rocks as they are formed tells how Earth's magnetic field was oriented at the time. These rocks can yield data on the location of Earth's magnetic poles at that time, from which the location of the geographic poles (the top and bottom of Earth's axis of rotation) can be estimated, enabling scientists to determine the latitude at which the rocks were formed. This information is important in tracking the wanderings of continents and in investigating ancient climates. For example, paleomagnetic evidence has established that North Africa lay at the South Pole in the late Ordovician, a fact that helps account for the geological evidence of glaciers in that region at that time. Although it appears from the evidence that the South Pole 'wandered' to North Africa, what actually wandered was Africa.

Earth's magnetic field is fickle, however, and sometimes reverses polarity: the north magnetic pole can turn into the south magnetic pole, and vice versa.

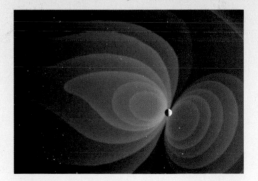

ABOVE *This illustration shows Earth's magnetosphere in the solar wind from the Sun. The field lines originate from the poles and form a magnetic shield that protects the Earth from the radiation of the solar wind, which is made up of energetic charged particles.*

Do Oceans Have Climates?

Oceans do not have climates, because climate is defined as a characteristic of the atmosphere. But oceans affect climate and are affected by it. On a global scale, a warmer atmosphere is usually accompanied by a warmer ocean surface. For example, both the atmosphere and the ocean surface become cooler farther from the equator. Ocean currents introduce modifications on smaller, regional scales. For instance, although Quebec, Canada, and France are at the same latitudes, the Gulf Stream brings warmer water and air temperatures to northern Europe than are found at the same latitude in northern America or Asia.

Climate can also affect marine organisms by altering the temperature and chemistry of the water, just as it affects land organisms by altering their environment. Marine organisms depend for life on a number of physical variables, such as temperature, light, salinity, dissolved oxygen levels and stratification (layering of water). Climate changes can alter these variables.

On land, climates are differentiated by climate zones, with characteristic levels of temperature, humidity, light and other variables. Similarly, the oceans can be divided into regions, also called provinces. Scientists also talk about ecoregions, large land or water areas distinguished by environmental or biotic aspects.

As life came out of the sea and onto land, directly exposing itself to the atmosphere, climate became increasingly important.

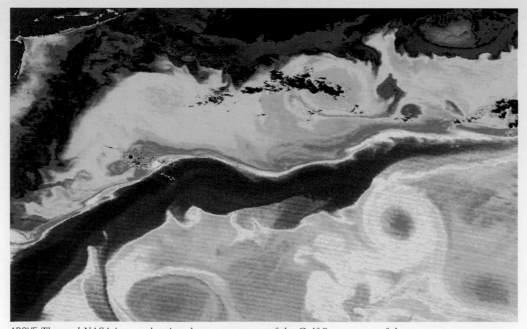

ABOVE *Thermal NASA image showing the temperatures of the Gulf Stream, one of the strong ocean currents which carry warm water from the tropics to higher latitudes. The coldest waters are shown as purple, with blue, green, yellow and red representing progressively warmer water.*

At the end of the Permian period came the largest mass extinction Earth has ever experienced. Scientists estimate that at least 80 to 90 per cent of all species were exterminated, including about 95 per cent of all marine species, more than 80 per cent of amphibian species, 90 per cent of reptile species and 50 per cent of insect species. Many plant species were wiped out. At sea, the trilobites died out completely after 300 million years of existence.

The causes of the Permian extinction are uncertain. The spread of deserts and the extreme seasonal fluctuations might have been too much for many organisms. The creation of the supercontinent may have changed the direction of ocean currents and wind patterns in ways that disrupted established ecosystems. Ocean chemistry might have changed. There is evidence that oceanic salinity fell, so that the water might not have been salty enough for many kinds of organisms. There is also evidence of anoxia, a state of low oxygen and high carbon dioxide in the water, which could have asphyxiated many sea creatures.

Volcanic activity is also a likely suspect as the cause of the mass extinctions. In the course of less than a million years, eruptions poured lava across Siberia, forming what are known today as the Siberian Traps. The eruptions might have produced enough debris to block sunlight, chilling the climate and contributing to the further spread of glaciers and lowering of sea levels. Because the percentage of atmospheric oxygen was high in this period, sulphur from the eruptions would have easily caught fire, producing sulphur dioxide that would have reached Earth's surface as acidic rain. Geologic evidence also exists that a sizable asteroid or comet hit Pangaea at this time.

THE MESOZOIC ERA

After the Permian extinctions, life was slow to recover, but in a few million years it did and what resulted was not just a new period, the Triassic, but a new era, the Mesozoic (248–65 million years ago). That era is commonly known as the Age of Reptiles because of the creatures that dominated its history. Most sensational of all the reptiles were the dinosaurs, which first appeared early in the Mesozoic and flourished until its cataclysmic end.

The Triassic Period

The climate of the Mesozoic was probably warmer than at present. The Paleozoic ice sheets disappeared. Crocodiles, which evolved in this era, could be found even in the high latitudes of North America.

During the Triassic, the continents were all joined together as the supercontinent Pangaea, which had formed completely about 220 million years ago. Since then, Pangaea has broken apart into smaller continents. This process is called continental fragmentation and has continued to the present day.

Swamp forests grew along the shores in the tropics, but further inland conditions were arid. Deserts and dry climates were widespread. In North America, the marine sedimentary record of the Triassic is the most meagre of any period.

Among animals, reptiles were best suited for the arid climate and they flourished and diversified. One group of reptiles, the diapsids, included the ancestors of dinosaurs, crocodiles and birds, as well as modern-day lizards and snakes. The first turtles and tortoises emerged in the Triassic and, late in the period, the first dinosaurs as well, including the

plant-eating, long-necked Massospondylus and the carnivorous Coelophysis and Eoraptor. But dinosaurs did not yet achieve the dominance they would later. Also appearing late in the Triassic were the first mammals.

In the ocean, oysters, lobsters and sea urchins evolved. Large reptiles took to the sea as predators, including dolphin-like ichthyosaurs and lizard-like nothosaurs.

The unity of the supercontinent meant that a successful species could spread across the entire world island without being stopped by an ocean. Originating in Gondwana, the dinosaurs were able to spread throughout Pangaea before tectonic forces split the supercontinent apart.

for many species, but a boon to the dinosaurs. The ecological niches for medium- and large-bodied terrestrial animals were emptied, creating room for new occupants. The dinosaurs were about to become the rulers of the Mesozoic.

The Jurassic Period

At the beginning of the Jurassic period (206–144 million years ago), the climate was still warm and dry. Then it grew rainier and wetter; the land became greener and deserts shrank. Temperatures rose and forests spread, thick with ground and tree ferns, cycads and gingkoes. The trees included the ancestors of present-day redwoods, pines and cypresses. Even high latitudes had significant plant cover.

THE PALEOZOIC ICE SHEETS DISAPPEARED. CROCODILES, WHICH EVOLVED IN THIS ERA, COULD BE FOUND EVEN IN THE HIGH LATITUDES OF NORTH AMERICA

At the end of the Triassic came another mass extinction. The cause is not known, although a comet impact that created the 97-km-wide (60-mile) Manicouagan impact structure in Quebec, Canada, may have played a part. The existence of smaller craters in Canada and France is evidence that the comet fragmented upon arrival, spraying the planet with a deadly rain of projectiles. The mass extinction at the end of the Triassic was a catastrophe

Tectonic shifts may have been behind the climate changes. During the Jurassic, the supercontinent Pangaea was splitting into pieces, which were slowly moving toward their present-day locations. North America separated from South America. A rift opened between North America and Eurasia and flooded with seawater, becoming the proto-North Atlantic Ocean. The South Atlantic appeared as South America began to divide from the African continent.

FOSSILS AND CLIMATE

Much of what we know about the history of life – and of Earth and its climate – comes from fossils, the remains or traces of living things found in rocks. Soft parts of animals and plants decay easily and are rarely preserved, but hard parts, such as bones, teeth, claws and shells, last longer. They can become fossilized when water containing minerals soaks into their pores, and the minerals replace the original material, duplicating its details.

Fossils can also form when a once-living plant or animal leaves behind an impression or mould in sediment that turns to rock, or when it is caught in sap that over time becomes amber. Relatively few fossils survive from Precambrian times, because Precambrian organisms had not yet evolved the hard parts that are more readily fossilized. The evolution of hard parts began to emerge in the Cambrian period and from then on the fossil record is rich and varied.

Fossils have uses beyond helping scientists learn about species that lived in the distant past. To nineteenth-century scientists, fossils were useful for ascertaining the relative age of rocks. Sedimentary rocks – those rocks which form from particles that are then cemented or compacted into stone – are laid down in layers, with a more recent layer on top of an earlier layer. Therefore, if a certain type of fossil were always found in a lower layer, that fossil could be judged to be relatively older than those in higher layers. If found elsewhere in the world, that same fossil could identify the rocks around it as being from that period. Using fossils in this way, geologists constructed a geologic timescale of eras, periods and epochs. Some of the names of these divisions refer to an interval's place in the chronology of life. For example, 'Paleozoic' comes from the Greek *palaios,* 'ancient', and *zoe,* 'life', while 'Mesozoic' and 'Cenozoic', also Greek-derived, mean respectively 'Middle Life' and 'Recent Life'. Other terms refer to a geographic location where fossil-bearing rock formations from that time were first found or were particularly well displayed. For example, the Cambrian period refers to Cambria, the Roman name for Wales, and the Jurassic period to the Jura Mountains of France and Switzerland. Still other names have other origins. The Carboniferous period refers to the coal-bearing rocks that are the remains of the forests that predominated during this time.

Determining Fossil Age

At first, this geological timescale, developed from studying fossils found in sedimentary rock layers, showed only relative dates: that the Silurian period was before the Devonian period, for example. In the twentieth century, radiometric dating made it possible to assign absolute dates to rocks and the fossils they contain.

LEFT *Fossilized head and neck of a Tarbosaurus bataar, a carnivorous dinosaur. This species inhabited Central Asia in the late Cretaceous period, about 75 million years ago.* UPPER LEFT *Mosquito embedded in amber, which is fossilized resin produced by now extinct coniferous trees during the Jurassic period.*

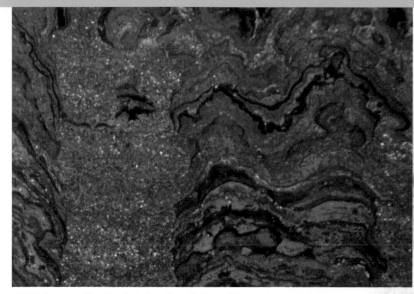

ABOVE *Stromatolite fossil dating from 2,400 million years ago. Among the oldest organic remains to be found, Stromatolites are large, stony masses composed of layers of blue-green algae.*

ABOVE *A paleontologist excavates the fossilized skeleton of an unidentified dinosaur. Photographed at the Dinosaur National Monument, Utah, USA.*

This kind of dating is based on the existence of radioactive isotopes, forms of elements that decay to produce other elements. Uranium-238, for example, breaks down to form lead-206. Scientists know how quickly certain radioactive isotopes decay and they use this knowledge in dating rocks. It takes 4,500 million years for half of a given quantity of uranium-238 to decay into lead-206; that is, uranium-238 has a half-life of 4,500 million years. With this radiometric clock, the age of a rock containing uranium-238 can be determined by finding out how many atoms of each material it contains. Other radioactive isotopes that are commonly used in dating rocks include uranium-235 and potassium-40. Having numerous isotopes available, with different half-lives, allows accurate dating over many timescales. For example, carbon-14, with its half-life of 5,730 years, is used in dating organic material.

Reading the DNA

Molecular biology offers another way of investigating life in the past. By comparing the DNA of organisms from two species, scientists can establish when the

ABOVE *A fossilized fern that dates from the early Cretaceous period, found in the Antarctic. The polar climate was warmer during this period than it is today.*

two organisms last shared a common ancestor. This is possible because random changes in a given DNA sequence generally happen at a regular rate, functioning as a molecular clock. By measuring differences in a shared DNA sequence in two organisms, scientists can discover how many changes have occurred since the organisms' ancestors began to diverge into distinct species and therefore how long ago that divergence happened.

Discoveries about prehistoric life, through whatever means they happen, are important to reconstructing prehistoric climates. Much of what we know about past climates depends on information gleaned from the types of plants and animals that predominated in a given period. For example, the presence of crocodile fossils even in high latitudes of North America indicates the environmental influence of the warm Mesozoic climate. Sometimes the fossilized organisms influenced climate themselves. The animals of the Cambrian period had hard parts, such as shells and exoskeletons, that were more likely than soft parts to survive in the fossil record. Made of calcium carbonate and other materials, these hard parts amounted to a new carbon sink, a reservoir for carbon storage that would help keep atmospheric carbon dioxide and climate in balance.

India splintered off from what had been Gondwana, the southern continent, and Antarctica and Australia, which were attached, split off from Africa.

The breakup of the supercontinent led to higher sea levels and the flooding of low-lying areas with inland seas. The new oceans came with new ocean currents and wind patterns, affecting regional climates. The increased tectonic movement was also linked to greater volcanic activity, which led to higher carbon dioxide levels in the atmosphere. The resulting boost to the greenhouse effect may have contributed to the elevated temperatures of the Mesozoic.

The moist, warm climate of the Jurassic was a spur to dinosaur evolution, encouraging the development of herbivores that could live off the lush plant life. The largest of these herbivores, known as sauropods, grew to enormous sizes – some, like Argentinosaurus, as big as 36.6 m (120 ft) long and 100 tonnes in weight. Standing on four column-like legs, they were the largest land animals the world has ever known. Carnivores, meanwhile, were evolving to take advantage of the feast represented by the new herbivores.

Hunting the herbivores were the theropods, relatively lean and fast two-legged dinosaurs. Equipped with fearsome teeth and claws, they came in many sizes. Some theropods, such as Allosaurus, were 10.1 m (33 ft) long. Others like the Coelurus were 6.8 m (6.5 ft) long. The giant dinosaurs get all the attention, but dinosaurs came in many sizes, down to some that were the same size as a crow. Some of the small theropods – those 20.3–50.8 cm (8 to 20 in) in length – were developing into a new kind of creature, the bird. It is now generally believed that birds evolved from dinosaurs. By the late Jurassic, when the primitive bird Archaeopteryx appeared, birds had emerged as animals covered with feathers and capable of flight.

The niches for small terrestrial vertebrates were occupied not just by birds and dinosaurs, but mammals as well. During the Jurassic, mammals were small, furry, mostly nocturnal creatures, similar to rats and shrews. They probably ate insects, lived in burrows and reproduced by laying eggs, like the dinosaurs.

At the end of the Jurassic, there was a minor extinction. A third of all marine species died and many land species were affected.

The Cretaceous Period

At the end of the Jurassic and the beginning of the next period, the Cretaceous (144–65 million years ago), a general cooling occurred, with cold winters at high latitudes. Yet by the mid-Cretaceous, around 100 million years ago, global climate was extraordinarily warm. The average global temperature was 6°C to 12°C (11°F to 22°F) warmer than at present. There were no icecaps at the poles, and in the Arctic and Antarctic circles forests grew and dinosaurs grazed. The oceans, too, were unusually warm.

The movement of the continents probably contributed to the hothouse climate. The continents were still spreading apart, moving more rapidly than today, with what may have been the most vigorous plate tectonics in Earth's history. The Rocky Mountains began to form during this period. An east-west seaway opened at the equator, bringing warmth and moisture to middle latitudes.

The volcanism associated with plate movements may also have been a major cause of the warming trend. Volcanoes were unusually active in this period, spawning, among other structures, an undersea lava plateau in the western pacific called the Ontong Java, which is about two-thirds the size of Australia. The volcanoes spewed carbon dioxide. The atmospheric level of that gas was from seven to ten times the present carbon dioxide level. The carbon dioxide trapped heat, keeping the Cretaceous warm.

Swamp forests proliferated—over time the dead plant material from these swamps formed extensive coal deposits. At sea, micro-organisms abounded, eventually settling to the bottom and turning into oil deposits; about 60 per cent of all known oil reserves are from the Cretaceous.

Many new varieties of dinosaurs evolved. Ornithopods were particularly abundant in the Cretaceous. These herbivores were represented by such creatures as the 5.5-tonne Iguanodon and the hadrosaurs, or duck-billed dinosaurs, which were found as far towards the poles as 80 degrees north and south. Even though the global climate was warm, there were still seasons and these would have been pronounced at such high latitudes. The hadrosaurs may have migrated like contemporary caribou, heading towards warmer climates in summer and winter.

Through much of the Cretaceous world, new herbivores such as the hadrosaur displaced the giant sauropods, which became rare. The new breed of herbivores led to the evolution of new predators. The greatest of these was Tyrannosaurus, which gave its name to a whole family of tyrannosaurs, such as Albertosaurus. Tyrannosaurus was 12.2 m (40 ft) long, with rows of 15 cm (6 in) teeth. Among the smaller carnivores were the dromaeosaurs, sickle-clawed, bird-like hunters that included Velociraptor.

RIGHT *An illustration of a comet hitting Earth. The clouds of dust and water vapour resulting from such an impact would lower global temperatures and cause mass extinctions. Such events are rare, but have occurred in the past and will almost certainly occur in the future.*

Angiosperms, or flowering plants, evolved in the middle of this period, along with insects that became specialized in carrying pollen from flower to flower, assisting the plants' reproduction.

The spread of the angiosperms may have been aided by forest fires, which were probably frequent in the Cretaceous, and by the havoc wreaked on local environments by huge herds of plant-eating dinosaurs. There were many opportunities for the fast-growing, weed-like angiosperms to re-populate scorched or trampled habitats.

At the end of the Cretaceous came the best known mass extinction in Earth's history, the one that killed off the dinosaurs. Although not as severe as the Permian extinction, it eliminated almost all the planet's large vertebrates, along with most plankton, many tropical invertebrates and many land plants. There have been numerous theories as to the cause of this calamity – known as the Cretaceous/Tertiary or K/T extinction – but the most widely held theory today assigns principal blame to an extraterrestrial impact.

At the close of the Cretaceous, 65 million years ago, an asteroid 9.7 km (6 miles) in diameter blasted into Earth, producing a crater more than 160 km (100 miles) wide. The Chicxulub crater remains in the northwest Yucatán peninsula of Mexico (see illustra-tion on pages 50–51). Other evidence, found in rocks of this age, includes a worldwide layer of iridium, a metal that is rare on Earth but abundant in meteorites, and which was probably scattered by the blast. The impact released 100 million megatons of energy which produced tsunamis, hurricanes, earthquakes, acid rain, toxic debris and possibly a global heat pulse that set off fires worldwide. But the most devastating effect would have been a cloud of dust suspended in the atmosphere, creating global darkness for several weeks or months, then twilight conditions for several more months. The cloud of dust would have brought on frigid conditions and shut down photosynthesis at sea and on land, causing the collapse of food chains. Global warming, lasting thousands or tens of thousands of years, followed. This warming, which raised the average global temperature about 10°C (18°F), resulted from excess carbon dioxide blown into the atmosphere by vaporized limestone and a burned biosphere.

This impact theory has the virtue of helping to explain why mammals survived while dinosaurs did not. Small and rat-like, late Cretaceous mammals were mostly omnivores, able to live off whatever food was available and not requiring it in large quantities. They were warm-blooded and, in many cases, nocturnal, preadapting them for the time of cold and darkness that followed the impact.

THE CENOZOIC ERA

With the destruction of the dinosaurs, a new era, the Cenozoic (65 million years ago to present), began. Mammals became the dominant land vertebrates, giving the era its common appellation, the Age of Mammals.

The Tertiary Period

The first period of the Cenozoic era was the Tertiary (65–1.8 million years ago). The direct effects of the asteroid impact at the end of the Cretaceous were relatively short-term and Earth's climate reverted fairly rapidly, in geologic terms, to its prior state. A million years later, the climate was similar to that which had prevailed before the impact – warm. The warmth persisted or increased in the early Cenozoic

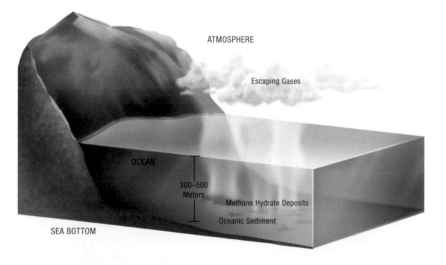

ABOVE *A methane explosion. Fifty-five million years ago, the movement of tectonic plates released methane from seafloor deposits that had been buried under the sea floor into the atmosphere, raising the global temperature.*

era. The poles were free of ice, and ferns grew in Antarctica and the forests north of the Arctic circle.

Mammals, small and five-toed at the beginning of the Tertiary period, diversified in many directions. Ancestors of many modern groups appeared, including elephants, primates, rabbits, shrews and whales. Overall, mammals became larger, their brains bigger, their teeth more specialized to suit specific diets and their limbs more specialized for certain ways of life.

During the first epoch of the Tertiary, called the Paleocene (65–54.8 million years ago), angiosperms became the dominant land plants. The evolution of plants moved in tandem with that of the birds and mammals that fed on them. Fruits, for example, evolved as a mechanism by which plants use birds and mammals to transport their seeds for them. Birds and mammals, attracted by the colours and aromas of fruits, eat them, pass the hard seeds through their digestive systems and drop the seeds, complete with fertilizer, in new places where they can sprout. Some birds and mammals also evolved specialized beaks or teeth to crack open seeds and nuts and they added them to their diets.

In the role of large carnivores, mammals could not compete with birds. These included large, flightless birds such as Diatryma, which stood 1.8 m (6 ft) tall and had a pow-

RIGHT *Artwork showing the Chicxulub crater on the Yucatan Peninsula, Mexico, shortly after impact. This crater was created by an asteroid 65 million years ago, and measures 180 km (112 miles) across. This calamity and its resulting global climate change brought on the end of the Cretaceous period, with the extinction of the dinosaurs and seventy per cent of all Earth's species. The Cenozoic era followed, also known as the Age of Mammals.*

erful beak and sharp claws. At sea, new kinds of sharks and bony fish replaced the large marine carnivores, such as ichthyosaurs and plesiosaurs, which had become extinct.

In the early to middle parts of the next epoch, the Eocene (54.8–33.7 million years ago), the global climate remained warm. In fact, the early Eocene, up to about 50 million years ago, was the warmest period in the Cenozoic era. Tropical conditions stretched 10 to 15 degrees of latitude further towards the poles than at present. Increased levels of methane, a greenhouse gas, were responsible for the warming trend. Fifty-five million years ago, tectonic activity released frozen methane deposits from beneath the sea floor into the atmosphere, raising the global temperature.

The continents moved closer toward their present arrangement. In the early Eocene, North America and Greenland split from Europe, opening a link between the Arctic and Atlantic oceans. Eurasia and Africa approached each other. About 45 million years ago, India collided with Eurasia. The crash turned India into a subcontinent and raised the Himalayas, the world's highest mountain range. Australia and Antarctica, still connected, broke away completely from South America, then drifted away from each other. Antarctica moved over the South Pole.

For more than 20 million years, Earth basked in the early Eocene warmth. Lush tropical forests spread. Mammals extended their dominance, venturing even beyond land. By the early Eocene, the first bats had taken to the sky and whales had gone to sea. An early whalelike animal was Ambulocetus natans, which had large hind legs that allowed it to function on both land and in water.

In Africa, the first proboscideans, ancestors of the elephant, evolved. Small ancestors of the horse roamed Europe and North America. Camels appeared in North America but had not yet spread to Asia and Africa. Rodents, which had probably existed since the Paleocene, increasingly became the principal group of small animals. Birds were evolving as well. Ducks, pelicans and herons were widespread.

Primates, which had existed in Africa as early as the Paleocene, had evolved into two groups by the early Eocene, one more closely related to modern monkeys and apes, the other to lemurs.

By the late Eocene, monkeys had evolved. The early primates probably ate fruit in tropical forests. The balmy climate of the Eocene was perfect for the evolution of the group that would lead to human beings.

From about 50 million years ago, a significant cooling trend began, one that would mark the rest of the Cenozoic era. Known as the Cenozoic climate decline, it was not a steady fall, but one that included both long periods of relative stability and short periods of rapid change. The temperature dropped especially sharply at 50 million years ago and again at 38 million years ago. The latter drop may have

contributed to an extinction event. By 38 million years ago, ice was developing at the South Pole.

The causes of the extended cooling were many and are still not completely understood. They included the movement of continents to their present positions, which affected ocean circulation and ice formation, and the raising of mountain ranges, which affected atmospheric circulation. Decreasing levels of carbon dioxide in the atmosphere also may have caused the cooling.

By the Oligocene epoch (33.7–23.8 million years ago), a large ice cap was forming over the South Pole, increasing the planet's albedo (see page 22) and advancing the cooling trend – though Antarctica was not yet completely covered by ice. Oceanic deep-water circulation became more vigorous, with the deep water becoming colder. The drifting apart of South America and Antarctica, opening the Drake Passage between them, and the northward movement of Australia allowed a circumpolar ocean current to develop, isolating Antarctica from warming winds and currents.

Since the glaciers were tying up water in the form of ice, sea levels fell, exposing more dry land. Grass first appeared in significant quantity in this period. The tropical forests of the previous period retreated, while woodland and grassland environments spread.

The cool climate of the Oligocene was uncongenial to primates and may have restricted their geographic distribution. Nevertheless it was probably during the Oligocene that the first apes appeared in Africa, as hominoids split off from monkeys. The first human-like, or hominid, creature would not evolve until after the Oligocene.

CLIMATE AND HUMANKIND

Human Evolution and Climate

The influence of climate on human evolution and history has been profound. Changing climate drove our ape ancestors out of the forests and onto the grasslands. Climate fluctuations during the Pleistocene ice ages (1.8 million to 10,000 years ago), shaped the development of larger brain size and the spread of human beings throughout the world. Once the ice ages ended, climate played a part in the history of human cultures and civilizations, from the invention of agriculture to the rise and fall of empires.

ONTO THE SAVANNAH

Throughout history, climate has been cyclical. It does not move forever in one direction, but instead fluctuates between greater warmth and greater cold. Such fluctuations have been common since the earliest hominids split from their ape-like ancestors; indeed, the past few million years have seen climatic oscillations probably as violent as at any time in Earth's history.

Climatic cycles occur in complex patterns. In fact, the grand glacial epoch known commonly as the Ice Age was actually a series of short ice ages alternating with warmer, interglacial periods. The scientific term for this period is the Pleistocene glaciations. These ice ages, which occurred during the Pleistocene epoch, lasted from 1.8 million to 10,000 years ago.

In more recent times, the Little Optimum, also called the Medieval Warm Period, was followed by the Little Ice Age, a cold period that lasted until the nineteenth century, when it was succeeded by the current period of increasing warmth. The main difference today is that industrial activity is contributing to the changing cycles of climate, so that the warming trend is much stronger than it would have been without human influence.

The story of climate and humankind begins with the evolution of the first hominids—the first human-like animals. This happened in the Tertiary period of the Cenozoic era, when the ancestral line that leads to modern humans split from the rest of the ape family. The split occurred eight to five million years ago, either late in the Miocene epoch (23.8–5.3 million years ago) or early in the Pliocene (5.3–1.8 million years ago), during a period of general cooling.

Climate and Evolution

About 13 million years ago, during the Miocene epoch, the southern ice cap spread rapidly, covering all or most of Antarctica. The deep water became colder, by 4.4°C to 5.5°C (8°F to 10°F). Ice developed at the North Pole. The previous epoch, the Oligocene, had been cool, but the global climate was now even colder.

At high latitudes, climates became more seasonal, with greater swings of temperature between seasons. Around the world, climates became drier. Prairies spread, with grasslands covering more of North and South America, Europe, Asia, Africa and Australia than during the last epoch. The Mediterranean Sea may have dried up and refilled several times between 7.25 and 5.5 million years ago. At work in the Mediterranean were a combination of tectonic forces closing and reopening the Strait of Gibraltar and glaciers lowering sea levels.

Tectonic shifts contributed to the drying of East Africa as well. At about this time, as part of the ongoing splitting of Africa, two crustal plates were moving apart, deepening the Rift Valley that runs north and south through Ethiopia, Kenya and Tanzania. Territory east of the rift climbed higher and therefore became cooler and drier. New mountains rose on either side of the Rift Valley, changing air circulation patterns. The result was greater dryness in eastern Africa, but persistent moistness to the west.

As climate changed five to six million years ago,

RIGHT *Ol Doinyo Lengai mountain, a volcano in Tanzania overlooking the Great Rift Valley. Many of the oldest anthropological finds have been made in this region.* PREVIOUS PAGE *Clockwise from upper left: Homo habilis skull found in Tanzania, believed to be about 1.8 million years old; Native American drawings showing a hunting scene; dust storm in Depression-era Oklahoma, 1936; ancient Egyptian art on papyrus.*

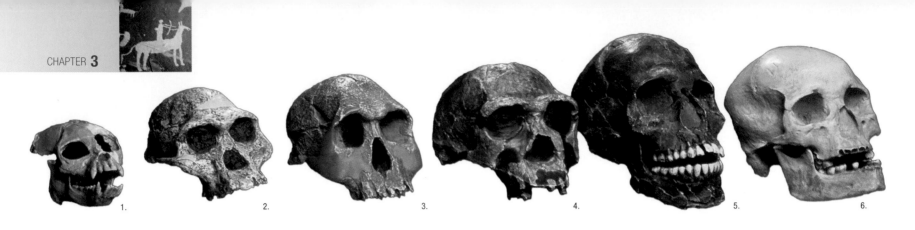

ABOVE *Progression of skulls belonging to ancestors and relatives of modern humans.*
Left to right: 1. Proconsul (a primate from 23 to 15 million years ago)
2. Australopithecus africanus (3 to 1.8 million years ago) 3. Homo habilis

(2.1 to 1.6 million years ago) 4. Homo erectus (1.8 to 0.3 million years ago)
5. Modern human (Homo sapiens sapiens) around 92,000 years old 6. French
Cro-Magnon human from around 22,000 years ago.

savannah, or flat grassland with scattered trees, spread in eastern Africa, reducing that region's dense forests. In the Sahara, plants adapted to arid conditions. Around the world, mammals grew larger and diversified. Grazing herbivores such as antelopes spread, taking advantage of the expanding grassland.

Theories differ as to why and how our first ancestors – the first hominids – evolved from ancient apes. However, they all point towards changes in climate and subsequent changes in vegetation as

that their knees and hips changed in ways that reflected this increasing bipedalism.

The bipedal stance offered many advantages to these hominids, such as allowing them to see predators from a distance over tall grass. Their hands were also freed to carry food or infants. Hominids were able to travel faster and more easily across the open, treeless savannah. The bipedal stance also reduced heating of the body by as much as 60 per cent. This change helped to decrease the risk of overheating and dehydration

Beyond the Savannah Hypothesis

In the 1990s, other views besides the savannah hypothesis became widely considered as ways of understanding the link between climate and human evolution. The American paleontologist Elizabeth Vrba of Yale University introduced the turnover pulse hypothesis, which connects critical events in human evolution to broader pulses of animal evolution that are driven by climate change. Stages in hominid evolution are linked to rapid change, or turnover, of antelopes, rodents and other animals. So far, some fossil evidence has supported Vrba's view, but there are many challengers.

Another view, the variability selection hypothesis, devised by the American paleoanthropologist Rick Potts of the Smithsonian, stresses the importance of large-scale fluctuations in climate and environments. According to Potts, hominids evolved when they were challenged to adapt to highly variable, unstable, inconsistent environments. During the Pleistocene ice ages, there were repeated changes in vegetation and landscape that led to the breakup of habitats and the separation of populations.

English anthropologist Robert A. Foley (UK) of Cambridge University argues that climate change influences evolution through extinction, not by

ACCORDING TO A THEORY KNOWN AS THE SAVANNAH HYPOTHESIS, SOME OF THE APES THAT LIVED IN THE AFRICAN FORESTS WERE FORCED ONTO THE OPEN SAVANNAH

driving the emergence of the first hominids.

According to a theory known as the savannah hypothesis, some of the apes that lived in the African forests were forced onto the open savannah. There they evolved an upright stance and began to walk on two legs. Fossil evidence shows

on the open savannah. By standing upright, hominids reduced the amount of their body surface that was directly exposed to the Sun. Also, more of the body was now kept away from the hot ground. Instead, their upright stance meant that hominids were exposed to cooling breezes.

Bonobos and Chimpanzees

Climate was involved not only in the evolution of humans, but also in that of our closest relatives, bonobos and chimpanzees. All three species evolved recently from a common ancestor, making them so genetically similar that British geographer and author Jared Diamond has called human beings 'the third chimpanzee' in this group.

The bonobos, or pygmy chimpanzees, live in a part of the African rainforest south of the Congo River in Congo-Kinshasa. Their social groups are dominated by females and are generally peaceful. Sexual contact – heterosexual and homosexual – is freely used to seal social bonds and resolve conflicts, as if their motto were 'Make love, not war'. Chimpanzee society, by contrast, is dominated by males who do make war and have less frequent sex. Both bonobos and chimpanzees eat fruit as a staple, but chimpanzees are much more likely to supplement it with meat obtained by hunting.

Behind the differences between the two primates may be climate. After hominids split off from the rest of the apes, bonobos stayed in a relatively moist forest habitat, while chimpanzees branched out, becoming adapted to more open, dry environments. Today, chimps live in a range of habitats, from humid rain forests to dry grasslands with few trees, in a broad swathe of tropical Africa. Bonobos, by keeping to their old climate, may have conserved much older characteristics – perhaps behaviours similar to those of the common ancestor of bonobos, chimpanzees and humans.

Some scientists conclude that bonobos and chimpanzees thought to be our two closest relatives, are distinct products of different climates.

ABOVE *Photograph of a bonobo, also known as a pygmy chimpanzee, in the Democratic Republic of Congo. These close relatives of both humans and chimpanzees live in female-dominated, generally peaceful social groups.*

LEFT
Chimpanzees in Kivu province, Democratic Republic of Congo. Chimpanzee society, which is male-dominated, is less peaceful than that of the bonobos. Some scientists believe the difference may result from the two species living in distinctly different climates. Chimpanzees have adapted to living in a wider range of habitats and may have changed certain behaviours as part of that adaptation.

driving the development of new species. He maintains speciation is more dependent on local competitive conditions (although competitive conditions are influenced by climate). Other scientists have argued the importance of bottleneck effects, in which the near extinction of a species leaves a reduced gene pool and a small population, prompting rapid evolution.

Whatever model best explains their evolution, both fossil and biochemical evidence – the latter based on measuring differences in the DNA of two species – show that the last common ancestor of hominids and apes lived between five and eight million years ago.

The first hominid was part of a larger story of ape evolution. During the Miocene epoch, before hominids developed, apes evolved into separate groups. One kind of ape was the common ancestor of chimpanzees, the bonobo (see sidebar, left) and humans. Living in Africa, it was probably an omnivorous, chimp-like creature able to walk upright for short periods, but usually walking on its hind legs and front knuckles.

About four million years ago, a group of hominids called australopithecines emerged, members of the genus Australopithecus. In most respects, they resembled apes. They were small herbivores with ape-like faces, and brains that were about one-third the size of a present-day human brain. But they walked on two legs, which made them hominids.

Though hominids had evolved upright locomotion, their brains were not significantly larger than those of apes. Their brains would evolve, however, because of a change in the climate.

THE ICE AGES

Twenty thousand years ago, ice filled the basin that now holds North America's Great Lakes. Sea levels were so low that the British Isles were part of the European mainland – and ice covered Britain, too, as far south as Oxford. In some places the great ice sheets were as much as 5 km (3 miles) thick. To the immediate south of the ice margins lay tundra – cold, dry, treeless areas with grass and other plants that were grazed by mammoths and giant deer. Winds howled around them, more vigorously than winds today. In the coldest periods, temperatures in Europe were 8°C to 11°C (14°F to 20°F) colder than at present.

A Frozen World

The Pleistocene glaciations, or ice ages, began between two and three million years ago, while the australopithecines still roamed the savannahs of Africa. Glaciers covered large regions of the planet during each ice age. These ice sheets largely receded during the interglacial periods, only to expand again to start a new ice age. It was not the first time Earth had gone through such a glacial epoch, but it was the first time human beings were there to witness it. In fact, it was during the Pleistocene ice ages that modern humans first evolved.

Paradoxically, the glaciations may well have been triggered by a development in the steamy tropical region of Panama. About 3.2 million years ago, during the Pliocene, the Isthmus of Panama formed between North and South America as a result of the collision of the Pacific and Caribbean tectonic plates. The pressure and heat of the collision raised

Types of Glaciers

There are two principal varieties of glacier. Ice sheets, or continental glaciers, are large-scale masses of ice that cover vast areas of land near the poles. Flowing outward from the centre in all directions, they conceal the whole landscape except for the highest peaks. At present, the only ice sheets are in Greenland and Antarctica. The Greenland ice sheet covers more than 1.6 million sq km (1 million sq miles) and has an average thickness of 1.6 km (1 mile). The Antarctic ice sheet occupies over 12,874,400 sq km (8 million sq miles) and has an average thickness of about 4 km (2.5 miles). Together, the areas covered by these continental glaciers represent nearly 10 per cent of Earth's land area.

Alpine or mountain glaciers are smaller-scale masses of ice that fill the valleys of high mountains, where streams once ran. Like streams, they are usually longer than they are wide. They occur in the Andes, Alps, Himalayas and other mountain ranges.

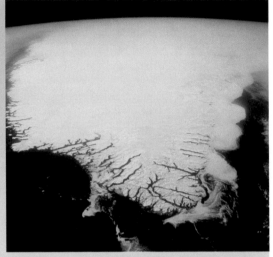

ABOVE *The southern tip of Greenland, showing a massive ice sheet glacier.* BELOW *Wright Glacier near Juneau, Alaska, USA, an example of an alpine, or valley glacier.*

underwater volcanoes, some of which thrust above the waves and formed volcanic islands. The sea floor was elevated and ocean currents brought in sediments, continuing the land-building task until the isthmus was complete.

The new land bridge allowed migration of plants and animals from one American continent to the other. More importantly, it altered global climate by changing ocean currents. The flow of water between the Atlantic and Pacific oceans stopped. Atlantic currents were forced northwards, eventually forming the Gulf Stream. This flow of warm waters from the Caribbean to the northeast Atlantic helped make the climate of northwestern Europe warmer than it otherwise would have been.

Most important for the start of an ice age, the newly formed Gulf Stream carried great quantities of moisture to the far north. This led to more snowfall in northern areas and the growth of glaciers in Eurasia and North America. Ice sheets as high as mountain ranges arose over Scandinavia and Canada. In a classic positive feedback effect, the ice sheets increased Earth's albedo. More solar radiation was reflected back into space, causing further glaciation. About this time, tectonic forces were forming the Tibetan Plateau and the mountains of western North America. Higher mountains tended to raise albedo levels, because the mountains held winter snow. Mountains also shifted

RIGHT *The world's fifth highest peak, called Makalu, 8,462 m (27,765 ft), in the Himalayas, Nepal. Mountain glaciers like those shown here on Makalu contributed to ice age conditions.*

storm tracks southwards, increasing winter snowfall in other areas. The result of these conditions was that glaciation accelerated from 3.1 to 2.5 million years ago. The ice ages were underway.

The ice ages consisted of a series of about 20 cycles of cooling and warming, of glacial and interglacial periods. In addition, during the glacial periods, there were occasional temperate intervals called interstadials. Each cycle of cooling and warming lasted for about 100,000 years. Most of that span occurred during the Pleistocene epoch (1.8 million years ago to 10,000 years ago), the first epoch of a new period, the Quaternary period.

The Milankovitch cycles (see page 30) were probably one cause of the alternation between glacial and interglacial episodes. But other factors were at work. Changing levels of carbon dioxide brought about greenhouse effects (high levels of carbon dioxide producing warming) and reverse-greenhouse effects (low levels, producing cooling).

During the ice ages, ice left its mark on about 30 per cent of Earth's land surface, with the Northern Hemisphere having about twice as much glacial ice as the Southern Hemisphere, where there was less land surface.

Though the glaciers were mostly in higher latitudes, they had effects worldwide. They cooled the global climate and, in most of the world, caused increasing dryness. With moisture locked up in the glaciers, sea levels fell worldwide and precipitation declined by about half in polar regions. In many places, deserts spread. There were polar deserts in Europe, very cold with little precipitation, and the Sahara engulfed the whole of North Africa. The rainforests of central and western Africa shrank, and the savannah of East Africa expanded again. That alteration triggered a momentous evolutionary change for hominids.

The Ice Ages and Hominids

By around 2.5 million years ago, near the start of the Pleistocene ice ages, hominid brain size began increasing. The new kinds of hominids were the first members of the genus Homo, our own genus, which emerged from australopithecine ancestors. Like earlier hominids, Homo was bipedal. But it also had a large cranial capacity, relatively small teeth and the ability to make tools.

One of the earliest examples of Homo was Homo habilis (skillful man), evidence of which was first found in Tanzania in 1960. At about 705 cubic cm (43 cubic in), the brain of Homo habilis was bigger than an australopithecine's but little more than half the size of a modern human's. Its tools included stone scrapers, cutters and choppers.

Scientists disagree about why hominids with larger brains emerged at just this time of changing climate. In all likelihood, it was not a coincidence. According to the variability selection hypothesis, the variable climate, with its alternation between long glacial and short interglacial periods, would have favoured individuals with the ability to adapt to the changing climate. In Africa, this would have meant adjusting every few thousand years from the cool, dry spells of the glacial episodes to the warm, wet spells of the interglacials. Greater intelligence affords just such an ability to adapt. More so than Australopithecus, Homo could handle the changing conditions that affected Africa as the glaciers advanced.

One of the more outstanding characteristics of Homo was that it became more carnivorous – finding meat either by scavenging or hunting. This provided nutrition even during dry spells. In turn, it speeded the evolution of bigger brains, because meat provided more calories for a mother nurturing the growth of her child. A larger brain proved useful in adapting to changing conditions, because it allowed the development of cooperation through culture – tools, strategies, symbols and beliefs.

ALTERNATION BETWEEN LONG GLACIAL AND SHORT INTERGLACIAL PERIODS WOULD HAVE FAVOURED INDIVIDUALS WITH THE ABILITY TO ADAPT TO THE CHANGING CLIMATE

Cyclical climate change during the ice ages probably forced or encouraged populations of Homo and other animals to migrate. By about two million years ago, for example, Africa was returning from its cool, dry climate to a slightly warmer, more humid state, making it easier for hominids and other fauna to expand away from the equator. Homo habilis expanded southwards into the southern Africa temperate area.

Homo Erectus and Homo Sapiens

During the Pleistocene epoch, a new species of hominid, Homo erectus (upright man), evolved in East Africa. It emerged about 1.8 million years ago and lasted until at least 300,000 years ago and perhaps later. At about 1.7 m tall (5.5 ft), it was taller than earlier hominids and had a bigger brain than Homo habilis, about 900 cubic cm (55 cubic in). It made tools, was social and may have used language. Its omnivorous diet included meat, though it is not clear whether hunting or scavenging was the pre-

dominant method of food gathering. This intrepid hominid moved out of Africa to colonize many parts of Europe and Asia, including China and Indonesia. By 500,000 years ago, it learned the use of fire – not necessarily how to create it, but how to find and

maintain it. This would have been a valuable skill in ice age conditions – a way of modifying the environment to make it warmer, as well as increasing the range of foods that could be eaten.

Despite the broad range of environments Homo erectus eventually called home, this species was biologically adapted to hot, arid conditions. This is clear from the fossil skeleton of a juvenile known as Turkana Boy, or the Nariokotome Boy. Found in 1984 near Lake Turkana, Kenya, it was either Homo erectus or a related species, Homo ergaster (workman man). The boy's skeleton was 1.6 m (5 ft, 3 in) tall, but he would have been over 1.8 m (6 ft) tall at maturity. His lanky body was adapted to hot, dry

RIGHT *Major hominid fossil sites. Hominid fossils have been found throughout Africa, Europe, Asia and Australia. Fossils of the most ancient human ancestors are found only in Africa, indicating that continent's status as the birthplace of humanity. Fossils of later hominid species, including H. erectus; H. heidelbergensis; H. neanderthalensis; and our own species H. sapiens, are more widely distributed.*

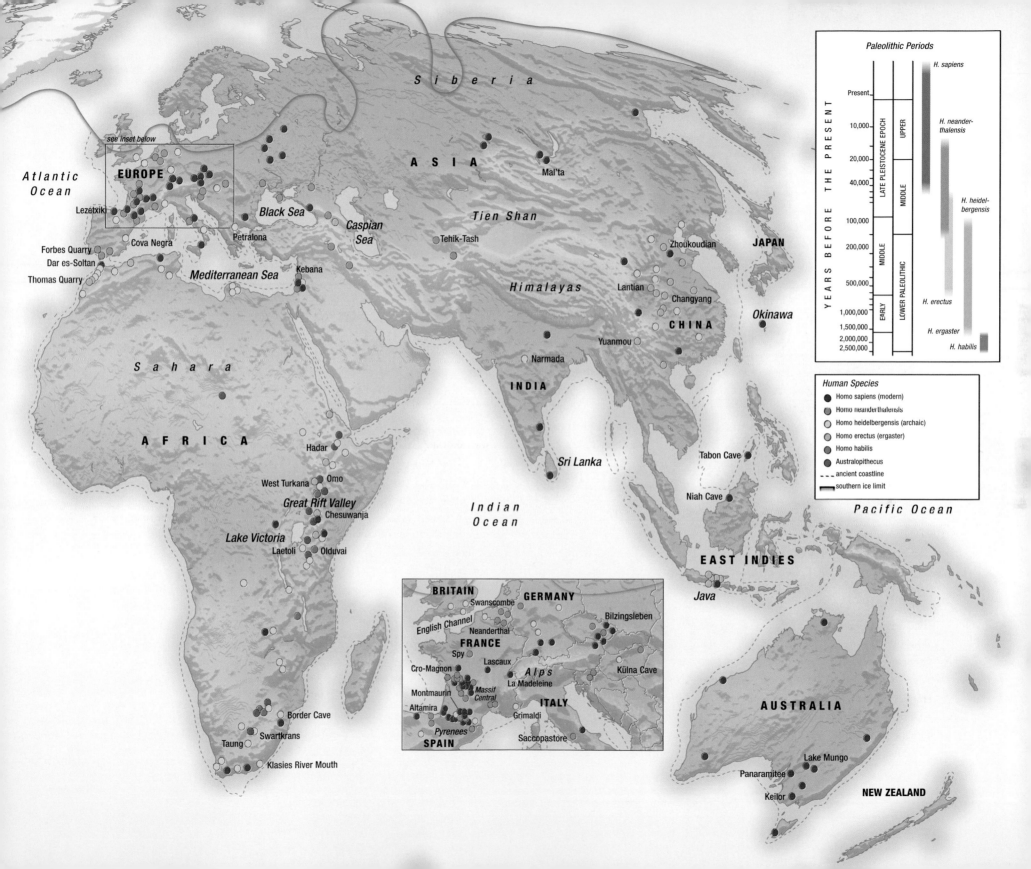

Siberia

ASIA

Mal'ta

Atlantic Ocean

EUROPE

see inset below

Lezetxiki

Black Sea

Caspian Sea

Petralona

Tien Shan

Zhoukoudian

JAPAN

Forbes Quarry

Cova Negra

Dar es-Soltan

Thomas Quarry

Kebana

Tehik-Tash

Himalayas

Lantian

Changyang

CHINA

Okinawa

Mediterranean Sea

Yuanmou

Narmada

INDIA

Sahara

AFRICA

Hadar

Omo

West Turkana

Sri Lanka

Indian Ocean

Tabon Cave

Niah Cave

Great Rift Valley

Chesuwanja

Lake Victoria

Laetoli

Olduvai

EAST INDIES

Java

Border Cave

Swartkrans

Taung

Klasies River Mouth

AUSTRALIA

Lake Mungo

Panaramitee

Keilor

NEW ZEALAND

Pacific Ocean

Paleolithic Periods

			H. sapiens

YEARS BEFORE THE PRESENT

Present
10,000
20,000
40,000
100,000
200,000
500,000
1,000,000
1,500,000
2,000,000
2,500,000

LATE PLEISTOCENE EPOCH — UPPER / MIDDLE

MIDDLE — LOWER PALEOLITHIC

EARLY

H. neander-thalensis

H. heidel-bergensis

H. erectus

H. ergaster

H. habilis

Human Species

- ● Homo sapiens (modern)
- ◑ Homo neanderthalensis
- ○ Homo heidelbergensis (archaic)
- ◐ Homo erectus (ergaster)
- ● Homo habilis
- ● Australopithecus
- - - - ancient coastline
- ▬▬ southern ice limit

Inset

BRITAIN

Swanscombe

GERMANY

Bilzingsleben

English Channel

Neanderthal

FRANCE

Spy

Lascaux

Alps

Külna Cave

Cro-Magnon

La Madeleine

Montmaurin

Massif Central

ITALY

Altamira

Grimaldi

Pyrenees

Saccopastore

SPAIN

conditions because it had more surface area for sweating and evaporation than a short, squat one. Turkana Boy had probably lost the thick coat of body hair of his ape ancestors, which afforded him more skin area for perspiration. Homo erectus also possessed a projecting bony nose, similar to that of modern humans and unlike the flat nose of apes and earlier hominids. In an arid climate where water loss through evaporation was a constant problem, the new nose was a water-retention system: it humidified the incoming air and retained moisture from the outgoing air.

Other hominid species followed Homo erectus out of Africa. In doing so, they took opportunities offered by the ice age pattern of glacial extension and retreat. These fluctuations opened and closed corridors for movement. One such corridor led north through the Sahara and over the Sinai Peninsula to the Levant (Syria, Lebanon, Israel and Jordan) and Europe. Another, a southern path, led east, across the mouth of the Red Sea to the Yemen, Oman and India. The northern path would have been practical only during an exceptionally warm, humid period, when the Sahara and Sinai were green and well-watered. During glacial times they would have been forbidding deserts. In contrast, the southern path would have been passable only during the most severe glacial times, when sea levels fell enough so that the mouth of the Red Sea became a narrow channel dotted with reefs and islands, accessible to rafting. Climate change determined when these routes were open and hominids took advantage of them when they were.

How early hominids gave rise to our own species, Homo sapiens (wise man), is a subject of controversy.

One widely accepted view is the 'Out of Africa' theory, which holds that all humans alive today are descended from one population of Homo sapiens that evolved in East Africa 100,000 to 200,000 years ago. The other principal theory, the multiregional model, contends that hominids in various parts of the world – Eurasia, Africa and perhaps Australia – evolved separately but in step to yield the human species as it now exists. Continuous contact between regions maintained the hominids as a single species.

Whatever theory is correct (and possibly there is some truth in both), fossil evidence shows that anatomically modern Homo sapiens had emerged by about 100,000 years ago. They gradually colonized the entire world, replacing earlier hominid populations, including Homo erectus and, in Europe, the Neanderthals.

Neanderthals and Cro-Magnons

By 120,000 years ago, the Neanderthals were living in Europe and the Middle East. Many scientists believe they were a subspecies of Homo sapiens, though they may have been a distinct species. Neanderthals were bigger than earlier hominids, with a large, rounded forehead. Their brain size was 1,500 cubic cm (92 cubic in), which was bigger than the modern human brain – 1,400 cubic cm

ABOVE *Cave paintings from the Cro-Magnon era at Lascaux, Dordogne, France. These paintings are believed to have been created between 15,000 and 10,000 BCE. The images of animals and of hunting demonstrate the Cro-Magnons' use of both sophisticated tools and symbols.*

(85 cubic in). They lived in a climate that was becoming colder as it approached the maximum frigidity of the last glacial episode. The polar ice cap reached down to southern England. The glaciers of the Alps stretched across central France and Germany.

Neanderthal bodies were adapted to that climate: short, squat and massive, with large nasal regions useful for warming and humidifying cold, dry air. They wore clothing made from animal skins, took shelter in caves, lived communally and may have believed in an afterlife, judging from evidence of funeral customs, including flowers placed in the graves of their dead.

A significant step in the evolution of human intellect may have occurred about 70,000 years ago, when the climate cooled globally and glaciation was

on the increase. The small size of the Homo sapien population could have facilitated its rapid evolution at this time. It has been speculated that modern humans' ability to use symbols dates to this point, though that is uncertain. What is certain is that a group of Homo sapiens called the Cro-Magnon people, who spread into Europe around 45,000 years ago, were able to use symbols. For a time they coexisted with the Neanderthals and may even have interbred.

About 40,000 years ago, the Cro-Magnons underwent a cultural explosion, during which their tools became much more sophisticated – including spearheads, harpoons and fishhooks – and they developed music, sculpture and painting. To cope with the ice age cold, they made sewn clothing and shelters. By 30,000 years ago, the Cro-Magnons had supplanted the Neanderthals.

End of the Pleistocene Ice Ages

Climate change is sometimes gradual, advancing imperceptibly from year to year, with a full-scale shift not tangible for hundreds, thousands or even millions of years. But sometimes climate can change with extreme abruptness, in a matter of decades or less. When that happens, it is called a Rapid Climate Change Event (RCCE), a profound reorganization of the climate system occurring in a very short time. Such an event happened as the Pleistocene ice ages came to an end.

About 14,500 years ago, a rapid warming trend began. Then, about 12,900 years ago, in what was called the Younger Dryas event, the climate cooled suddenly and dramatically. The reversal was dramatic, with the climate switching from temperatures much like today's to a full-blown ice age state in a matter of several decades or even less than a decade. The climate turned cold, dry and windy. Glaciers advanced; lakes in Africa shrank.

The Younger Dryas happened because the great ocean conveyor belt (see page 26) stopped. Melting ice from glaciers may have poured too much fresh water into the ocean, upsetting the salt balance on which the thermohaline circulation depends. This type of phenomenon, in which ice is deposited rapidly and massively in the ocean, causing climate change, is called a Heinrich event. Heinrich events occur about every 7,000 years during glacial periods. The cold that accompanies them is followed by a warm period.

The climate remained in a glacial state until about 11,600 years ago. Then the Younger Dryas ended as abruptly as it had begun. Temperatures rose again, by about 7°C (13°F), and glaciers receded. Forests spread across northern Europe. Moist, westerly winds blew from the Atlantic to the eastern Mediterranean, superseding the cold, dry, north-easterly winds of centuries past. The Pleistocene ice ages – or at least their latest episode – were over and the world settled into more or less the climate of historic times.

Sea Level and Migrations

When much of the world's water was locked in glaciers, at the peak of the last glacial episode, sea levels were nearly 112.8 m (370 ft) lower than at present. This exposed a land corridor, now called the

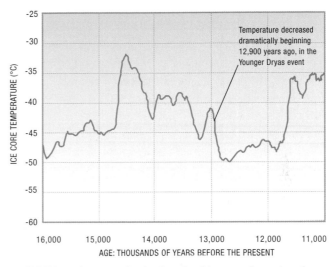

THE YOUNGER DRYAS EVENT

Temperature decreased dramatically beginning 12,900 years ago, in the Younger Dryas event

ABOVE *Data from a study of a Greenland ice core shows that about 14,500 years ago, after the ice ages, Greenland's temperatures rose rapidly. In the Younger Dryas event, temperatures dropped abruptly about 12,900 years ago, bringing back glacial conditions.*

Bering Land Bridge, nearly 1,600 km (1,000 miles) wide from Siberia to Alaska that lasted from 25,000 to 14,000 years ago. At various times during that period, Asian hunters walked eastward across Beringia (the name given to that land bridge and adjacent areas of Siberia and North America), pursuing grass-eating mammals such as mammoths, mastodons, giant bison and musk ox. Herbivores on the steppes or tundra of Beringia also included horses, caribou and camels, while carnivores included sabertooth cats, American lions, grizzly bears and wolves. The hunters would not have known they were on a land bridge, or that they were migrating from one continent to another. But when they reached Alaska and stayed in the Americas, they became the first Americans.

Evidence of the Ice Ages

Sea floor sediments (deep sea cores), ice cores, tree rings, pollen accumulations, lake bed sediments, fossils and paleomagnetism yield information used to investigate past climates. More specifically, scientists study the following evidence to increase their knowledge of the ice ages:

ISOTOPES: Shellfish use oxygen to build their calcium carbonate shells. Studying these shells in deep sea cores, scientists look for the relative abundance of two isotopes of oxygen, O_{16} and O_{18}. The latter, which is the heavier of the two, is more abundant if the seawater is cold. If more of it is found, that is evidence of colder temperatures. It also provides an index to past ice volumes, and therefore to glaciations. Seawater may be relatively enriched with O_{18} as a result of local temperatures or as a result of global shifts in temperature that lead to high levels of O_{16} ending up in glaciers. Oxygen isotopes are measured in ice cores as well as shells.

CARBON DIOXIDE: Air bubbles trapped in polar ice sheets will contain less carbon dioxide if formed at a time of widespread glaciation. Scientists are able to analyse the air bubbles to determine the concentration of carbon dioxide.

TREE RINGS: Every year, the new wood in the trunk of a tree forms a distinct ring. If growing conditions are favourable, the ring will be broad. If the tree is under stress – as from cold or drought – the ring will be narrow.

ROCKS: Glaciers move slowly, flowing downslope at a rate of anywhere from a few to many centimetres per day. As they move, they erode the land beneath them and shape it in distinct ways. The geological data left behind by a glacier that has since retreated can give evidence about it. For example, glacial striations – long grooves and scratches in bedrock – are formed by rocky debris carried by a glacier as it moves. By mapping the striations, scientists can reconstruct glacial flow patterns and estimate the past extent of glaciers.

POLLEN: Certain plants live only in certain climates, so their pollen is a clue to climate at the time the plant lived. For example, the cold period known as the Younger Dryas is named for the dryas, a mountain flower that today is found only in cold areas at high altitudes or high latitudes. Dryas pollen in a layer of sediment dating to the time of the Younger Dryas is evidence that the location was cold during that period.

ABOVE, FROM LEFT TO RIGHT *Foraminifera, single-celled organisms found in the sea or salt water lakes. Their shells provide evidence of climate change; glacial striations in rock, carved by rocky debris carried by glaciers as they move; Dryas octopetala, or mountain avens, photographed in Denali National Park, Alaska, USA. Dryas pollen found in a layer of sediment indicates a period of cold in that location.*

There may have been other migration routes. Some early Americans may have come along the southern shore of Beringia by boat, then followed the coast down to South America. Some may have travelled by watercraft across the Bering Strait, which was created when Beringia was flooded as the last glacial episode ended. Those who did cross the Bering land bridge would have faced an obstacle to further migration: kilometre-thick sheets of glacial ice blocking the path east or south from Alaska. But once the glaciers melted back, an ice-free, north-south corridor opened between western and eastern Canada. That may have been used for southward migration, along with another potential route along the mountainous northwestern coast. However they arrived, humans spread across both North and South America, developing the panoply of Native American cultures during millennia of isolation from the Old World.

Low sea levels assisted other migrations. About 40,000 years ago, humans reached Australia when stretches of dry land reduced the gap between Asia and Australia, although even the shortest route between continents would have required sea passage of 48 to 97 km (30 to 60 miles). At the last glacial maximum, land bridges existed between Australia and New Guinea, Asia and Indonesia, and continental Europe and Britain. All these pathways, like the Bering Land Bridge, were submerged when meltwater from glaciers made sea levels rise. That rise was not sudden: it progressed in some places at the rate of 4 cm (1.6 in) per year, or about 4 m (13 ft) per century. Nevertheless, it must have forced coastal communities to relocate inland and created barriers between peoples who were once connected.

AFTER THE ICE AGES

By the time the ice ages relented, modern humans were more or less the only hominid species left on Earth and many animals had become extinct. From about 16,000 to 11,500 years ago, a wave of extinctions, including that of mammoths, woolly rhinoceros, giant deer and many smaller animals, occurred. About 80 genera (species groups) disappeared from northern Eurasia alone. Different causes have been proposed for the extinctions of the megafauna (large animals), such as overhunting by humans – a view known as the prehistoric or Pleistocene overkill hypothesis. But a strong candidate for these losses is climatic change. The warming trend during this period may have been intolerable to many large animals. It would have transformed their environments and food supplies, changing, for example, the steppes and tundras on which mammoths grazed into forests. Possibly overhunting and climate combined to cause the extinctions.

Even as some wild animals became extinct, a new kind of creature came into being – the domesticated animal. Around 14,000 years ago, dogs were domesticated from the wolf in Mesopotamia (Iraq). Domestication of herd animals began with goats in Persia (Iran) about 12,000 years ago. Human beings were making their first steps towards a way of life in which they shaped not only tools but also living things to meet their needs.

The Dawn of Agriculture

Since the end of the last glacial period, Earth's climate has been warmer and more stable than in previous eras. The climate shifts have not been as large as the dramatic revolutions during and at the end of the ice ages and the shifts are often difficult to interpret. Nevertheless, there have been fluctuations large enough to influence the course of civilization.

One such shift may have influenced the development of agriculture, which coincided with the beginning of a new age in geologic time: the Holocene epoch (10,000 years ago to the present). In northern Mesopotamia about 8000 BCE (10,000 years ago), agriculture was born when wheat and barley were first cultivated. In the New World, the invention of farming was independently duplicated at about 7000 BCE, when pumpkins and gourds began to be cultivated in Mexico. In China and Japan, rice, millet and gourds were first grown, probably independently, about 6000 BCE. The development was momentous. Agriculture would transform many peoples from nomadic hunter-gatherers to settled societies. Food surpluses permitted larger populations, more pronounced social stratification and job specialization, the building of cities and the growth of empires.

The engine behind the birth and spread of agriculture may have been climate change. Since the end of the Younger Dryas, there had been a warming trend. Because of the Milankovitch cycles, Earth reached a peak of summer heating in the Northern Hemisphere about 8000 BCE. There was a lag of several millennia in general temperature rise as the glaciers melted and the seas warmed to a sufficient depth. The last remnants of the Laurentide Ice Sheet (the continental glacier that covered North America during the Pleistocene epoch) disappeared around 5000 BCE. Then the present interglacial period reached its Climatic Optimum, about 5000 to 4000 BCE, with global average temperatures at their peak and considerably

SINCE THE END OF THE LAST GLACIAL PERIOD, THE EARTH'S CLIMATE HAS BEEN WARMER AND MORE STABLE THAN IN PREVIOUS ERAS

more rainfall in most of the world. Indeed, no large deserts are known to have existed at this time. In most parts of the world, the climate was 1°C to 3°C (2°F to 5°F) warmer than today. In the lower latitudes of Asia, the greater warmth strengthened the summer monsoon, bringing heavier rainfall. The tropics were also warmer, and Africa was much wetter.

The mild, moist climate would have sped the development and spread of agriculture. By providing food for large, stable populations, agriculture made possible numerous other innovations, including bronze, which first appeared in Sumer, in Mesopotamia, about 3500 BCE. In Mesopotamia, this development marked the beginning of the

Bronze Age and the end of the Stone Age.

China was particularly warm relative to its present-day climate. Temperatures in the middle of winter were probably 5°C (10°F) warmer than today. Rice was sown a month earlier than it usually is today and bamboo grew five degrees of latitude further north than it does today. Even northern Europe's climate was milder. Trees grew in parts of Iceland that are today covered with ice. During this period, people of the British Isles and in Europe raised megalithic monuments, including Stonehenge (2800–1500 BCE).

Agriculture led to a new kind of dependence on climate, as farmers anxiously counted on rain and Sun to deliver the conditions needed for a good harvest. Farmers learned early how fragile this dependence was. About 6200 BCE, a period known as the Mini Ice Age began, lasting about 400 years. The cause, as may have been the case with the Younger Dryas event, was freshwater from melting glacial ice that slowed or even stopped the ocean conveyor belt. Cold, dry conditions, with widespread drought, ensued from Europe to the Middle East. As far south as Indonesia, the sea surface cooled by about 3°C (6°F). Then, about 5800 BCE, as the ocean conveyor belt resumed its work, the warm climate returned.

The Black Sea Flood

A different kind of ecological calamity struck shortly after the Mini Ice Age. About 5600 BCE, the Mediterranean Sea, its waters raised because of the melting glaciers, overflowed into a basin of river valleys and deltas. There, farming and fishing people lived on the shores of Euxine Lake. As torrents gushed in, the lake waters rose at the rate of 15 cm

ABOVE *Photograph of stone relief depiction of Gilgamesh, of Babylonian legend. Gilgamesh is the hero of the Gilgamesh epic, a work of some 3,000 lines, written on 12 tablets, c. 2000 BCE and discovered among the ruins at Nineveh, in Iraq. These tablets describe a powerful deluge and flood, thought by some to be the same flood mentioned in the Bible and texts from other cultures. The flood is thought to have been caused by melting glaciers after the Mini Ice Age.*

(6 in) a day. The people's villages were inundated; their farms destroyed. An unknown number of lives were lost. In two years, Euxine Lake had reached the same level as the Mediterranean. It had been transformed into the Black Sea.

This cataclysm, some speculate, could have been the source of the legends that grew into the story of Noah's flood in the Bible. Many ancient cultures in the region had stories of world-destroying floods that were survived by only a few individuals who had escaped in some sort of ark. The Greeks told of

a flood in which the survivors were Deucalion and his wife, Pyrrha. The Mesopotamian Epic of Gilgamesh describes a deluge so powerful it frightened even the gods. Conceivably, all of these cultures could have been looking back to the same dimly remembered event in their region.

On the other hand, cultures even farther away – including American Indian tribes and peoples in India, Southeast Asia, China and Australia – have flood legends of their own. Not all of them could have known about the Black Sea. Possibly they are each remembering the continued rise in sea levels that followed the end of the Pleistocene ice ages and which may have caused catastrophic losses for many peoples. Perhaps they are only recording the impact that floods have for people in every region. What historical events, if any, lie behind the flood stories remain a mystery.

Cooling, Drying and the Suez Canal

After the Climatic Optimum, conditions gradually became harsher. Beginning around 3500 BCE, the climate began to cool and become drier. This trend is known as the Neoglaciation and, with fluctuations, it has continued to our own time. As this episode began, the ancient civilizations of Egypt, the Middle East and China were founded. Fortunately for paleoclimatologists, from this point on, indirect evidence of climate change, such as tree rings and ice cores, is supplemented with historical records.

Rainfall declined during this period. The monsoon of southwestern Asia weakened and moved southward, with the rains beginning later and ending earlier each season. Droughts became more frequent. In what is now the southwestern United States, between 4000 and 3000 BCE, a movement towards agriculture came to a halt because of the recurring droughts. Between 1500 and 1300 BCE, glaciers advanced in Alaska, the Alps and elsewhere, and they have yet to retreat. Most or all of the glaciers in the Rocky Mountains south of Canada are thought to have formed since that time.

The condition of the Sahara region of northern Africa was a dramatic example of the drying trend. Now the world's largest desert, the Sahara's climate had been wet during the Pleistocene epoch. It had numerous lakes and streams, and elephants and giraffes populated its grasslands and forests. About 3500 BCE, the region's climate became drier and the Sahara began to become a desert. That desert has expanded ever since, with people contributing to the desertification cutting down trees and brush along the region's boundaries and overgrazing their livestock.

There were fluctuations in the cooling and drying trend. From about 2500 to 2000 BCE, mountain glaciers receded, marking a time of greater warmth. Sea levels rose, so much so that Egyptian pharaoh Sesostris I, in the twentieth century BCE, built the Suez Canal that linked the Nile Delta and the Red Sea. The high sea level probably made the project seem workable, since it would have reduced the land barrier.

After 2000 BCE, the glaciers again expanded. The expansion coincided with a decline in rainfall in the Middle East and North Africa. In the Canadian Arctic and Siberia, the tree line, the altitude above which forests do not grow, receded. Sea levels fell by 1.8 m (6 ft) or more from 2000 to 500 BCE. The long-term cooling and drying trend was once again felt. However, the trend remained variable, with repeated episodes of glacial expansion – for example, around 200 BCE and from 500 to 800 CE – alternating with warmer periods.

El Niño and Ancient Egypt

The drying trend spelled disaster, or near-disaster, for many societies. In 2184 BCE, Egyptian civilization almost collapsed because of droughts that drastically reduced the flow of the River Nile and

RIGHT *Rock paintings in the Tassili N'Ajjer mountains, Sahara Desert, Algeria. Dating back some 6,000 years, the Tassili N'Ajjer paintings include depictions of animals which are now confined to more southerly temperate zones of Africa. This indicates that the climate of the Sahara was then less arid.*

brought on widespread starvation. The droughts may well have been connected to a now well-known climate fluctuation called El Niño, which affects the planet at more or less regular intervals.

In the 1800s, the name El Niño was attached to a local ocean current originating far from the Nile Valley, off the coast of South America. Today, scientists use it to describe a recurrent phenomenon that results from the interaction of the atmosphere and the tropical Pacific Ocean. Trade winds in the tropical Pacific generally blow east to west: from a high-pressure centre off the coast of South America, where upwelling keeps water relatively cool, to a low-pressure centre near Indonesia, where warm water piles up as a result of the easterly winds. But in an El Niño, which occurs about every two to seven years and typically lasts about 18 months, an unusually strong and persistent southward current of warm water develops off

Ecuador and Peru. Air pressure drops in the eastern Pacific and rises in the west, so that the east-to-west winds weaken, or even reverse. Widespread climatic changes result, affecting patterns of rainfall and dryness around the world. For example, western South America experiences more rainfall than usual, while Indonesia becomes drier. A reverse pattern, La Niña, often follows; it occurs when the water in the western Pacific is warmer than usual. The alternation between El Niño and a period without an El Niño is called the Southern Oscillation, or the El Niño Southern Oscillation (ENSO). The droughts that plagued Egypt in the twenty-second century BCE were probably related to El Niño, which brought drought again to northern Africa as recently as 1997 and 1998.

No one knows when ENSO started – whether it is a phenomenon only of the Holocene or dates back to more remote prehistory. But it appears to

have had a strong influence on societies for at least the last 5,000 years. Egyptian civilization depended on the annual floods of the Nile, which in turn depended on heavy monsoon rains over the Ethiopian highlands. Probably because of ENSO events, the monsoon weakened or failed, and the Nile's flow all but halted. Crops failed, famine spread and social disorder erupted. Pharaohs succeeded each other rapidly but were unable to prevent the empire from disintegrating. Only after a long civil war did King Mentuhotep I reunite Egypt in 2046 BCE and establish the Middle Kingdom. During the following 250 years, there was a period of prosperity, the rulers learning from the lessons of climatic instability. They prepared grain storage facilities for times of crisis, developed new irrigation techniques and instituted a centralized system for feeding their people.

Climate as a Catalyst in History

The long-term cooling and drying trend of the Neoglaciation had widespread effects on human culture. In the previous three millennia BCE, the trend led to widespread droughts and failing harvests, sparking social collapse and great migrations as people searched for more fertile lands. The effects are difficult to measure; rarely can a particular episode of climate change be assigned as the cause of the rise or fall of a civilization.

LEFT *Ancient Egyptian art on papyrus depicting men in a boat on the Nile. The prosperity of the Egyptian civilization was closely linked to the flood cycles of the Nile and monsoon conditions in Ethiopia. Climate change brought on by El Niño may have contributed to near disintegration of the empire prior to 2046 BCE.*

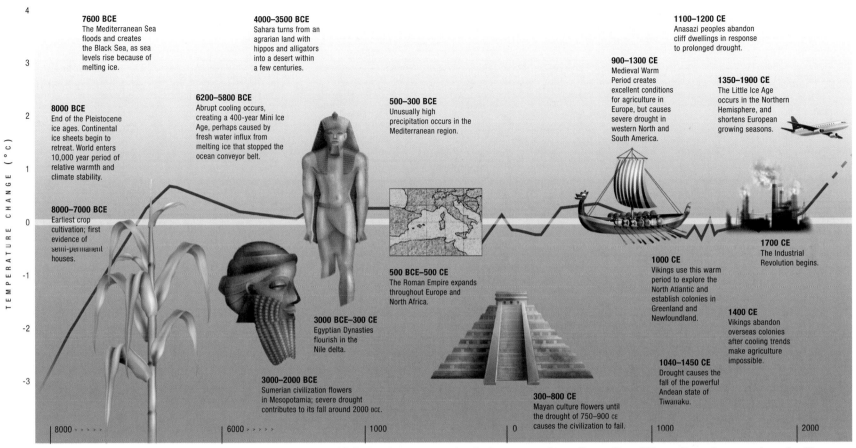

4

7600 BCE
The Mediterranean Sea floods and creates the Black Sea, as sea levels rise because of melting ice.

4000–3500 BCE
Sahara turns from an agrarian land with hippos and alligators into a desert within a few centuries.

1100–1200 CE
Anasazi peoples abandon cliff dwellings in response to prolonged drought.

900–1300 CE
Medieval Warm Period creates excellent conditions for agriculture in Europe, but causes severe drought in western North and South America.

3

6200–5800 BCE
Abrupt cooling occurs, creating a 400-year Mini Ice Age, perhaps caused by fresh water influx from melting ice that stopped the ocean conveyor belt.

500–300 BCE
Unusually high precipitation occurs in the Mediterranean region.

1350–1900 CE
The Little Ice Age occurs in the Northern Hemisphere, and shortens European growing seasons.

2

8000 BCE
End of the Pleistocene ice ages. Continental ice sheets begin to retreat. World enters 10,000 year period of relative warmth and climate stability.

1

8000–7000 BCE
Earliest crop cultivation; first evidence of semi-permanent houses.

0

500 BCE–500 CE
The Roman Empire expands throughout Europe and North Africa.

1000 CE
Vikings use this warm period to explore the North Atlantic and establish colonies in Greenland and Newfoundland.

1700 CE
The Industrial Revolution begins.

-1

3000 BCE–300 CE
Egyptian Dynasties flourish in the Nile delta.

1400 CE
Vikings abandon overseas colonies after cooling trends make agriculture impossible.

-2

1040–1450 CE
Drought causes the fall of the powerful Andean state of Tiwanaku.

-3

3000–2000 BCE
Sumerian civilization flowers in Mesopotamia; severe drought contributes to its fall around 2000 BCE.

300–800 CE
Mayan culture flowers until the drought of 750–900 CE causes the civilization to fail.

T E M P E R A T U R E   C H A N G E   (°C)

8000 › › › › › 6000 › › › › › 1000 0 1000 2000

ABOVE A timeline of world history. Throughout history, the rise and fall of civilizations has been linked to climate change. About 8000–7000 BCE, the warming trend that followed the Pleistocene ice ages encouraged the invention of agriculture. But climate is volatile and a 400-year Mini Ice Age occurred from 6200–5800 BCE. The Roman Empire benefited from a mild climate. The Mayan civilization was wracked with drought, and harsh climate caused its collapse. Since 900 CE, climate has gone through still more dramatic shifts, from the Medieval Warm Period to the Little Ice Age to the rising global temperatures of the 20th and 21st centuries. Today's shift is distinctive because it is strongly influenced by human activity: by the Industrial Revolution that began in the eighteenth century and the greenhouse gases released by the burning of fossil fuels.

Many civilizations managed to adapt to climate change. From about 3500 BCE to 2000 BCE, the ancient civilization of Sumer flourished in the fertile plain formed by Mesopotamia's Tigris and Euphrates rivers. This was a time of generally drier climate, which may have spelled an end to its civilization. But the Sumerians adapted by building canals to irrigate their fields. Their civilization thrived; in fact, the Sumerians made extensive advances in agriculture and architecture. In addition they invented the world's first writing system, using cuneiform tablets.

In the 2300s BCE, Sargon of Akkad conquered Sumer. The fate of the Akkadian empire that he founded shows how climate can interact with politics to change – and in this case end – a society. According to some archaeologists, the northern parts of the Akkadian empire began to suffer a prolonged drought that lasted about 300 years and was probably connected to a series of more frequent ENSO events. Agriculture in the north declined rapidly, and much of the northern population migrated south. There the Tigris and Euphrates rivers provided more stable water supplies, but the presence of the refugees overburdened the government. About 2200 BCE, under the weight of climate change and inadequate policies, the Akkadian empire collapsed.

Tradition and Climate

Climate and weather are facts of life, and people throughout history, in all parts of the world, have tried to make sense of them. Long before the advent of modern science, cultures around the globe did their best to grasp the cycles and unpredictability of the atmosphere, through religion, mythology, legend, folklore, or other traditions.

Religion and mythology offered explanations for climate and weather phenomena that would have otherwise seemed inexplicable. The ancient Greeks believed that the winds, for example, were kept locked up in the cave of Aeolus, king of the winds. When it was time for one of the winds to blow – Boreas the north wind, for example – Aeolus released it from its cave and it went howling out to do its work. Similarly, the Iroquois of North America gave control of the winds to Ga-oh, a giant spirit who kept the reins on four animal spirits that personified the four winds. The Inuits explained the thick mists of spring by saying that the Creator raised them because he did not like to see the Inuits killing his favourite animals, the giant bowhead whales, in their spring hunts.

Religion and myth not only explained causes, but offered rituals through which people believed they could exert some control over the weather. The Chinese, for instance, offered sacrifices to Longwang, the Dragon King and god of rain, for well-timed rainfall and good harvests.

THE GOLDEN AGE AND RAGNARÖK

It is possible that some of the myths and legends record, in symbolic form, actual climatic episodes. Many cultures have legends of a Golden Age, a time when the world was better, greener and more bountiful, and which ended with humankind being thrust out into a more wretched world. In the well-known biblical story, 'Every tree that is pleasant to the sight and good for food' grew in the Garden of Eden, before Adam and Eve were expelled for sinning. The Greeks and Romans also had Golden Age stories. Possibly these traditions record an ancestral memory of the Climatic Optimum, a time when the climate was warmer, wetter and milder than it became later. Elements of the Norse legend of Ragnarök, the battle of the end of the world, may have been inspired by folk memories of the climate change that took place in Norway around 800 to 700 BCE. At that time, glaciers advanced dramatically and trees that were adapted to warmth were replaced by spruce. When it describes Ragnarök, Snorri Sturluson's Edda poem of about 1220 CE depicts a time when 'the snow drives from all quarters with a biting wind; three such winters follow one another and there is no summer in between'.

PREDICTING WEATHER, MAKING RAIN

While some traditional approaches to weather may now seem fanciful, in cultures where survival depended on foreknowledge of the weather, accurate forecasting has not been uncommon. Through oral tradition, many peoples

ABOVE *The Andreas Stone with a relief depicting the final battle of the god Odin from the legendary Norse poem Ragnarök, or 'Doomsday of the Gods'.*

accumulated detailed knowledge about climate and weather in their region. In Kenya, traditional rainmakers became skilled at monitoring and predicting rainstorms, windstorms, droughts and seasonal rains based on observation of the environment and knowledge of natural signs of changing weather, from clouds to trees. The Maori of New Zealand learned to predict shifts in temperature, wind direction and rainfall based on observation of such factors as the blooming of trees and flowers, the appearance of stars and the movements of migratory birds.

On the other hand, much of the world's folkloric wisdom about climate and weather leaves something to be desired. The familiar proverb 'Red sky at night, sailor's delight/Red sky at morning, sailor take warning' has a certain degree of truth. But 'Lightning never strikes the same place twice' is false. New York's Empire State Building, for example, is struck 23 times a year on average.

LEFT *Incan gold artifact depicting Apu Inti, the Sun god. The Incas referred to themselves as 'children of the sun'.*

Far away on the Indian subcontinent, societies struggled with climate change just as the Mesopotamians did. Beginning around 2500 BCE, the Harrapan culture flourished in the Indus Valley of western India and Pakistan. (Archaeologists have come to know it from the ruins of the cities of Harappa and Mohenjo-Daro.) Sustained by varied agricultural crops, from wheat to melons, the civilization developed systems of writing, counting, measuring and weighing. It collapsed around 1700 BCE, probably

Lava and ash destroyed the community living there. Nearby Crete was overwhelmed as well. The Minoan civilization shared by Crete and Thira never recovered from the calamity.

Migrations and hardships, brought on by changes to climate, continued all over the world from 1200 to 800 BCE and from 600 to 200 BCE. In ninth-century Phoenicia, an ancient land that corresponds roughly to coastal Syria, Lebanon and Israel, a prolonged drought so afflicted the people

IN CARTHAGE, THE HISTORIAN PROCOPIUS WROTE, 'THE SUN GAVE FORTH ITS LIGHT WITHOUT BRIGHTNESS, LIKE THE MOON DURING THIS WHOLE YEAR, AND IT SEEMED EXCEEDINGLY LIKE THE SUN IN ECLIPSE'

because of increasing drought. Changing river patterns, including a series of floods, may also have played a part.

In China around 2000 BCE, severe floods on the lower Yangtze and Yellow rivers inundated farmland and may have contributed to the decline of the Longshan and Liangzhu cultures. The Xia culture rose in central China to take their place, leading to the founding of China's first dynasty about this time.

Between 1470 and 1450 BCE, a massive volcanic eruption produced a dense cloud of debris that must have lasted several years. By blocking sunlight, the cloud had a widespread cooling effect. However, the most direct effects were felt on the Greek island Thira, or Santorini, where the eruption occurred.

that Ethbaal, king of Tyre, made ritual sacrifices of children in an attempt to stop it.

The period from 900 to 300 BCE was generally cooler and wetter in higher and middle latitudes and the last two centuries of that period were particularly cold. Changing levels of solar activity may have been at least partially responsible. A reduction in the number of sunspots (see p. 23) occurred in 850 BCE, and carbon-14 levels rose. This isotope of carbon is formed through the action of cosmic rays, which are affected by solar activity; when the Sun is more active, fewer cosmic rays reach earth and so less carbon-14 is preserved in organic material. The rise in carbon-14 and the reduction in sunspots indicate a fall in solar radiation.

Climatic and Political Chaos

Until 900 CE, climate in Europe remained generally colder and more variable than the previous era. To the south and east, North Africa and western Asia suffered through dryness and drought. This was especially so in Arabia in about 600 CE, when many farmers abandoned their land because of drought.

A catastrophic climatic event took place in 535 or 536 CE. The cause was probably an enormous volcanic eruption – perhaps Rabaul, in New Guinea, or a volcano in Sumatra or Java in Indonesia. Enough debris was blown into the atmosphere to blot out much of the Sun's radiation for 18 months. Another possible source of the cloud was a meteorite impact. It would have had to strike one of the oceans; had it struck land, the meteor would have left a tremendous crater, and no such formation exists from that time.

Whatever the cause, a dense, dry fog covered Europe, China and southwest Asia. In Carthage, the historian Procopius wrote, 'The Sun gave forth its light without brightness, like the moon during this whole year, and it seemed exceedingly like the Sun in eclipse.' Worldwide climatic chaos ensued, with accompanying disasters: severe cold, storms, droughts, floods, famine, bubonic plague.

In China, snow fell in August, destroying the harvest. Tree rings from western Europe show that tree growth slowed for a decade, from 536 to 545 CE. The territory of the Moche civilization of Peru became arid and was wracked by dust storms.

Many political and cultural changes of the following century may be partially attributable to the climatic chaos. They include the decline of the powerful city of Teotihuacan in what is now Mexico.

CLIMATE AND CIVILIZATION

The early centuries CE saw the rise of two empires in different hemispheres: the Roman in the Old World and the Mayan in the New. The story of both civilizations is intimately linked to that of climate. The ascent of the Roman empire coincided with a general warming. The global climate from 200 BCE through 100 CE, when the Romans were building their empire, was similar to our own, though it was somewhat wetter in North Africa and the Near East.

ABOVE *Pont du Gard Bridge, a Roman-built aqueduct in Languedoc, France. Through skilled engineering, the Romans controlled water resources to further the spread of their empire.* UPPER LEFT *Apollo, the mythological Greek god of the Sun.*

The wetness of the time allowed the Romans to grow crops extensively in North Africa and allowed for settlements in places like Petra, now in Jordan. Once an important centre of trade, Petra later fell due to the expanding desert.

The Romans had the advantage of an exquisitely fertile site for their domain. Starting from their home city of Rome in Italy – founded, according to legend,

in 753 BCE – the Romans built their empire around the Mediterranean Sea, which they called *Mare Nostrum* (Our Sea). The warm water of the Mediterranean gives most of the surrounding lands a subtropical climate, with hot, dry summers and mild, rainy winters conducive to agriculture. (Similar conditions, wherever in the world, are often referred to as Mediterranean climates.) Many ancient cultures, from the Egyptians to the Greeks, developed on the Mediterranean's shores. Several, including the Phoenicians and the Carthaginians, had tried to control the sea. But it was the Romans, with their formidable army and navy, who achieved full mastery of the Mediterranean. By the first century CE they governed all the surrounding lands.

Changing Climates, Changing Fortunes

During the Roman empire, the Mediterranean climate extended north into much of western Europe, allowing the Romans to practice their style of agriculture across a vast domain. That style involved extensive production of cereal crops such as wheat and millet for large populations.

As the Caesars ruled and time passed, the climate warmed, reaching its apex around 400 CE. The trend was accompanied by rising waters: sea levels also reached their peak – at or slightly above present-day levels – around 400 CE. During the Roman empire, sea levels rose about 0.9 m (3 ft) altogether, enough to submerge some Mediterranean harbour installations and depopulate the coasts of the Netherlands and Flanders.

Factors other than climate were involved in Rome's decline, but climate played a part. As the climate warmed, it became drier as well, and the period 300–400 CE was one of drought in the Mediterranean, North Africa and western Asia. Then, in the fifth century, the climate grew cooler, wetter and more variable, with still more serious consequences for Rome. To sustain its western European empire, Rome depended on a climate that allowed crops to be grown fairly far north – grain bought loyalty from subject peoples. But as the climate cooled, cereal production in much of the Roman territory of Gaul (France and some of its adjoining areas) suffered. Rome's influence in Gaul declined and invading Franks and Goths overran much of the land Rome had formerly held. The Roman empire collapsed during that century – at least in the west. In the east, it was able to survive as the Byzantine empire.

Mayan Civilization

Across an ocean, another empire was at its height even while the Romans still reigned. In 300 CE, the Mayan civilization, based in Mexico's Yucatán peninsula, stretched across modern-day Guatemala, Honduras and Belize. The Mayans made great advances in agriculture, astronomy, mathematics, architecture and writing; they raised pyramids and monuments and made pottery and ornaments of copper and iron. The Mayans experienced periodic droughts but developed irrigation techniques to manage them. Their accurate calendar and extensive record-keeping allow scholars to chart their rise and decline. The Mayan civilization reached its apex in the eighth century.

Then came a drought that was so severe they were unable to cope with it.

North America suffered a number of droughts in the first millennium CE. In the northern Great Plains, for example, extreme droughts occurred from 200 CE to 370, 700 to 850 and 1000 to 1200. But sediment measurements from the bed of Lake Chichancanab, Mexico, show that the period from 750 to 900 was the driest in that region in the last eight millennia. The exceptional dry spell may have stressed the Mayan empire's governing institutions, prompting social unrest. Some Mayan cities revolted and, by about 900, the civilization collapsed. According to the archaeologist Richardson Benedict Gill, the collapse happened in three stages. A first drought in 810 affected the cities of Palenque and Yaxchilán. Another in 860 overthrew the cities of Caracol and Copán. Finally, from 890 to 910, Tikal, Uaxactún and other cities fell to a third drought. In Tikal people resorted to cannibalism to survive. Across the Mayan empire, cities were abandoned as people died or dispersed into scattered villages. Mayan culture endured in the Yucatán, but it never returned to its former glory.

LEFT El Castillo, part of the Mayan city of Chichen Itza, Yucatán, Mexico. These pyramids were temples to the gods of the Sun and the rain. Worship of these gods was vital to the Mayan civilization, which prospered or suffered at the whims of weather. RIGHT A stone carving of Chac, the Mayan god of rain.

Teotihuacan fell to famine, epidemics, revolution and collapse. During this period also, the nomadic Avars left drought-ridden Asia to invade and build a powerful empire as far west as Hungary and Germany.

The Little Optimum

Starting in about 900 CE, the North Atlantic climate warmed. It remained mild for 400 years, until about 1300. This episode is known as the Little Optimum, the Medieval Warm Period, or the Medieval Climatic Anomaly. Summer temperatures in western and central Europe were probably 1.1°C (2°F) higher than today. European agriculture and population boomed and lush vineyards grew in England. Grain was grown further north in Norway than is currently possible.

The bountiful harvests and benign climate helped cultures to flourish in many parts of Europe. The High Middle Ages, the period when medieval civilization reached its peak, occurred during the Little Optimum, which lasted from the eleventh to thirteenth centuries. This was a time of burgeoning trade and expansiveness. Medieval towns increased in size and number. Gothic cathedrals rose throughout Europe.

The warm period influenced exploration as the Scandinavian adventurers known as the Vikings or Norsemen settled parts of Greenland. Situated mostly north of the Arctic Circle, Greenland today is largely a frozen island – 80 per cent of its surface is covered by ice. But at the time, Greenland's conditions made

RIGHT *Aerial view of the Greenland ice sheet. At the time the Vikings settled Greenland, the country was actually green, not ice-covered as it is today.*

it more worthy of its name, a title bestowed upon it to attract settlers. From Greenland, taking advantage of seas that were relatively free of ice, the Vikings sailed to North America. There, in about 1000 in what is now Newfoundland, Canada, they founded the colony of Vinland, reaching the New World nearly five centuries before Columbus.

In the western Americas, however, the Little Optimum brought not bounty but drought. Five centuries of aridity strained the agricultural and hunter-gatherer societies of western North and South America. At the same time, off the coast of what is now California, upwelling – the phenomenon by which cold water rises from the depths to replace warm surface water driven away by the wind – increased.

The rising water is rich in nutrients that encourage the growth of seaweed and phytoplankton as well as the flourishing of the animals – from fish to seabirds –

that depend on them. The Chumash Indians of California exploited the increased marine productivity, building more settlements and accumulating more wealth in the form of beads, shells and canoes. The growth of their fisheries compensated in part for the droughts on shore.

Another Native American society did not fare as well. From 750 to 1300, the Anasazis had developed an elaborate civilization in the southwest, particularly in the Four Corners region where Arizona, Colorado, New Mexico and Utah now meet. They lived in multi-family pueblos, or towns

built like apartment complexes, with various levels connected by ladders. Some pueblos, called cliff dwellings, were built in canyon recesses and under rock overhangs. At places such as Chaco Canyon, New Mexico, the Anasazis made strides in agriculture, irrigation, trade and crafts, and maintained a spiritual life that centred on underground ceremonial structures called kivas. While rain was plentiful, the population grew. But in the twelfth century, rain was not plentiful and drought plagued the land. The Anasazis were also beset by invading Athapascan peoples, the ancestors of the Navajos and Apaches. The Anasazi moved to the wetter parts of their domain, but their large populations soon overtaxed the land.

Drought struck again in the thirteenth century. Around 1300, unable to endure the drying out of the land, the Anasazis abandoned their cliff dwellings – which today remain as ruins – and dispersed to more fertile parts of the southwest, including the Rio Grande valley.

Tiwanaku, a powerful state in South America, also suffered catastrophic drought. In the first millennium CE, it raised a magnificent Andean city of palaces, temples and plazas and came to dominate an area the size of California. Its economy was based on a system of intensive agriculture designed to cope with the seasonal contrasts of the *altiplano*, the high plains of Peru and Bolivia. Crops grew on raised platforms of earth, separated by canals that limited the danger of drought and frost. The system supported a growing population, but it could not handle the horrendous drought that began in about 1040. The raised fields dried up and Tiwanaku fell. Its cities were abandoned and

Epidemics, Famine and Strife in the Middle Ages

By the fourteenth century, the pendulum of climate was swinging back toward lower temperatures and a less predictable climate was affecting much of the North Atlantic region. From 1314 to 1317 in Europe, a series of cold, wet summers, connected to the general cooling trend, led to devastating harvest failures. Heavy rains ruined fields, eroding soil and leaving deep gullies.

Crops failed from Scotland to Italy to Russia. Grain prices soared and famine and disease stalked the countryside. Horses, oxen and sheep perished from cold, hunger and illness in such large numbers that people spoke of the 'great dying of beasts'.

Disease epidemics are rarely traceable purely to climate, but climate may have been a factor. The Black Death of the Middle Ages is an example. From 1347 to 1352, this epidemic of bubonic plague killed a quarter or perhaps as much as a third of the European population. The disease is caused by a bacterium, *Yersinia pestis,* which is transmitted to humans principally by the bite of fleas from infected rats. The rats that brought the plague to Europe, perhaps by hitching a ride on a caravan, probably originated in the steppes of central Asia, or further east in China. In 1332, China was wracked with huge floods that killed millions of people. These floods, tied to a generally wetter climate, caused migrations of many wild animals, including rats, some of which may have been carrying the virulent form of bubonic plague. However indirectly, climate may well have played a role in the Black Death.

In addition to plague, the fourteenth century brought floods, droughts and famine to Europe. It was a time of shrinking population, the breakdown of the feudal system, peasants' uprisings and widespread war.

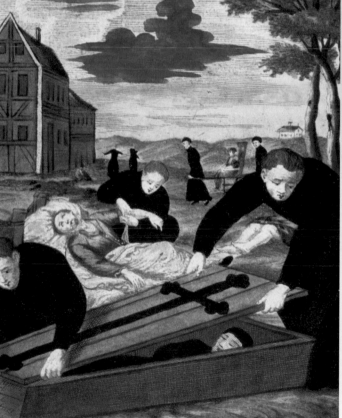

ABOVE *Spanish Jesuits tending to victims of the plague, which may have been carried by rats escaping flood conditions in China caused by climate change.*

its people turned to more sustainable activities, such as raising llamas and alpacas.

Had the mild influence of the Little Optimum persisted, there is no telling what effects it might have had in Greenland, Europe or on the California coast. But the climatic episode came to an end. By the late fourteenth century, the North Atlantic climate had become frigid enough that contact between Scandinavia and the Norse settlements in Greenland drastically diminished. The settlers in Greenland were forced to eat all their farm animals and dogs before dying themselves. The colony of Vinland in Canada was also lost to history.

Winds and the Age of Exploration

In 1492, Christopher Columbus established lasting contact between the Old World and the New World for the first time since the Pleistocene ice ages. The cultures of the two hemispheres had developed essentially in isolation from each other for 14,000 years, since climate change caused the inundation of the Bering Land Bridge. Columbus's trip was made possible by another element of climate: a system of winds that developed as Earth's continents moved into their present-day positions and that still prevails today.

A Genoese mariner sailing under the flag of Spain, Columbus was searching for a westward route to Asia, not a new world. After sailing first to the Canary Islands, a Spanish possession off the African coast, he took advantage of the trade winds, which blow from the northeast at that latitude (about 30 degrees north). These dependable winds earned their name from an obsolete meaning of trade as 'regular course'. The winds took him to

ABOVE *Columbus's ships sailed for the New World in 1492. This expedition took advantage of the trade winds off the coast of Africa, a dependable system that still prevails today.*

the Bahamas, where he made landfall.

The fifteenth to nineteenth centuries are known as the Age of Exploration, because of the daring voyages of Columbus, Ferdinand Magellan, Vasco da Gama, John Cabot, Jacques Cartier, Francis Drake, James Cook and others. Their feats of navigation depended on winds and surface ocean currents, which in turn depended on climate. But the winds could fail explorers as well as transport them. In 1519, having embarked from Spain on his

round-the-world voyage, Magellan found himself in the doldrums, a region near the equator with little wind. It is now also known as the intertropical convergence zone, because the trade winds from both hemispheres converge there. For several weeks he and his fleet lingered there, low on water and with signs of scurvy appearing. Eventually, the South Equatorial Current drifted the ships to the edge of the trade winds, where they took advantage of a light north wind and moved on towards Brazil.

THE LITTLE ICE AGE

From about 1450 to 1850, Earth passed through what is called the Little Ice Age. (Some scientists place the start date as early as 1300 and the end date as late as 1890.) During this period of renewed cold, alpine glaciers advanced in virtually all the world's mountainous areas and the Arctic islands' ice caps grew larger. Winters became colder and summers cooler, though the effect on winters was generally greater.

The effects of the Little Ice Age varied by region and from year to year. Climatic volatility was the only constant: very cold periods alternated with warmer ones, dry ones with wet ones. In Switzerland in the 1590s, for example, winters were exceptionally harsh and summer harvests were poor, but afterwards only the winters were unusually cold – summers were closer to normal. From 1730 to 1739, European winters were the mildest for a generation – but cold returned with a vengeance from 1739 to 1742. Despite these regional and time-specific fluctuations, the general trend during these centuries was toward cold.

Worldwide, the climate change damaged many eco-systems. Floods, plague and famine devastated Europe. Crops failed, especially in northern regions. The Baltic Sea froze. England's Thames River developed ice several centimetres thick. In higher latitudes, great storms increasingly roiled the skies. A storm that hit southern England on December 7–8, 1703, blew down a lighthouse, wrecked houses, tossed ships onto land and killed 8,000 people. In North America, the Little Ice Age forced southwards the northern limit of the maize belt, the area where corn can be cultivated. In about 1600, corn growing was abandoned at sites near Winnipeg, Canada. The corn farmers from that area, ancestors of today's Cree, returned to the traditional hunter-gatherer diet they had eaten before corn cultivation was introduced.

Drought and flood often besieged the same areas. Timbuktu, a trading city near the southern edge of the Sahara in what is now Mali, suffered great winter floods in 1592, 1602–1603, 1616–1619, and several more times into the eighteenth century. The floods exceeded all previous bounds, spilling into the citadel, the ancient royal palace and scattering the population. Yet Timbuktu itself was not subject to heavy rains at this time. The floods were caused by drenching summer rains far away over the upper basin of the River Niger. The area around Timbuktu suffered drought and famine, even while the city was periodically being flooded.

The hard times of the Little Ice Age influenced political decisions. In the late sixteenth century in

ABOVE *Pieter Brueghel's* Hunters in the Snow, *1565. The painting shows the effects of the harsh winters which prevailed in Europe during the Little Ice Age, which lasted from about 1450 to 1850.*

England, poor harvests and high grain prices led to social unrest. Parliament responded with what was called the Great Act, aimed at poverty relief. From 1693 to 1700, poor harvests in Scotland led to widespread famine – one of the factors that prompted Scotland to unite with England in the 1707 Act of Union. In France, the climatic chaos of the eighteenth century brought bitterly cold winters, floods, droughts, crop failures, hunger and rising grain prices.

IN FRANCE, THE CLIMATE CHAOS OF THE EIGHTEENTH CENTURY BROUGHT BITTERLY COLD WINTERS, FLOODS, DROUGHTS, CROP FAILURES, HUNGER AND RISING GRAIN PRICES

All of these contributed to the French Revolution of 1789. Before the Bastille fell that year, mobs rioted seeking bread and grain, commodities in short supply because of the climatic changes.

Phases of Freezing

Evidence exists that suggests the Little Ice Age consisted of two principal phases, the first from 1450 to 1700 and the second from 1700 to 1850. In the first phase, precipitation was generally heavier than in the second period, with particularly intense episodes around 1470 and in the late 1600s. The 1690s were extremely cold in Europe, but from then until the 1730s came a strong warming trend, with temperatures comparable to those of the warmest parts of the twentieth century. Then, in 1740, exceptional cold returned, especially in winter. The chillier

conditions prevailed until about 1850, when winters began to become warmer again.

Nineteenth-century North America suffered the coldest conditions. In the Southern Hemisphere, the most extreme episodes of cooling occurred in the sixteenth and seventeenth centuries. In Asia, the seventeenth century was the coldest period, with several more cold decades around 1800. For China's Ming dynasty, the seventeenth century brought widespread drought and calamitous floods, resulting in epidemics, famine and economic troubles. Revolt spread against Ming rule, making the government vulnerable to growing attacks by Manchu invaders from the north. In 1644, the Ming dynasty fell to the Manchus.

The causes of the Little Ice Age are not well understood. Sunspot cycles may have contributed (see page 23). The coldest period of the Little Ice Age in Europe, from the mid-seventeenth to the early eighteenth centuries, corresponds to the time of very low sunspot activity known as the Maunder minimum (1645–1715). The total number of sunspots observed throughout those 70 years was fewer than the number typically seen in a single year at present. In addition, carbon-14 levels (as measured from tree rings) were high

during those years, indicating a jump in cosmic rays and a fall in solar activity. Both the sunspot and carbon-14 data may indicate reduced energy output from the Sun.

Changes in ocean circulation may also have played a part in the Little Ice Age, especially changes in the flow of warm water into the North Atlantic from the south. This kind of variation may be characteristic of an interglacial period such as ours. Evidence from ice cores suggests that the previous interglacial, which ended about 130,000 years ago, may have been even more volatile with more short-term climatic variability than this one.

Another factor possibly connected to the Little Ice Age was a change in a phenomenon known as the North Atlantic Oscillation (NAO), an approximately ten-year cycle that may affect climate in Europe and Africa. A seesaw of atmospheric pressure between a low-pressure centre near Iceland and a high-pressure centre around the Azores results in an oscillation, or periodic alternation, between strong and weak westerly winds. Strong winds accompany a steep pressure gradient, or difference, between the two pressure centres; weak winds result when the gradient is low. The strong westerly winds, when they occur, push mild air across Europe and into Russia and bring mild winters to much of North America, while pulling cold air over western Greenland. They also bring more summer rain in Europe. The reverse, weaker pattern pulls cold Arctic air and heavy snows into Europe, while directing mild air toward Greenland. At present, the NAO accounts for about half the variability in winter temperatures in northern Europe. Since 1870, the NAO has fluctuated on a scale of time ranging from several years to a few

decades, bringing a period of warm winters in western Greenland and severe winters in northern Europe, then the reverse.

The NAO has apparently affected the European climate for millennia. The Little Optimum of 900 to 1300 was a period when mild westerlies predominated. The cold, unpredictable climate of the Little Ice Age was probably a period when the NAO fluctuated rapidly from one extreme to the other. The low periods, when westerly winds are weak, seem to have been related to cold periods during the Little Ice Age, such as in the late seventeenth century. Warmer periods, such as the 1730s in Europe, were probably related to high periods of the NAO, when westerly winds are strong.

Volcanism and Variations

Volcanism may also have been involved in some phases of the Little Ice Age. The eruption of Mount Tambora in Indonesia in April 1815 shot five to ten times more debris into the stratosphere than the Philippine volcano Mount Pinatubo did in 1991. Tambora's eruption was linked to exceptionally cold conditions worldwide over the next two years and may have contributed to lower temperatures for longer than that. In Canada and New England, 1816 was dubbed the 'year without a summer'. In India, the volcano-induced climate shift led to crop failure and famine, triggering a cholera pandemic that spread across the globe.

Artists and writers were inspired by the cold of the Little Ice Age. Flemish artist Pieter Brueghel the Elder painted his *Hunters in the Snow* in 1565, recording for posterity the unusual severity of that winter. In general, the skies depicted in European paintings from 1550 to 1849 tended to be darker and cloudier than in the period 1400–1550 or 1849–1967. Mary Shelley conceived the novel *Frankenstein* during the chilly summer of 1816 in Geneva, Switzerland, where the bad weather convinced Lord Byron and his houseguests to stay indoors and make up scary stories. The white Christmas of Charles Dickens's 1843 story *A Christmas Carol* may owe much to the snowy winters in early nineteenth century England.

In Ireland, the summer of 1845 was cool, grey and wet, while the next summer was warm and dry. However, excessive dampness in even one summer could be disastrous. Irish farmers depended heavily on the potato as a staple crop. The potato blight fungus spreads rapidly in moist conditions, and it ruined the Irish potato crop of 1845, triggering a famine. In the ensuing years, poor government relief policies and a typhus outbreak made things worse. About one million people in Ireland died of starvation or disease during the famine (1845–1849), and about 1.25 million people left the country.

Even after the Little Ice Age, volcanic activity still brought on occasional cold spells. The 1883 eruption of Krakatau, in Indonesia, which killed about 36,000 people, resulted in a huge volcanic dust cloud that may have caused a worldwide drop in temperature that lasted for five years.

ABOVE *Irish tenants are shown after being evicted as a result of the economic troubles brought on by the potato famine. An estimated one million people in Ireland died as a result of the famine. The potato blight fungus was able to spread rapidly in the unusually cool, wet summer of 1845.*

THE TWENTIETH CENTURY

In general, since the Little Ice Age, climate has been growing warmer. Temperatures reached a high in the 1940s, became cooler until the mid-1960s, and since have been rising again, setting records for warmth in the 1990s. Despite this trend of global warming, some researchers believe that the Little Ice Age may not be over, but that it is being counteracted by rising carbon dioxide levels generated since the Industrial Revolution. It is possible that atmospheric circulation patterns characteristic of the Little Ice Age may still be in place. And if the Little Ice Age is a Rapid Climate Change Event like the Younger Dryas, it would be expected to last about a thousand years, not four hundred.

The story of climate is complex – never more so than in how it affects human history – as a few examples from the twentieth century will illustrate.

From the 1880s to the 1940s, a warming trend occurred in the Atlantic region of the Arctic and in northern Siberia. On the night of 14-15 April, 1912, however, some 640 km (400 miles) southeast of Newfoundland, cold was at work. That night, the British passenger ship *Titanic* collided with an iceberg and sank to the bottom of the North Atlantic. The disaster, which occurred on the ship's maiden voyage from Southampton, England, to New York City, killed 1,517 people. Icebergs form when large pieces of ice break off from the front of a glacier and splash into the ocean. The edges of the Greenland Ice Sheet produce thousands of icebergs a year, many of which drift into the North Atlantic, endangering shipping, particularly in the months from April to June. The Greenland Ice Sheet had existed since before there were ships, and had grown and thickened when

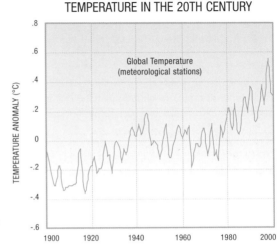
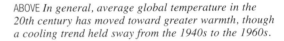

TEMPERATURE IN THE 20TH CENTURY

Global Temperature (meteorological stations)

ABOVE *In general, average global temperature in the 20th century has moved toward greater warmth, though a cooling trend held sway from the 1940s to the 1960s.*

glaciers worldwide were expanding. What sank the *Titanic* was a stray chunk of the ice ages.

Less than two decades later, dust and wind proved as potent as ice. The Dust Bowl of the 1930s brought economic disaster to the southern Great Plains of the United States. The causes – a drought and high winds – were aggravated by poor agricultural practices. Farmers in the early twentieth century had converted much of the natural grassland of the Great Plains to wheat fields, and the

LEFT *A painting by Willy Stöwer of the sinking of the* Titanic *during the night of 14 April, 1912. The disaster occurred when the ship collided with a giant iceberg in the North Atlantic. Such icebergs form when large pieces of ice break off from the front of a glacier, in this case the Greenland ice sheet.*

ABOVE *A Depression-era photograph showing an Oklahoma home in the USA covered in vast mounds of dust, as a result of a great dust storm in the 1930s. The dust storms were brought on by a drought on the Great Plains that lasted from 1931 to 1938. The effects were aggravated by poor agricultural practices; native grasslands had been largely cleared and converted into wheat fields and livestock were allowed to overgraze the remaining prairies.*

remaining grassland was overgrazed. By the early 1930s, these practices had left the soil dry and loose, and it was further damaged by drought from 1931 to 1938. From 1934 to 1938, strong winds swept the region, blowing away the soil in huge dust storms. Such storms had wracked the Great Plains before, but never on such a scale. About 40 large storms afflicted the area in 1935 alone, with dust frequently reducing visibility to about a kilometre. Newly planted fields were blown away, dirt had to be shovelled out of houses and vehicles were

disabled. Their farms ruined, many farmers went bankrupt and thousands of them abandoned the land, especially in the hardest-hit parts of Kansas and Oklahoma, where more than half the population left. Many of these 'Okies', as they were known, migrated to California.

Winter weather changed history as much as dust storms. In the 1940s, the North Atlantic Oscillation entered the phase of its cycle in which Europe becomes severely cold. In 1941, German dictator Adolf Hitler invaded the Soviet Union, but his

advance was impaired by the winter weather of 1941–42. That year, winter was unusually early and bitter, with temperatures that fell to -34°C (-30°F). It was, in fact, the coldest winter since the Russians began keeping temperature records in 1752. The German troops lacked proper clothes, supplies and equipment. Snowstorms blocked resupply trains. With the aid of the climatic shift, Soviet forces defeated the Germans in the Battle of Stalingrad (1942–43), a major turning point in the war.

Climate and weather events can also impact the founding of nations. On 12-13 November, 1970, a tropical cyclone, or hurricane-force storm, in the Bay of Bengal caused a storm surge, or rapid rise in sea level. The coast of East Pakistan, then part of Pakistan, was flooded, killing an estimated 300,000 people. East Pakistanis were already dissatisfied with the Pakistani government's treatment of their region. Public discontent boiled over when the government did not provide timely relief after the disaster. East Pakistanis accused the government of delaying relief shipments to stricken areas. Civil discord between East Pakistanis and the government escalated quickly in the ensuing months, erupting into civil war from March to December 1971. With assistance from India, East Pakistan won its independence as Bangladesh.

The effects climate will have on civilization in the future remain, as ever, unknown. What has become increasingly clear today, however, is that people are directly affecting climate and that the climate changes of the coming centuries may be brought on by humanity itself.

CLIMATE TODAY AND TOMORROW

HUMAN IMPACT ON CLIMATE

The Facts Behind the Theory

From prehistoric times until the relatively recent past, all people had at least one thing in common. Whether they lived in ancient China or eighteenth-century Europe, they could not affect the climate on a large scale; climate affected them. Today, something fundamental has changed. Through their actions – particularly the burning of fossil fuels and the release of greenhouse gases into the atmosphere – people are now influencing the world's climate.

HUMAN-GENERATED CLIMATE CHANGE

Before the Industrial Revolution, human beings affected only their local or regional climate when they cut down or burned forests, drained swampland or built cities. Then, for good or for ill, they lived with the consequences, most of which were local. But with accelerating advances in technology and the world's rapidly increasing population, people are now causing changes in the global climate – with global consequences.

Some scientists argue that humankind has entered a new geological interval in which the human influence on the global climate and ecosystem has become dominant. This new interval has been dubbed the Anthropocene epoch (from the Greek *anthropo,* human being, and *cene,* recent). The name originated with Nobel Prize-winning Dutch atmospheric chemist Paul Crutzen.

According to Crutzen, climate change is only the most obvious of the human-generated effects that are altering the planet physically, chemically and biologically. In his view, the Holocene epoch ended and the Anthropocene began in the late eighteenth century, around the time of James Watts's invention of the steam engine and the beginning of the Industrial Revolution, when global concentrations of carbon dioxide began to rise.

Other scientists have proposed that the Anthropocene epoch started as early as remote antiquity. Regardless of how geological time is divided and labelled, however, the growing influence of humankind on the environment is undeniable. Since the Industrial Revolution, a new, complex link has been forged between human activity and climate – rising levels of carbon dioxide.

Over the past two hundred years, the emission of greenhouse gases such as carbon dioxide – through the extraction and burning of fossil fuels and other activities – has caused climate changes that are often oversimplified and placed under the umbrella term 'global warming'. Over the last century, Earth has been growing warmer overall at an accelerating pace, yet the changes are complex and vary from place to place: some regions are growing warmer, others cooler; some drier, others wetter. Anthropogenic, or human-induced, climate change is poorly understood and not very predictable. Although a small minority of scientists are sceptical about anthropogenic climate change, most scientists are convinced that it is real. Certainly, there is enormous evidence of its existence in a variety of forms, from records of carbon dioxide and temperature levels down through the centuries, to highly sophisticated climate models – computer simulations that show what the global climate was like in the past or will be like in the future.

Heat Islands

A barefoot walk across asphalt in midsummer demonstrates how human beings have transformed their local climate. Replacing light-hued vegetation with dark asphalt alters the land's reflectivity, or albedo (see page 22), and changes how much heat the land absorbs and retains. Covering soil with asphalt also reduces the amount of water that the ground absorbs.

Changing the albedo is not a new phenomenon. Deforestation and agriculture, occurring over thousands of years, can affect regional climate by altering the land's albedo and moisture retention, through changed rates of absorption and evaporation, changed water levels and other factors.

RIGHT *Steam and smoke rise from an industrial plant in British Columbia, Canada.* PREVIOUS PAGE *Clockwise from upper left: building goes on in Beijing, China, even as dust storms, an effect of drought, exacerbate smog; slash-and-burn practices clear Brazilian rain forests for cattle grazing; drilling rigs look for oil beneath the ocean; seabirds are coated with oil from a tanker wreck – potentially a more common sight if melting Arctic ice opens new shipping routes.* PAGES 82–83 *Rubbish chokes a canal in Beijing. Drought diminishes access to safe drinking water, already limited by pollution.*

Artificial Weather: Showers and Clouds

Not all human influences on weather and climate have been inadvertent: deliberate attempts to change weather have been common as well.

In cloud seeding, particles of a substance such as silver iodide or dry ice are introduced by aircraft into clouds to act as freezing nuclei. If the technique works, cloud droplets stick to the introduced particles and shower to the ground as precipitation. The droplets must be supercooled, existing in the liquid state at subfreezing temperatures, and for that reason the method works only on cold clouds tall enough to reach into high, cold regions of the atmosphere. Cloud seeding has also been used to suppress hail and to dissipate cold fog, a kind of fog consisting of supercooled water droplets.

Cloud seeding has been practised for more than 50 years in 43 countries, mainly to increase precipitation for agriculture or for hydroelectric uses, yet there are still disputes about its effectiveness. Some studies indicate that the technique does not increase precipitation, while others suggest that, under the right conditions, it can boost precipitation by 5 to 20 per cent.

ABOVE *This cylinder of silver iodide is used in cloud seeding. Aircraft introduce the substance into high clouds in order to cause precipitation.*

ABOVE *Russia's Myasishchev M-17 was designed for map-making photography and hail control. It seeds thunderclouds with small crystals that cause the release of hail before it reaches crops, and damages them. Conceivably, the M-17 could carry ozone generators to 'patch' holes in the ozone layer.*

Cities are called heat islands because their temperatures are a few degrees higher than the surrounding region. The heat island effect is just one easily understood example of human-generated climate change.

Asphalt and other materials with low albedos are concentrated within cities, trapping more heat than less developed outlying areas. Urban 'canyons' are full of surfaces that reflect incoming sunlight, with some absorption happening at each point. The canyon-like topography also tends to keep the long waves of infrared radiation from escaping. Industrial activity, air conditioning and vehicle exhausts also generate heat. And because drainage systems are usually underground, the streets tend to be drier than the vegetated rural areas, so less surface moisture exists to moderate the heat. As a result of all these influences, cities tend to be considerably warmer than their environs – problematic during heat waves, advantageous during cold spells.

The phenomenon of urban heat islands influences local precipitation as well. The built-up heat increases convection; more warm, moist air rises and condenses into clouds. Because cities also tend to block winds, the rising air potentially contributes to the cloud cover. Industrial, vehicular and other emissions may introduce large concentrations of particles into the atmosphere; these serve as nuclei for the condensation of clouds and the increase of precipitation. Given the right regional wind and moisture conditions, cities can generate greater quantities of rain and snow.

Some city planners have taken steps to mitigate the heat island effect. Introducing more vegetation in parks and elsewhere is one tool in that effort. In Germany, the city of Stuttgart has deliberately added plant materials to the urban landscape to limit the heat island effect. In the United States, the community of Dayton, Ohio, has tried building 'green' parking lots – car parks that intermix grass and tarmac. Such efforts have to be citywide, however, or their effects will be limited to their immediate area.

Fossil Fuels, Greenhouse Gases

Local climate changes are just one part of the global picture. Scientists now estimate that carbon dioxide in Earth's atmosphere has increased by 31 per cent since pre-industrial times and is higher than it has been in 420,000 years and probably in the past 20 million years. Other greenhouse gases – methane, nitrous oxide and tropospheric ozone – are also at their highest recorded levels. Air bubbles trapped in ice cores taken from glaciers provide a continuous record of atmospheric carbon dioxide levels stretching back 1,000 years; they confirm that carbon dioxide levels began to rise during the Industrial Revolution.

The primary cause of the rise of these greenhouse gases is the combustion of fossil fuels – that is, coal, oil and natural gas. Fossil fuels are so named because they were formed from the fossilized remains of prehistoric plants and animals. They have been extracted from the earth and have been burned in increasing quantities since the early nineteenth century, first to power factories and trains and later to drive cars, planes and utilities.

These fossil fuels now provide 90 per cent of the world's commercial energy. All plants and animals are predominantly composed of molecules of carbon

RIGHT *Creating more grazing land, often by clearing forests, puts more greenhouse gases in the air: cows release methane, one of the greenhouse gases.*

and hydrogen. When fossil fuels are burned in air, their molecules react with atmospheric oxygen to form oxides of hydrogen – that is, water – and oxides of carbon – carbon dioxide and carbon monoxide. On average, burning one gallon of petrol creates 22 pounds of carbon dioxide (or 10 kg of carbon dioxide for every litre of petrol). Seventy per cent of the world's fossil fuels are burned by industrialized nations, even though they comprise less than 25 per cent of the world's people.

Agriculture and land-use changes are another cause of the rise in greenhouse gas levels. Forests are carbon sinks – natural storage areas for carbon – as they absorb carbon dioxide from the atmosphere and turn it into wood, leaves and other organic matter. Especially during the last two centuries, great swaths of forest have been cut for timber and to make way for fields and towns. This land's capacity to store or capture carbon may be significantly reduced. In addition, the cut wood, when burned or left to decay, releases the carbon that had been stored in the living forest, increasing the atmospheric carbon dioxide concentration.

Carbon dioxide is not the only greenhouse gas originating from agricultural activity. Nitrous oxide, a greenhouse gas 300 times more powerful than carbon dioxide, is given off when farmers use nitrogen-based fertilizers. Methane is released by rice paddies and cattle, which have proliferated as food sources as the size of the population increases.

HOW WE UNDERSTAND CLIMATE

Scientists have used many methods to determine whether and how human activity is changing Earth's climate. They begin with careful observations of climate conditions, past and present, around the world. Pursuing further knowledge, they also pore over historical records, study ice cores, run statistical analyses, evaluate climate models and look at classifications of large geographic regions on biome maps. Eventually, the facts emerge.

Climate Observations

To learn more about Earth's climate, scientists deploy specialized instruments on land, on or under the sea, suspended from balloons in the upper air and aboard satellites in space. These instruments record a host of data: temperature, humidity, precipitation, wind speed, energy fluxes entering and leaving the atmosphere and the changing extent of sea ice in the polar regions, to name a few. One of these instruments, the radiosonde, has been used since the 1930s to measure air temperature, humidity and air pressure from a balloon at altitudes of up to 29 km (18 miles). Since 1985, the Advanced Very High Resolution Radiometer (AVHRR) carried by the United States' National Oceanic and Atmospheric Administration (NOAA) and National Aeronautics and Space Administration (NASA) satellites have been providing daily measurements of sea surface temperatures worldwide.

Instrumental records of climate typically extend back only to the middle of the nineteenth century. For reliable evidence of past weather and climates, scientists draw on evidence from sea floor sediments (deep sea cores), ice cores, tree rings,

ABOVE *Magnetic fields that conduct electrified gases over the surface of the Sun are measured using a Solar Vector Magnetograph. This data helps predict solar activity that affects weather.*

LEFT *A weather balloon held by a scientist at the European Space Agency's (ESA) Ariane launch site in French Guyana carries instruments for measuring temperature and humidity.* UPPER LEFT *The MetOp satellite, seen in an artist's view, is now in the planning stages at the ESA. This will be one of three operational meteorological polar orbiting satellites. The first launch is planned for 2006.*

pollen accumulations, lake bed sediments, fossils and paleomagnetism.

The instruments observing the climate from surface, air and space generate a vast amount of data. This information is gathered, compiled, processed by computers and subjected to sophisticated analyses. Statistical techniques are applied as climatologists search for cycles, trends and shifts. The challenge is to distinguish changes indicating real trends from natural climate fluctuations. For further understanding, comparisons are made with historical records.

Climate Models

Scientists use computer-based tools called climate models to understand the key processes that govern climate and, ultimately, to predict climate change. Atmosphere-ocean general circulation models (AOGCMs) are the most sophisticated variety to date. They are similar to the models used to predict the everyday weather, in which the physical laws governing the motion of the atmosphere are merely systems of equations to be solved on supercomputers. However, climate models must also include equations representing the behaviour of the oceans, vegetation on land and the cryosphere (including sea ice, glaciers and ice caps). Many climate models have been developed by scientists all over the world. For the 2001 report of the United Nations-sponsored Intergovernmental Panel on Climate Change (IPCC), 34 models from 18 research laboratories were used.

Among these were models that were run by the Hadley Centre of the British Meteorological Office and by the NOAA Geophysical Fluid Dynamics Laboratory in the United States.

Climate models are only approximate representations of very complex systems. They cannot yet simulate all aspects of climate. However, since their modest beginnings in the mid-1950s, they have constantly improved and confidence in the ability of models to project climate is growing. Climate models can now provide credible climate simulations, at least down to subcontinental scales. For instance, they have been able to reproduce the twentieth century's warming trends as well as some aspects of the El Niño–Southern Oscillation and of climates in ancient times. Climate models can never be 100 per cent accurate, but at the moment they are the best tools available.

Biomes and Climate

Scientists often divide the globe into biomes, large geographic regions that are distinguished from one another by their climate and dominant vegetation. Maps of these biomes provide climatologists with valuable information on climate and climate trends.

One particularly useful system for the study of climate is that of Holdridge life zones, in which terrestrial biomes are differentiated by biotemperatures (a combined measure of temperature and growing season), annual precipitation and evapotranspiration ratio (a measure of how much moisture is available in the environment).

Holdridge life zones span all the ecosystems of the globe, from the polar desert to cool temperate steppe to tropical rain forest. Changes in the size and the location of a particular biome may reflect changes in the climate patterns of the region as well as changes in its flora, fauna and agricultural potential that may affect the livelihood of the local population.

ABOVE *An ESA satellite photo shows the Ubari and Murzuq sand seas in Libya. Monitoring desert weather and dust storms is vital to climate observation.*

GLOBAL WARMING AND OTHER CLIMATE CHANGES

Evidence that an increased concentration of greenhouse gases in the atmosphere contributes to global warming is abundant. In Earth's prehistoric past, higher concentrations of greenhouse gases have been associated with warmer periods. The tale of our climate over the past 400,000 years is told by an ice core more than 3.2 km (2 miles) long, taken from Vostok Station in East Antarctica. The bubbles inside the core – the longest ever recovered on Earth – reveal an extremely close and consistent correlation between the concentration of carbon dioxide in

records began nearly 150 years ago – and all were within the past decade. The IPCC (Intergovernmental Panel on Climate Change) reported in 2001 that global average surface temperature increased during the twentieth century by 0.6°C (1.1°F), and that the increase in temperature during the twentieth century in the Northern Hemisphere was probably greater than for any other century in the last thousand years. The upward trend was not continuous: temperatures increased from the turn of the century to the 1940s, then cooled slightly until the

be measured from the ocean surface all the way down to a depth of 2.9 km (1.8 miles).

The amount of floating ice in the polar oceans is rapidly diminishing. This includes perennial ice – ice that remains year-round, even after seasonal ice has melted in the summer. Studies show that ice thickness may have decreased by 40 per cent over the past 50 years, while the extent of perennial sea ice is diminishing faster than previously thought, at a rate of nine per cent per decade.

Ice at higher altitudes is also disappearing. From Mount Kilimanjaro in Africa to the Himalayas in Asia, mountain glaciers worldwide have melted significantly since the 1960s. There has been an estimated net loss of more than 400,000 cubic km (96,000 cubic miles) of water – more than the annual discharge of the Orinoco, Congo, Yangtze and Mississippi rivers combined. This loss occurred more than twice as fast in the 1990s than during previous decades.

THE TALE OF OUR CLIMATE OVER THE PAST 400,000 YEARS IS TOLD BY AN ICE CORE MORE THAN 3.2 KM (2 MILES) LONG, TAKEN FROM VOSTOK STATION IN EAST ANTARCTICA

the atmosphere and air temperature over the past 400 millennia: carbon dioxide rises, the air temperature rises; carbon dioxide falls, the air temperature falls. Because of this evidence, there is reason to believe not only that global temperatures will rise in the future as a result of the burgeoning quantities of greenhouse gases, but also that they have already been rising with a vengeance.

Global temperatures are now rising faster than at any time in the past thousand years. At the end of 2004, the World Meteorological Organization announced that 1998, 2002, 2003 and 2004 were the four hottest years since the use of instrumental

1980s, when the warming trend resumed; it has continued ever since. Temperature changes have not been uniform across the globe but have varied from region to region and even from one part of the lower atmosphere to another.

Signs of Climate Change

Both measurable and visible signs of climate change are all around us. Besides the recorded rise in terrestrial surface temperatures, the world's oceans, our global heat sink, have warmed at least ten times more than other parts of Earth for the last half century. The warming can

Changing climates have also spurred numerous alterations in living things, from the rhythms of their biological processes to the patterns of their migrations. Many birds are reproducing earlier and flowers are blooming earlier than they have in the past. In some cases, the changes can be deadly: coral reefs have been destroyed by the water becoming too warm; polar bears can be threatened by the loss of sea ice on which to hunt.

Other elements in Earth's climate system have also changed significantly over the past century. Precipitation has probably increased by five to ten per cent over the Northern Hemisphere at the middle and high latitudes since the late 1960s, although snow cover has declined by ten per cent. Average sea levels globally rose by between 10 and 20 cm (4 and 8 in) during the twentieth century. Records dating back 300 years from Amsterdam, the Netherlands; Brest, France; Swinoujscie, Poland; and other sites in Europe confirm that sea levels rose more rapidly during the twentieth century than during the nineteenth.

The IPCC has determined, based on climate data for the last thousand years, that the twentieth-century climate change was indeed unusual and unlikely to be the result of entirely natural causes. Natural factors, including volcanic eruptions and variations in solar radiation, may have been played a part and been the primary contributors to warming during the first half of the century. They do not, however, explain the warming since 1950. When climate models exclude the natural effects of volcanic eruptions and variations in solar radiation and include only the anthropogenic effects of greenhouse gases and sulphate aerosols – an industrial pollutant – a distinctive pattern of temperature change is revealed, one that closely fits the temperature observations of the last 50 years. The evidence is strong: the IPCC panel of experts concluded that most of the warming observed during that period has been the result of the increase in greenhouse gas concentrations generated by human activities.

The IPCC predicts that if carbon dioxide concentrations continue to increase, the globally averaged surface temperature will rise between 1.4°C and 5.8°C (2.5°F and 10.4°F) by 2100. The effects on the world's societies and ecosystems would be powerful and widespread.

Climate Contrarians

In science, dissent is essential to the process of discovery. Scientists are trained to scrutinize and test explanations and results – whether their own or those of colleagues – and to regard conclusions as perpetually revisable in the light of new evidence. Absolute certainty is not regarded as possible. As a result of their scepticism, scientists rarely achieve unanimity, especially on complex subjects that are still under

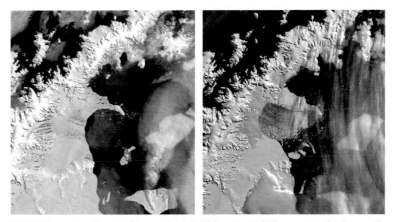

ABOVE *The epic collapse of Antarctica's 3,237-sq.km (1,250-square mile) Larsen B Ice Shelf is shown in early 2002, in satellite photos. The dark striations on the ice shelf, left, are melt ponds.* RIGHT *The break-up of the shelf is shown.*

BELOW *Spring melt of sea ice in Admiralty Inlet, Nunavut, Canada. Floating ice in the polar oceans is rapidly diminishing due to global warming.*

Advice on the Issues

In 1988, the United Nations Environment Programme (UNEP) and the World Meteorological Organization (WMO) established the Intergovernmental Panel on Climate Change (IPCC) to provide independent scientific advice, 'on a comprehensive, objective, open and transparent basis', on the issue of climate change. The panel was asked to prepare a report on climate change and the impacts of such change, based on available scientific information and to formulate response strategies.

The IPCC's First Assessment Report appeared in 1990. Since then, two more multi-volume assessment reports have been published, with the fourth report due in 2007. To date, about 1,000 experts from more than 100 countries have been involved in the drafting of the reports and an additional 2,500 experts have worked on the review process. IPCC authors are nominated by governments and international organizations and are drawn from universities and research centres, as well as business and environmental associations. The IPCC does not conduct new research.

The teams of authors prepare their drafts based in large part upon peer-reviewed literature and on the most recent scientific findings and reports from national academies of sciences, industry and UN bodies. These drafts then undergo a two-stage review and the final reports are accepted at the annual plenary meeting.

The IPCC's reports have often been controversial, drawing criticism from climate contrarians. In 2002, the controversy was linked to the removal of the IPCC chairman, Robert T. Watson, a US citizen who was widely believed to have been voted out of office because of his outspokenness about the dangers of climate change. His successor, India's Rajendra Pachauri, affirmed the urgency of the issue, saying, 'Climate change is for real. We have just a small window of opportunity and it is closing rather rapidly. There is not a moment to lose'.

investigation. Consequently, the existence of climate contrarians – as the dissenters on the issue of human-generated climate change are sometimes known – is not surprising. Some scientists continue to disagree or to remain sceptical about one or more aspects of climate change. What is more significant is how broad the scientific consensus has become on these issues.

For example, there is widespread agreement that carbon dioxide levels are rising. And there is agreement, even among most contrarians, that global average surface temperature has increased during the course of the twentieth century. But debate remains about the degree to which humans play a role in climate change. Some scientists argue that natural variability can account for the observed climate changes. And opinion varies on how climate will change in the future.

The scientific issue now rests largely on how important certain environmental feedbacks will be – for example, whether the water vapour produced by increased evaporation in a warmer climate will reinforce warming or counteract it. It also rests on what happens to human-generated carbon dioxide emissions once they enter the atmosphere, how long they stay there and where they go afterwards.

Answering the Contrarians

Dissenting views in this debate certainly have not been suppressed; most of the contrarians' objections have been considered and taken into account in IPCC documents and the climatological papers that support them. Indeed, those works are sometimes the source of the contrarians' data. The US-based climatologist Stephen H. Schneider, formerly of the National Centre for Atmospheric Research and now a professor

of interdisciplinary environmental studies at Stanford University, supports the existence of human-generated climate change and said in a 1999 interview, 'The part I've resented most about the contrarian attack is the pretence that they're the guys with the caveats. Almost all the uncertainties and caveats they raise are in the papers we write'.

A great deal of evidence has been published showing that human activity is the cause of the current observed climate change and that this climate change will continue. Some of this evidence comes from climate models developed by climatologists, and dissenters raise doubts by arguing that such models may be erroneous and are unverifiable.

In fact, some climatologists agree that a certain amount of scepticism is healthy and has helped to ensure the development of better and more reliable models during the past ten to fifteen years. Ronald Miller of the Goddard Institute for Space Studies of the US National Aeronautics and Space Administration (NASA) put the problem of answering the sceptics this way: 'It's hard to know if we will ever 'win' over the climate change sceptics. The models ... give us a chance to see what effect our actions will have before we have to live with the consequences of our actions. The only other way to find out what will happen is to wait around and see. That would be risky, risky business'.

Some critics say that climate models are inadequate because the models disagree about changes on local or regional scales. There is no doubt that different global climate models do make somewhat different predictions about changes on those scales, yet all model simulations agree on the key elements of global changes.

Some dissenters have argued that even in the event that the global temperature does rise, the effects will be largely beneficial. Who would not welcome the idea of milder winters and the possibility of longer growing seasons?

But human-induced climate change is not simply a matter of global warming. Many far less benign conditions may develop as a consequence. Drought, flooding, stronger hurricanes and even the extinction of species may follow, and each of these disasters has its own perils for humanity.

Business Interests

Not all contrarians are scientists. Energy corporations, and groups and individuals economically invested in fossil fuel production and use – as most of us are to some degree – are active in denying the existence or importance of human-induced climate change. Some energy corporations have paid for contrarian research and lobbied politicians in Washington, DC, to oppose efforts to limit climate change. This is not surprising: for anyone who enjoys the fruits of fossil fuel use, it is difficult to come to grips with the fact that fossil fuel consumption will have to be reduced. Yet that does not alter the evidence that climate change generated by the various activities of human beings is already well under way.

Finally, there is the frequently heard opinion that nothing should be done about climate change until there is 100 per cent agreement about it, because the actions that must be taken to slow climate change, or to mitigate it, are far too expensive to undertake without certainty. That position illustrates the misconception that unanimity can be expected in science and ignores the possibility that it may cost even more to deal with the consequences of climate change in the future. Absolute certainty is no more available on this subject than on any other scientific issue. In fact, lack of certainty may indicate the wisdom of taking an aggressive approach to limiting climate change. It could well be that the effects of climate change will be worse than anticipated.

Even if the certainty of climate change is in doubt, preventive actions can be of benefit whatever the future outcome. For example, changing to a more fuel-efficient vehicle or improving energy efficiency in the home reduces fuel costs, even as less carbon dioxide is released into the atmosphere.

Based on current trends, the United Nations Environmental Programme Finance Initiative estimates that natural disasters driven by climate change will cost the global economy $150,000 million a year within the next decade. What matters in preparing for the future is not what is known for certain (which is bound to be little) but what there is probable cause to think will happen.

LEFT *Controlled burning emits greenhouse gases as it clears a forest in Guatemala. Combustion releases carbon and puts more carbon dioxide into the air; the bare land may be less able to capture and store carbon.*

TOO HOT TO HANDLE

Effects of Warming

Climate change will affect different regions in different ways. Many places will become hotter. Some already are – Earth's global average surface temperature increased during the twentieth century by 0.6°C (1.1°F). Research shows that Earth is now out of energy balance, with more solar energy coming to the planet than terrestrial heat radiating into space. As this heating-up process continues, the effects will become more severe and if the temperature rises high enough, the damage to species may increase beyond the point of no return.

HOW HEAT AFFECTS LIVING THINGS

Although not necessarily the most striking consequence of the changing climate, rising heat is the consequence most readily associated with the term 'global warming'. It also has a direct effect on many aspects of life on Earth, from agriculture to forestry, from the health of the human population to the welfare of the animal kingdom. In considering the future of a warmer world, one question is key: how much heat can we handle?

Rising temperatures have a direct impact on living things of all kinds, from bacteria to plants to animals. Heat affects many aspects of biology: the chemical reactions that are the basis of life; individual growth rates; availability of food and moisture; the timing of biological events such as flowering, spawning and birth. Even after an organism dies, heat is still important, because the decomposition process is also temperature-dependent. The very morphology, or form and structure, of organisms can change as temperatures rise. A study of red deer in Norway found that males have become larger as the climate grew warmer over the past 32 years. Similarly, the eggs of European pied flycatchers have become bigger as warmth has increased.

A 2003 report by biologist Camille Parmesan, of the University of Texas, USA, and economist Gary Yohe, of Wesleyan University, USA, showed that a 'climate change fingerprint' is already visible in living systems. Parmesan and Yohe found effects of rising temperatures on 84 per cent of the 334 species they studied. These included signs that rising temperatures were having an effect – signs such as species migration to new locations or changes in the timing of flowering or hatching. Some of these changes appear benign (longer growing seasons, for example). Others clearly are not (greater heat causing a decline in coral populations). Though optimal temperature – the temperature range within which an organism thrives – varies from species to species, all living things will be affected by climate change, sometimes in complex ways.

In the boreal forest, or taiga, of northern Canada and Alaska, temperatures have risen 2°C (3.6°F) during the last 90 years. A study of growth rings showed that the trees grew faster during the 1930s and 1940s as temperatures and carbon dioxide levels increased. But after this period, the growth rate levelled off, even as temperatures and carbon dioxide levels continued to increase. The study found that the warmer temperatures that encouraged tree growth also caused moisture to evaporate faster and enabled harmful insects to spread and mature more quickly. As a result, trees grew more slowly.

Climate change is also expected to bring more fires. Boreal forests benefit from some degree of fire, but too many fires can be damaging and fires have been on the increase since the late twentieth century.

It is difficult to know exactly how much heat an ecosystem can handle. Some studies forecast that the boreal forest will expand northwards as the Arctic warms. Others predict that the combination of warmer temperatures, longer droughts and more harmful insects will cause increasing damage to the boreal forest. Either way, the result will be significant, because boreal forests make up one third of the planet's forests.

Heat's Effects on Human Health

Increasing warmth can have positive and negative effects on human physiology. Winter feels more bearable on a mild day than a cold day. But in summer, an increase in temperature that is coupled with high humidity can produce a heatwave that subjects people to heat stress and other heat-related injuries.

RIGHT *A wildfire burns rapidly through a forest near Leuk, Switzerland, in 2003.* PREVIOUS PAGE *Clockwise from upper left: a dry lake bed in Bhopal, India, where hundreds died from a heatwave in 2002; industrial emissions billow into the sky; corn dries and withers on the stalk during a drought in central Texas, in the United States; a man tries to control a brush fire on the island of Kalimantan (formerly Borneo).*

CLIMATE AND MALARIA RISK
— Current medium-/high-risk areas of malarial transmission
— Projected areas of medium-/high-risk malarial transmission
Current Biomes:
☐ Ice cap
☐ Tundra
☐ Subarctic
☐ Cool continental
☐ Warm and humid
☐ Mediterranean
☐ Semi arid
☐ Arid
☐ Tropical
☐ Humid equatorial

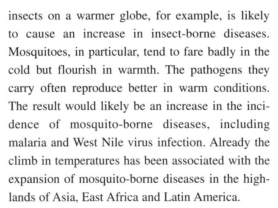

The human body can keep its temperature stable in varying conditions, but only up to a point. People who become overheated may suffer heat exhaustion, also called heat prostration; they cannot sweat enough to regulate their body temperature and feel dizzy, nauseous, tired or weak. Worse, heatstroke may occur: the body's heat-regulating mechanism is disturbed, sweating stops and body temperature rises high enough to damage the brain and other organs. Untreated, heatstroke can be fatal.

Heat strains human health in other ways. Under hot conditions, the heart must work harder to pump more blood to the capillaries in the upper layers of the skin, where these tiny blood vessels can shed some of the excess heat to the atmosphere. Thus, high temperatures put a strain on the heart. Although metabolic rate initially increases as temperatures rise, metabolism may fall as the body limits how much heat is produced by internal chemical processes. As metabolism falls, ability to resist infections falls as well. When global temperatures rise, severe heatwaves are likely to become more frequent, sometimes with fatal consequences for the very old, the very young and the frail.

Warmer weather may also lead to an increase in the occurrence of diseases. The proliferation of insects on a warmer globe, for example, is likely to cause an increase in insect-borne diseases. Mosquitoes, in particular, tend to fare badly in the cold but flourish in warmth. The pathogens they carry often reproduce better in warm conditions. The result would likely be an increase in the incidence of mosquito-borne diseases, including malaria and West Nile virus infection. Already the climb in temperatures has been associated with the expansion of mosquito-borne diseases in the highlands of Asia, East Africa and Latin America.

Even the simple act of breathing may become harder as the global climate grows warmer. Allergies will probably increase as rising temperatures and carbon dioxide levels speed plant growth. Physician Paul Epstein of Harvard Medical School, USA, showed in 2002 that when ragweed – the source of one common allergen – grows in an atmosphere with double the present-day carbon dioxide levels, it produces 61 per cent more pollen. Allergy-caused strain on the human respiratory system is likely to be

ABOVE *Risk of malarial transmission today is largely limited to the non-desertified, humid, tropical climes and, within these, high-altitude mountain areas are also relatively risk free. However, projected climatic changes over the next century are likely to extend this high-risk area far beyond the tropics, into hitherto negligible risk areas, some of them densely populated. These include the eastern seaboard of the United States, much of mainland Europe and southern Scandinavia and a wide area of increased risk across Central Asia.*

aggravated by air pollution, which tends to grow worse during heat waves. Especially over urban areas, stagnant air masses permit the accumulation of such contaminants as carbon monoxide, sulphur dioxide and particulates (particles found in the air, including dust, soot and smoke).

Extreme Heat and Its Effects

In spring 2002, daytime temperatures in India climbed above 46°C (115°F). This led to more than a thousand heat-related deaths, mostly among the poor and elderly, who were more vulnerable.

In summer 2003, Europe experienced a series of extraordinary heat waves that killed thousands of people. Scientists linked the devastating mortality to a curious fact about Earth's present climate: night-time low temperatures have been rising almost twice as fast as daytime highs. As carbon dioxide levels have risen, they have tended to trap heat near Earth's surface at night rather than allowing it to radiate into space. The result was that Europeans in 2003 had little chance to cool off in the evenings; heat stress increased, leading to more deaths, especially among the elderly. The World Meteorological Organization has projected that heat-related deaths in the world's cities will double within twenty years.

Forest fires are another consequence of extreme heat when it is accompanied by drought. Forest fires have been increasing for decades throughout Europe's Mediterranean region, from Greece and Italy to France and Spain. In 2003, the worst forest fires in 23 years raged through Portugal, threatening species such as the Iberian lynx, which is one of the most endangered cat species in the world, and the Bonelli eagle, of which there are currently fewer than 1,000 breeding pairs left in Europe.

In North America, too, the frequency of forest fires has increased. In California, nearly 303,500 hectares (three-quarters of a million acres) burned in 2003 alone. In Canada's Yukon, fires razed forests covering nearly 1,011,000 hectares (2.5 million acres) in 2004, darkening skies for days in some places. In a vicious cycle, fires stoked by climate change contribute to further climate change as the burning of the trees releases heat-trapping carbon dioxide.

As the number of extreme heat events increases, there will be an economic price. Stress on human health and an increasing number of forest fires both carry large economic costs. On a more mundane level, the cost to tourism in resort areas will also be significant. For example, by the year 2050, if global warming continues unabated, the eastern Mediterranean will experience twice as many days per year above 35°C (95°F) as at present. And as temperatures increase, air pollution is also likely to get worse. Even without heatwaves, hotter temperatures are likely to affect holiday destinations around the world. Areas known for their scuba diving – the Maldives, the Pacific Islands and Australia – may see fewer visitors as corals die off due to the increase in ocean temperatures.

DEATH IN PARIS

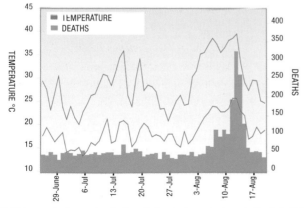

ABOVE *This graph shows deaths in relation to temperatures in Paris in 2003. The lower red line shows the daily low, the upper line the daily high.*

Lost to the Heat

In July and August 2003, during the big European heatwave, more than 25,000 fires were recorded in Portugal, Spain, Italy, France, Austria, Finland, Denmark and Ireland. Just under 650,000 hectares (1.6 million acres) of forest were destroyed. Portugal was the worst hit, with nearly 400,000 hectares (988,000 acres) burned, amounting to 5.6 per cent of its forest area, with a financial impact estimated at more than 1,000 million euros (£700 million).

Throughout Europe, the main economic impacts of the heat were on agriculture, livestock and forestry. In Switzerland, fodder had to be imported from as far away as Ukraine. Wine-making was seriously affected. Cereal production in the European Union dropped more than 23 million tonnes from 2002 levels. Crop development was accelerated by 10 to 20 days. More water had to be used for crops because of the high temperatures and solar radiation. Together with the summer dry spell, this acutely depleted the water in the soil and caused a reduction in crop yields.

ABOVE *A Swiss army helicopter fights a forest fire near the village of Leuk, Switzerland. Many wildfires in the area were linked to the 2003 drought, which caused dangerously dry conditions.*

CORAL BLEACHING

Coral reefs, the hard limestone structures built by tiny animals, appear brilliant in colour because of the microscopic algae that make their home inside the coral's tissues. Habitats for about 25 per cent of marine species, reefs support more biodiversity than most other marine ecosystems. But rising ocean temperatures are destroying reefs in places as far apart as Australia, the South Pacific nation of Fiji and the Central American nation of Belize.

Coral reefs provide natural breakwaters that protect shorelines from storms and also benefit humanity through fisheries and tourism. The shallow-water reefs that are familiar to most people occur principally in warm tropical and subtropical seas and they can easily be damaged by heat.

Rising ocean temperatures make coral reefs susceptible to a harmful phenomenon called coral bleaching. In this process, corals turn white because they lose the microscopic algae that live inside their cells and nourish them.

(Other marine organisms, such as giant clams, are also susceptible to bleaching when rising temperatures disrupt the relationship between host and symbiotic algae.) Without microalgae, the corals' survival is threatened and the reefs no longer provide a good habitat for fish. The disappearance of those fish, in turn, affects species higher on the food chain that depend on them, including human beings.

According to a 2001 study by the United Nations Environmental Programme, 58 per cent of the planet's coral reefs are threatened by human activities or by human-induced climate change. During mass bleaching episodes, the damage can be severe: in 2002, 60 to 95 per cent of Australia's Great Barrier Reef was affected by bleaching, with some areas experiencing 90 per cent mortality. During a worldwide bleaching episode in 1998, several reefs in the Indian Ocean suffered 95 per cent mortality.

Some coral populations are able to recover from bleaching episodes, but not when the episodes become more frequent and severe. Long-term or frequent bleaching may severely weaken the corals and make them more vulnerable to other stresses, such as overfishing, pollution, disease, predation and natural climate fluctuations such as those that are regularly caused by El Niño.

The threat to coral reefs is tied to climate change in an even more basic way. Increasing levels of atmospheric carbon dioxide not only drive climate

LEFT *Rising ocean temperatures can kill the microscopic algae that live inside coral and give it such vivid colorations.* UPPER LEFT *Orange cup coral.*

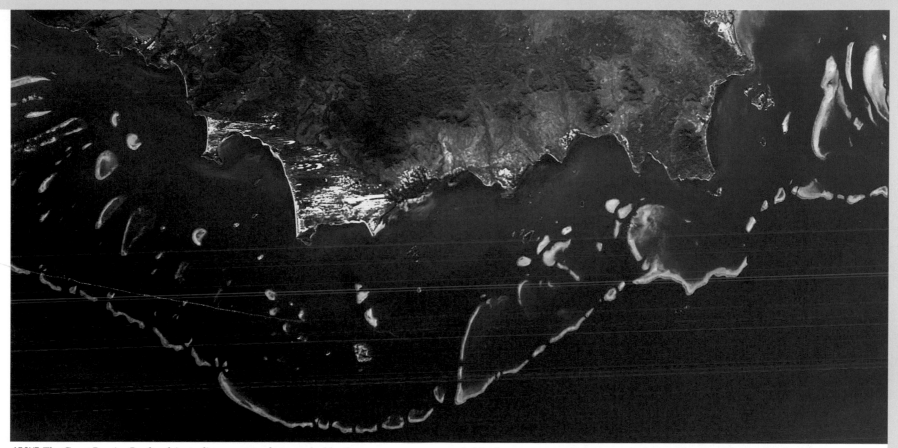

ABOVE *The Great Barrier Reef and Australian coast as photographed by the International Space Station (ISS). The smaller reefs that make up the Great Barrier Reef form shallow areas and small islands, which appear white and green. These are surrounded by deeper waters, which appear dark blue. The Great Barrier Reef is home to 400 species of coral and 1,500 species of fish. It is the world's largest reef, extending for 2,000 km (1,240 miles), and is now in danger as ocean temperatures rise steadily.*

change but also affect ocean chemistry. The higher the concentration of carbon dioxide in the atmosphere, the more carbon dioxide dissolves into the ocean surface. That raises ocean acidity and lowers the concentration of carbonate that corals and other marine organisms use to build their skeletons. As a result, rising carbon dioxide levels affect corals in two ways: by increasing global temperatures, causing potentially lethal bleaching and by slowing coral growth, so that it is harder to keep up with rising sea levels.

Corals will not necessarily become extinct. The rises in sea levels expected to accompany climate change within the next centuries are unlikely to drown coral reefs: reef expansion can keep up with them. Deep-water coral reefs are plentiful, reef organisms are adaptable and climate change varies greatly from place to place. But if warming is not curtailed, few shallow-water reef communities may survive, if any, and those that do will be quite different from now – with costs that are high, both ecologically and economically.

CORAL BLEACHING THRESHOLD

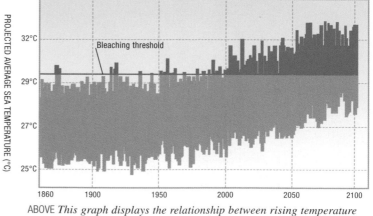

ABOVE *This graph displays the relationship between rising temperature and coral bleaching. At the bleaching threshold, corals begin to die.*

Dry, forested countries such as Australia and Spain could conceivably become less attractive to tourists as forest fires and insect-borne diseases that accompany warming decimate the forests that attract many vistors to these countries. On the other hand, even minor increases in temperature levels may boost the popularity of destinations that were once considerably cooler. In Scotland, during Europe's 2003 heatwave, temperatures hovered at an unusually pleasant 27°C to 32°C (80°F to 90°F).

As climate changes, tourists may react in different ways, some visiting a favourite destination earlier or later in the year and others travelling to entirely different destinations.

Ironically, tourism – which will be so directly affected by the changing climate – is also a major

Death Toll

During Europe's 2003 summer heat wave, temperatures ran 20 to 30 per cent higher than the seasonal average. The UK's record high of 38.1°C (100.6°F) hit in August of that year. In France, temperatures rose to 40°C (104°F) and stayed high for two weeks. In Geneva, Switzerland, temperatures exceeded the average by 5.4°C (9.7°F). July was dry in France, Spain, Germany and Italy; drought spread eastward by August.

The death toll – calculated by counting the number of deaths above the usual August average – was estimated at 30,000. The elderly were most affected. France was hardest hit, with 14,000 heat-related deaths. In Italy's 21 biggest cities, more than 4,000 elderly people died.

reduce Earth's biodiversity, which is already threatened by other factors, from overdevelopment to the introduction of invasive species. A study by Chris D. Thomas of the University of Leeds, UK, found that middle-range warming scenarios for 2050 would result in the extinction of 15 to 37 per cent of species – a rate unprecedented in history.

It is difficult to quantify such matters as the disappearance of iconic species such as polar bears; the loss of unique ecosystems; and the value of human lives taken by famine, heat waves and flooding. Different gradations and aspects of climate change, some irreversible, will have different costs. Some will be intolerable and it is best to assess them clearly now, while there may still be time to avert them.

There may be no one critical point for all species. The danger varies by region, because living systems vary and because different regions have warmed more quickly than others (the Arctic faster than the tropics, for example). Still, consensus is growing among ecologists and policy makers that every effort should be made to limit the global average temperature increase to 2°C (3.6°F). According to Lara Hansen, chief scientist of the Climate Change Programme of the World Wildlife Fund (WWF), even if change is limited to that figure, some species will be lost and others can be saved. But if the temperature increases by 4°C (7.2°F), many species will be lost and management options will be costly and not necessarily successful. With a 6°C (10.8°F) increase, the situation will be dire.

A WWF study found that a 2°C (3.6°F) rise in global temperature would result in the ecological collapse of Australia's Great Barrier Reef by 2100. Wilfried Thuiller of the National Botanical Institute

AIR TRAVEL IS THE FASTEST GROWING SOURCE OF GREENHOUSE GAS EMISSIONS – AND THUS IS CONTRIBUTING TO THE INCREASE IN GLOBAL TEMPERATURES THAT MAY DOOM TOURISM IN MANY DESTINATIONS THAT DEPEND ON IT

contributor to that change. That is because air travel is the fastest growing source of greenhouse gas emissions – and thus is contributing to the increase in global temperatures that may doom tourism in many destinations that depend on it.

LEFT Boys employed as cart-pullers in New Delhi rest at midday during the capital's 2002 heatwave. Early summer temperatures climbed to 49°C (120°F) in some areas of India.

How Much Heat Can We Handle?

Depending on how humanity reacts, the effects of climate change may range from a slight problem to truly dangerous. Dangerous climate change is that which passes a point of no return, beyond which many species cannot recover. Many species can adapt to some warming, but beyond a critical global average temperature, most species will no longer be able to adapt. The extinctions that result will

ABOVE *High surf from Hurricane Felix pounds against a seawall on the mid-Atlantic coast of the United States. Homes on barrier island communities are regularly lost to the sea despite sandbags and seawalls, incurring high costs for both homeowners and their communities. Rising sea levels will exacerbate this problem.*

in Capetown, South Africa, showed that the same temperature increase could put 22 per cent of Europe's 1,350 species of plants into the 'critically endangered' category and render 2 per cent extinct.

If global temperatures rise more than 3°C (5.4°F), the Greenland ice sheet is likely to begin melting, according to climatologist Jonathan Gregory of the University of Reading, UK. Such a melting would be virtually unstoppable. Over the course of one thousand years, the melt-off would

cause sea levels to rise approximately 7 m (23 ft). If warming does not rise above 3°C (5.4°F), the ice sheet could survive for several thousand more years.

Another way climate change can become dangerous is if the costs – both ecological and economic – grow higher than society is willing or able to handle. One study estimates that it will require £60,000 million to maintain current function and stability for 1,000 Japanese ports if sea level rises just 1 m (about 3 ft). Another study estimates that a rise in sea levels

of only about 0.5 m (1.5 ft) could cause cumulative impacts to US coastal property of £11,500 million to £86,500 million. Swiss Re, the global insurance and risk management company, determined in 2004 that the costs of natural disasters such as flooding, storms and heat waves, aggravated by climate change, threaten to double to £29,000 million per year in ten years. The IPCC predicts that if the global temperature rises by more than 2°C (3.6°F), the global food production system is unlikely to be able to adapt without price increases.

No certainty exists now about how hot the world will get in the next century. Ironically, it is up to human beings to decide how much and how fast the Earth warms. The amount of greenhouse gas in the atmosphere and the associated global warming, reflects population growth, socio-economic development and technological change. According to alternative IPCC scenarios, a world with a higher population growth rate and lower rate of technological change is likely to lead to a higher level of greenhouse gases in the atmosphere and more global warming. How world societies approach the issue will determine whether the global rise is closer to the 1.4°C (2.5°F) end of the spectrum forecast by the IPCC, or to 5.8°C (10.4°F).

While ideas vary as to what level of heat is dangerous, some environmental and international groups, such as the European Union and the Alliance of Small Island States (AOSIS), are committed to keeping the rise under 2°C (3.6°F). Some have more at stake than others. AOSIS is particularly concerned about the threat from rising sea levels. According to the US ambassador for St Lucia, Charles Fleming, 'If we wait for the proof, the proof will kill us.'

A SLOW BOIL: EFFECTS OF CHRONIC WARMING

Around the world, a major ecological change is underway: animal and plant populations are shifting towards the poles or to higher altitudes. Parmesan and Yohe's 2003 review of reported range changes found that animals had shifted toward the poles by an average of just under 6 km (4 miles) per decade. The range of certain butterflies has shifted northwards in Europe by 48 to 97 km (30 to 60 miles) or more. In the United States, species such as the red fox and Edith's checkerspot butterfly have moved north or to higher altitudes. The tree line at high latitudes in the Northern Hemisphere shifted polewards or upwards in elevation during the twentieth century. Pacific salmon species such as sockeye and pink salmon have been found much farther north than they used to be. All of these animals and plants are reacting in their own ways to

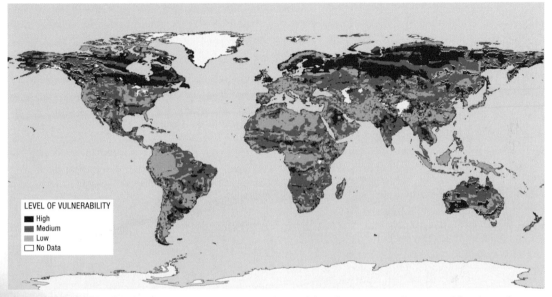

ABOVE *This World Wildlife Fund map, representing a synthesis of data from current climate models, shows that many of the world's terrestrial ecoregions could experience a significant shift in vegetation patterns if current global warming trends continue. In many regions, particularly in the savannahs of southern Africa and the great boreal forests of Russia and Canada, it may not be possible for the distribution of vegetation to shift fast enough to keep pace with the changing climate – in this case, temperature change.*

the slow boil of chronic temperature change.

According to what is known as the 'climate envelope model of species ranges', the geographic range of many species is determined largely by temperature and other climate variables. Therefore, species' ranges can often be expected to shift in response to changing temperatures. This can result from the migration of individuals to escape changing environmental conditions or to take advantage

LEFT *The Arctic fox, shown here, is in danger of losing hunting grounds to red foxes moving farther north as temperatures warm. The species could face extinction.*

of new opportunities, or from different survival and reproduction rates in populations at northern and southern edges of the existing species ranges.

The ability of a species to colonize new areas depends on how fast a population can migrate relative to how fast the climate changes. Many factors may interfere with the ability of a species to relocate. A study by Jay R. Malcolm of the University of Toronto, Canada, has found that climate change may require animals to migrate at rates much faster than those observed since the ice ages. Potentially, that could reduce biodiversity by selecting in favour of the species that proves itself the most mobile.

An ecological domino effect may also come into play. The invasion of species better adapted to a region's new climate might make it even harder for existing species to persist in a region. For example, the red fox, having migrated north, now competes with the Arctic fox, which had previously dominated the region and the Arctic fox may not survive.

Human civilization offers many barriers to migration, fragmenting and limiting natural habitats with cities, roads and farms. And some species – notably alpine and arctic ones – do not even have the option of migrating to colder places. If a species is unable to shift its range before its existing habitat is no longer viable, that species will become extinct.

A study of 179 South African animal species by Barend F. N. Erasmus of the University of Pretoria in South Africa showed that the effect of predicted climate change on ranges of species varied from little impact to local extinction. The researchers found that 78 per cent of species contracted their ranges and 2 per cent became locally extinct. A study of 1,870 species of Mexican fauna by Townsend Peterson of the University of Kansas, USA, found that by 2055, changing climates will have caused high species turnover in some local communities, affecting over 40 per cent of species.

The warming of the globe permits some species to migrate to regions where they were never before found. In Antarctica, for example, mosses and

BELOW *The endangered golden-striped salamander* (Chioglossa lusitanica) *is native to the north-west Iberian Peninsula, where its cool, humid habitat is growing increasingly dry and warm.*

other plants have colonized formerly bare ground. In the North Sea, rising water temperatures have prompted cold-water species of plankton to move north, while subtropical species have replaced them in the south. Such transitions can seriously disrupt a region's native species. The movement of North Sea plankton has damaged species in the food chain that depend on those plankton, including kittiwakes. Between 1999 and 2004, the population of this gull species declined by 50 per cent and 2004 saw the worst breeding season in the 19 years since monitoring began, apparently the result of a severe food shortage. Commercially important species have also suffered – among them the sand eel, which is an important food source for cod and is also processed for use in animal feed.

Timing of Ecological Events

Another sign of chronic warming is the altered timing, or phenology, of biological events worldwide. Driven by temperature change, birds are laying eggs, bearing young and migrating five days earlier on average than in the early twentieth century. At the University of Hannover, Germany, Gian-Reto Walther reported in 2002 that spring had been coming earlier in Europe and North America every year since the 1960s; plants are flowering earlier, trees budding earlier, butterflies appearing earlier and amphibians spawning earlier. Growing seasons are longer as well. In Fiji, octopuses have changed their spawning patterns, appearing in March rather than November.

An earlier spring might seem like good news, but the disruption in the timing of biological events can play havoc with many species.

Species at Risk: Marine Turtles

Sea turtles, also known as marine turtles, spend most of their lives in the sea. But females return to land when they are ready to lay eggs, dragging themselves onto the beach and burying their eggs in holes that they have dug in the sand. Sea turtles are fast swimmers and many migrate thousands of miles to return to the shores where they hatched in order to start the next generation.

Though found throughout the world's warm and temperate seas, sea turtles face an unusual problem brought on by climate change: a shift in the ratio of males to females. In sea turtles, as is true of several reptile species, temperature during incubation determines the sex of developing embryos. In some species, females are

ABOVE *A clutch of eggs from a green sea turtle. These turtles are found throughout the world's warm and temperate seas.*

produced when temperatures are warmer, males by colder temperatures. In other species, both cold and hot temperatures produce females while males are produced by intermediate temperatures. A five-year study of hatchling loggerhead turtles in Florida showed that rising temperatures skewed populations towards females, in some cases making them 87 to 99.9 per cent of the

population. As the climate grows warmer at sea turtle hatching sites, more females than males may continue to be produced, substantially altering the sex ratio. If this trend continues, the result could be the production of virtually nothing but females and likely extinction. By some estimates, if average global temperatures increase 4°C (7.2°F), male turtle offspring may be eliminated altogether, while increases of less than 2°C (3.6°F) may seriously skew the sex ratios.

Climate change can have other negative effects on sea turtles. In Malaysia, according to research by Joan Whittier, a biologist at the University of Queensland, Australia, a decades-long decline in green turtle numbers has resulted in part because beaches have grown too hot for effective incubation of turtle eggs. Hatchlings incubated in cooler nests are fitter and less susceptible to abnormalities. Sand temperatures are now too high, so hatchlings start out life unhealthy and are less able to scuttle down to the surf. Left to fend for themselves immediately after they hatch, baby sea turtles must move quickly to avoid predators, so a reduction in mobility can prove fatal.

Although most sea turtle species lay eggs during the summer, some turtles may adapt to warmer temperatures by laying when the weather is cooler. In some parts of the world, such as Malaysia, volunteers are helping to remedy the climate-caused skewing of the sex ratio by monitoring sand temperatures and relocating turtle nests as needed. But the long-term problem may defy such efforts by volunteers.

ABOVE *An adult ridley turtle. Sea turtles grow large: even the smallest species, the ridleys, grow to 71 cm (28 in), while the largest, the leatherbacks, may reach 2.4 m (8 ft).*

RIGHT *Newly hatched leatherback turtles make their way to the sea on the coast of Sri Lanka.*

Turmoil in the Water

Growing heat is proving deadly to many marine organisms. Some cannot tolerate the warmth; others are felled by pathogens that thrive in it. Disease that has spread because of the warming waters has caused the apparent extinction of a species of Caribbean sea urchin and killed many bottlenose dolphins in the Gulf of Mexico and the Atlantic off the US coast.

Some fish have suffered increased mortality from warming waters. In Scotland, in 2003, hundreds of salmon died as the rivers grew warmer. The fish could not tolerate temperatures above 20°C (68°F). In Portugal's Tejo River estuary, cold-water species such as flounder and red mullet have almost disappeared in the last two decades, being replaced by warm-water fish such as dogfish and Senegal sea bream.

The year 1997 saw a mass die-off in the Bering Sea of a species of bird called the short-tailed shear-

water. Migrants from Australia, the shearwaters suffered high mortality for several years when warmer waters caused a coccolithophore, a kind of plankton, to bloom in large numbers. The bloom turned the water an opaque turquoise colour, making it difficult for the birds to see their prey.

Death of Trees

Chronic warming has also had an impact on silviculture, or the tree-farming industry, affecting the ability of tree farmers to grow the species of trees they have in the past. In North America, as a result

debated. In the short term, growing plants can act as carbon sinks as they remove carbon dioxide from the atmosphere. In the long term, however, the same carbon will return to the atmosphere when the plant decays or is burned. In any case, finding ways to remove carbon dioxide from the air does not address the source of climate change – the emission of greenhouse gases from the use of fossil fuels. To curtail climate change, instead of looking for ways to get rid of these gases once they have been generated, a more direct and effective approach would be to focus on curbing emissions altogether.

IN THE SHORT TERM, NOT ALL THE EFFECTS OF CLIMATE CHANGE ARE BAD NEWS. SOME INDUSTRIES AND REGIONS OF THE WORLD MAY BENEFIT FROM RISING TEMPERATURES

of warming, bark beetles are attacking Douglas firs and tree growers are having to replace them with mixed conifers at higher altitudes. Although the range of the Norway spruce has expanded in some areas as temperature and carbon dioxide levels have increased, there are indications that because of climate change, it may no longer grow in many of the regions where it now makes up a major portion of the tree crop. Much farther south, in plantation states of Brazil's Amazonia region, the climate is expected to become drier, reducing silvicultural yields.

Silviculture has been proposed by some as a means of sequestering carbon to mitigate global warming. The value of this strategy is still being

Some May Like it Hot

In the short term, not all the effects of climate change are bad news. Some industries and regions of the world may benefit from rising temperatures. The wine business is one. According to research by Gregory Jones from Southern Oregon University, USA, the quality of vintages has improved over the past 50 years, a period during which growing season temperatures have increased by an average of 2°C (3.6°F) for most of the world's top wine regions. At least some of the improvement in quality is probably due to the change in climate, with the greatest effects detectable in cooler regions such as Germany's Mosel Valley. Jones predicts that the 27 wine regions he analyzed will have an average

ABOVE *The short-tailed shearwater* (Puffinus tenuirostris), *photographed off the California coast, is one of many migratory bird species that several nations' environmental agencies list as endangered.*

growing season temperature increase of just over 2°C (3.7°F) by 2049. This could be very welcome news to vintners in cool regions.

But for every region that benefits from climate change, another suffers. Warmer wine regions are likely to have grapes that mature to a 'sugar ripe' condition but lack flavours that are slower in developing. In California's Napa and Sonoma valleys, for example, retaining acidity and developing flavour have already become difficult because of the increasingly warm growing seasons. Some regions may have to switch to varieties of grapes that manage better in a warmer climate. And wineries in many parts of the world may have to deal with other unwanted effects of climate change, such as water scarcity and the need for increased pest control.

The melting of sea ice may also provide some benefits. Ice in the Arctic Sea has already receded by about 40 per cent since 1979 and the region could be ice-free in summer by the end of the twenty-first century. Trans-Arctic shipping lanes would cut oceanic travel time between the United Kingdom and Japan by up to 12 days. In the nearer future, commercial fishing fleets may be able to take advantage of the receding ice; energy companies will have easier access to oil and gas deposits in the region. However, in the Arctic as elsewhere, the benefits come with problems. The region's wildlife, including polar bears and seals, will be threatened when their habitats undergo fundamental changes. Shipping and mineral exploitation seriously endanger fragile ecosystems. Native communities are likely to find it harder, if not impossible, to pursue traditional ways of life. And increased access to oil and gas deposits will aggravate the fossil fuel

ABOVE *A container ship is unloaded in the docks of wintry Dudinka, in Russia's northern Siberia. Melting sea ice in northern regions will speed shipping of goods and enable fishing of previously inaccessible waters. But for every region that benefits from climate change, another region suffers.*

dependence that is driving climate change.

Colder regions will benefit from milder winters. In northern Europe, according to a study commissioned by the European Union, cold winters will be half as frequent by 2020 and non-existent by 2080. Energy needed for heating will drop, along with the number of work days lost to cold weather. Climate change should also benefit northern agriculture, fisheries, forestry and tourism. At the same time, warming may result in problems in the south, including flooding – counterbalancing the advantages to the more northerly parts of Europe.

Human beings and other living things are adapted to the climate they have known, and it takes effort to adapt to a new one. Rising heat will have some positive effects, but as temperatures climb, the costs may well outweigh the benefits.

MELTING SNOW AND ICE

Living in a Thawing World

Around the globe, snow and ice are melting rapidly at the poles, high in the mountains and at sea. This thawing, caused in part by global climate change, will have far-reaching effects on ecosystems and human lives. Organisms ranging from polar bears and penguins to tundra lichens and sedges are likely to be affected. So are the native peoples of the Arctic. Around the world, other populations will be threatened by the diminished supplies of drinking water, as well as the rising sea levels that result from glacial melting.

CHANGES IN ICE

For evidence that Earth's climate is changing profoundly, we need only consider the planet's extreme places: its highest mountains, the Arctic and the Antarctic. In places as far apart as Alaska and Tanzania, scientists are observing the effects of a warming Earth. Paradoxically, although warming reduces Arctic ice, the Antarctic ice sheet is gaining mass as rising temperatures bring more snow.

The Northern Hemisphere holds 39 per cent of the Earth's land surface as well as 90 per cent of its human population, almost all of its industrialized nations and all the native people who have long inhabited its perpetually icy regions. It is not surprising, therefore, that the most closely watched and studied effects of climate change are in the Northern Hemisphere. In addition, the southern polar region has been historically far less accessible.

With the help of satellite data, scientists have only recently been able to chart some startling changes in the ice in Antarctica. However, much remains uncertain about the effects of climate change on this vast continent.

In the Arctic, by contrast, the trend is clear: over the past few decades, temperatures have increased twice as fast as in the rest of the world. Sea ice in the Arctic Ocean has been thinning; the Greenland ice sheet is melting. Species are appearing further north than ever before and Arctic species are threatened.

The picture is similar further south, below the Arctic Circle. Glaciers in the world's high places are melting. The effects of climate change ripple through the physical and living elements of polar and mountain regions faster than in any other places on Earth. Indigenous subsistence cultures based on fishing and hunting are being threatened in many regions, as are local traditions tied to natural cycles.

The inhabitants of the Arctic and of the world's mountain regions will feel the impact of climate change most sharply and most rapidly. Arctic peoples thrive on hunting, fishing and herding cold-climate species. Cold temperatures are vital to the maintenance of even the firmness of the soil beneath their feet; many plant and wildlife species in polar and mountain regions rely on snowy habitats and permafrost – land that stays frozen year-round. Frozen water is essential to people and ecosystems around the world. Millions of people depend on glacially fed streams for water and agriculture.

The Arctic Climate Impact Assessment (ACIA), commissioned by eight Arctic nations, reported its findings in 2004, after four years of evaluating existing research on Arctic climate change and its effects.

The report concluded, 'The Arctic is now experiencing some of the most rapid and severe climate change on Earth. Over the next 100 years, climate change is expected to accelerate, contributing to major physical, ecological, social and economic changes, many of which have already begun. Changes in Arctic climate will also affect the rest of the world through increased global warming and rising sea levels.' It seems after all that no one lives too far away from the ends of the world.

Many changes have been well documented in sea ice and ice sheets in the Arctic and the Antarctic, as well as in glaciers worldwide; all have had serious effects on people and ecosystems around the globe. Scientists predict that still other changes lie ahead.

Worldwide Impact

While the North Pole and the summits of the Himalayas may seem remote and irrelevant to lower latitudes and altitudes, all the world has a stake in

RIGHT *Glacial retreat can be linked to climate change. It is the downhill flow of the 1,100 sq km (425 sq mile) Columbia Glacier, however, that leads to the loss of ice chunks, or calving, into Alaska's Prince William Sound. From 1992 to 2002, the glacier's annual shrinkage was measured at about 7 m (24 ft).* PREVIOUS PAGE *Clockwise from upper left: Adelie penguins in Antarctica; the glacier at Puna Arenas in Chile's Patagonia region; Inuit hunter, Baffin Island, Nunavut, Canada; an airplane surveys melting sea ice near Victoria Island, Canada.*

Snow and ice masses help keep the Earth cool because they have a high albedo, reflecting 90 per cent of the Sun's rays back into space and absorbing 10 per cent.

Old and dirty ice, as on glaciers, reflects about 65 per cent of the Sun's rays. This is not as much as is reflected by pure white snow, but still much more than is reflected by water or vegetation.

Sea water, because its colour is so dark, has a low albedo. It reflects only about 10 per cent of the Sun's radiation and absorbs the rest.

Tundra grass and other vegetation, such as forests have a low albedo, reflecting only about 20 per cent of the Sun's rays.

what is happening in these seemingly faraway places. This is because ice and snow play a crucial role in regulating the temperature of the Earth, reflecting radiation from the Sun, moderating global humidity, affecting ocean–atmosphere heat exchange and maintaining saline levels in the ocean. Without the planet's white surfaces of snow and ice, Earth would be even warmer than it is now.

Signs of Melting

Evidence is now abundant that the melting of ice and snow around the world is on the increase. According to David Robinson, a climatologist at Rutgers University, USA, the area covered with snow during summer in the Northern Hemisphere is about 10 per cent less than it was in 1966; winter measurements have remained approximately the same. Although seasonal and regional differences do exist, and some areas have received even more snow as a result of the changing climate, there has been a net loss of snow overall.

This loss of snow taken together with warming temperatures leads to the shrinking glaciers and ice sheets worldwide. Under normal conditions in some alpine areas, the amount of snow that falls during a season exceeds the amount that melts or evaporates. Over hundreds and hundreds of years, these snows build up, new snow layers compressing older ones,

LEFT *Albedo is the percentage of light, or radiation, reflected from a surface. Snow and ice help to keep Earth's climate cool because they have a high albedo; they reflect a majority of the incoming solar radiation. As ice melts, land or water is uncovered that has a lower albedo, so that less radiation is reflected and more absorbed, amplifying the effect of the warming.*

until they are compacted into glacial ice. Over thousands of years, as more snow accumulates than melts, glaciers grow into ice masses so enormous that they become known as ice sheets – sometimes measuring more than 50,000 sq km (19,000 sq miles). These rivers of glacial ice move downhill

GLACIERS ARE NOT IMMUNE TO WARMING TEMPERATURES. GLACIERS IN THE ALPS ARE 30 TO 40 PER CENT SMALLER THAN IN 1850

extremely slowly. Some of them eventually flow into the sea, where huge chunks break off and float away, during a process known as calving.

Glaciers and ice sheets are affected by warming temperatures. In fact, all around the world, mountain glaciers thinned during the latter half of the twentieth century, decreasing in thickness by close to 8 m (26 ft), an amount equivalent to more than 6,000 cubic km (3,700 cubic miles) of water. Glaciers in the Alps are 30 to 40 per cent smaller than in 1850. Glacier National Park in the United States has lost more than 120 of its glaciers since 1910. The Quelccaya ice cap in Peru is shrinking by 183 m (600 ft) a year at some of its edges. In Tanzania, Kilimanjaro's glaciers are 20 per cent of their 1912 size. Even the mighty Greenland ice sheet, which contains nearly 15 times as much ice as all non-Antarctic mountain glaciers combined, is losing 51 cubic km (12 cubic miles) of ice every year.

In the Arctic, sea ice extent declined by about 8 per cent between 1979 and 2004. New record lows in sea ice extent have been set every year since 2002.

Ice thickness has also been waning. At the end of the summer, Arctic ice is 1.3 m (4.3 ft) thinner than it was in the 1970s, or as much as 40 per cent less thick. Sea ice comes in different forms. First-year ice is usually smooth and relatively fragile and typically has a thickness of just over 1 m (3 ft). Multiyear ice, which

persists over the course of two or more winters, can range in thickness from about 3 m to 5 m (10 ft to 16 ft). Each year, more of the multiyear sea ice that covers the north polar region disappears. It is expected that towards the end of the century, the Arctic Ocean may be free of sea ice in summer.

Earth's ice, in all its forms, does not just melt – it also erodes with the force of the wind and it sublimates, or vaporizes, into thin air without first melting into a liquid. All the processes that reduce the size of glaciers and sea ice – together known as ablation – yield serious consequences for global climate, ecosystems and people.

Albedo and Sea Levels

Snow and ice are produced by low temperatures, but they also serve to keep the climate cold. That is because snow and ice reflect 20 to 90 per cent of solar radiation, while land and open water rarely reflect more than 30 per cent. Because of its reflective power, or albedo, ice insulates oceans and slows the warming generated by the Sun's radiation. Just as people wear white to keep cool in the hot summer sunlight, the whiteness of ice and snow helps the Earth stay cooler. Conversely, Earth absorbs more heat when it lacks white, reflective ice. And as Earth absorbs more heat, more ice and snow melt and the problem escalates, exemplifying a process called positive feedback, or feedback that speeds up change.

In addition to changes in the Earth's reflectivity, melting ice and snow also contribute to the rise of sea levels. Sea ice does not change sea levels when it melts. This is because floating sea ice displaces the same volume of water as its meltwater would add. However, since glaciers trap and store freshwater on land, their retention or loss of that water does affect sea levels. Large-scale glacial melting would cause a substantial rise in sea levels: water that has not been part of the water cycle

would suddenly be dumped into the oceans. In the past 100 years, the global sea level has risen between 10 cm and 25 cm (4 and 10 in). Within the next 100 years, sea levels could rise between 4 and 35 in (10 cm and 88 cm), enough to erode 2.5 m (8 ft) of sandy shore, contaminate island aquifers with seawater and flood low-lying islands and coastal areas.

A rise in sea levels of only about 1 m (3 ft) may not seem significant. Yet, worldwide, more than 100 million people live on coasts within that range. From low-lying Pacific islands such as those in Micronesia to sprawling seaside metropolitan areas such as Miami and Boston in the United States, as well as Calcutta in India, populations will have to adapt to and cope with the increased risks of flooding and erosion that global warming has brought.

Ocean Currents

Melting snow and ice do not simply add water volume to the oceans. They also contribute freshwater – and this is of key importance. A change in the salinity of the ocean disturbs thermohaline circulation, the patterns of deep-water currents that help control the Earth's climate (see page 26). Without this circulation of warmer surface water from the equator northward and the simultaneous movement of colder, deeper, saltier water towards the equator from the polar regions, Earth's climate would be radically different than it is at present. The water near the equator would grow progressively warmer, while colder water would remain fixed near the poles.

As an example, the warmer waters in the Atlantic Ocean travel from the tropics northwards along the eastern seaboard of the United States in an ocean current known as the Gulf Stream. After reaching the northeastern US coast, the main portion of the Gulf Stream continues to travel northeastward towards Britain and northern Europe. The warmer water in this current is crucial in providing the heat that maintains Europe's climate at more temperate levels in winter than is found in regions of North America at the same latitude.

Scientists have considered whether this vital North Atlantic current might suddenly disappear, but the present belief is that this is unlikely. However, during the 1990s, the National Aeronautics and Space Administration (NASA), using satellite technology, recorded a weakening of the current, compared to its strength during the previous 20 years. Although no scientific organizations or studies have directly identified global warming as the cause, the change has signs of being linked to warming ocean temperatures.

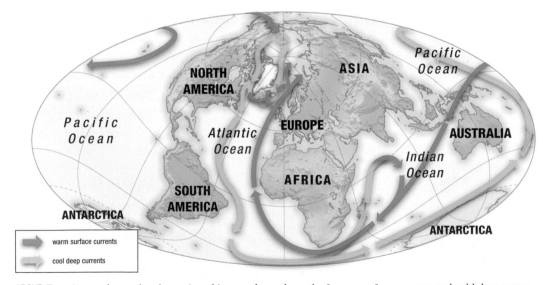

ABOVE *Focusing on the north polar region, this map shows the path of warm surface currents and cold deep water currents, or thermohaline circulation (see also page 26). From the poles, cold, salty currents move toward the equator. Melting ice in polar regions causes an influx of freshwater that threatens the patterns of the great ocean conveyor belt.*

THE WORLD'S THAWING PLACES

At the northernmost end of the world, ice and snow span the Arctic Ocean and reach down to the surrounding lands of Russia, Alaska, Canada, Greenland, Norway, Sweden and Finland. At and around the North Pole, multiyear sea ice covers the ocean year-round. In autumn, first-year sea ice starts to form and, throughout winter, extends its white reach southward, towards the surrounding lands. The frozen ocean prevents evaporation and creates a relatively dry atmosphere. At the edges of the Arctic Ocean, frozen land, or permafrost, allows meltwater to stay near the surface instead of seeping deep underground. Treeless tundra and northern forests thrive on this frozen land, which supports life ranging from bacteria to insects to polar bears to human beings.

Beginning in the nineteenth century, Arctic temperatures have risen to their highest level in four centuries. Just over the past few decades, they have been increasing twice as fast as in the rest of the world. An increase in solar radiation in accordance with the Milankovitch cycle (see page 30), together with less volcanic activity, likely initiated the warming trend that began in the 1800s. After 1920, however, greenhouse gases such as carbon dioxide became the more probable culprits.

Satellite data show that the extent of spring and summer sea ice is now 10 to 15 per cent less than it was in the 1950s. Changes in ice thickness are much harder to detect, but estimates hover at an average 10 to 15 per cent decrease since the 1960s. Most climate models, based on historical data, predict that 50 to 60 per cent of Arctic sea ice will vanish by 2100. Some models envision a much faster reduction, leading to an Arctic Ocean that will be ice-free in the summer by that same year. The differences in the models reflect the extreme difficulties in recreating the complex dynamics of sea ice.

Most Arctic glaciers and ice caps have been shrinking since the early 1960s, and the melting has been accelerating since the 1990s. Melting from the Greenland ice sheet and Alaskan mountain glaciers is estimated to account for over 20 per cent of the observed global sea level rise. Climate models forecast that Arctic glaciers will contribute even more to the rise in global sea levels over the next 100 years.

Permafrost and Tundra

Permafrost has dominated much of the Arctic landscape for more than 1,000 years. Permafrost is any kind of land – soil, rock, sand, gravel – whose temperature is permanently below the freezing level.

Permafrost is an important carbon sink, meaning it stores carbon that might otherwise circulate in the atmosphere as a greenhouse gas. Some scientists estimate that up to 14 per cent of the Earth's sequestered carbon is stored in the Arctic. Thawing permafrost releases this carbon in the form of two important greenhouse gases, carbon dioxide and methane, thereby accelerating climate change. One study determined that Arctic methane emissions increased from 22 to 66 per cent between 1970 and 2000, owing to permafrost and vegetation changes.

In some areas of the Arctic, only a few centimetres of the permafrost surfacethaw each summer. This active layer is what supports life.

LEFT *Aerial photo of a 2004 landslide in Prince Rupert, British Columbia, Canada, on the west bank of the Khyex River, which severed a natural gas pipeline. Further north in the Arctic, pipelines are also increasingly in danger of structural damage, as the permafrost that they rest on begins to thaw due to global warming.*

EVEREST AND KILIMANJARO

Rising temperatures worldwide are melting glaciers around the world and those on Mount Kilimanjaro and Mount Everest, two of the most fabled high peaks on Earth, are no exception. The American writer Ernest Hemingway immortalized the 'Snows of Kilimanjaro' in his short story of that name, and British explorer George Mallory called Everest the 'lord of all'.

Straddling the border between Tibet and Nepal, Mount Everest is the tallest mountain in the world, at a lofty 8,850 m (29,035 ft). Every year, mountaineers head for the summit, trekking over millions of tonnes of ice, snow and glaciers. In Tibet, on the great mountain's northern side, climbers cross the Rongbuk glacier to reach base camp. The ice beneath their feet is not what it used to be. Since 1966, the edge of the Rongbuk glacier has retreated up to 270 m (886 ft) in some places.

The Rongbuk glacier is not alone. In the eastern Himalayas, snow and ice cover is 30 per cent less than it was 30 years ago and some climate models predict that the Himalayan glaciers could melt completely by 2035.

Concern for the state of Everest's glaciers is mounting. One diverse coalition, made up of non-profit associations, in addition to a group of Nepalese climbers and British naturalist and broadcaster David Attenborough, has petitioned the United Nations Educational, Scientific and Cultural Organization (UNESCO) to add Everest National Park to its World Heritage in Danger List.

Having the world's tallest peak added to the UNESCO list could help focus international attention on serious hazards that are created as glaciers melt. Such meltwater often collects and forms lakes, but the extra water is no blessing. The boundaries of the lakes

LEFT A sherpa on his way to Everest base camp navigates Khumbu icefall, one of the most precarious parts of the mountain. Everest's melting glaciers endanger residents and climbers. Lakes filled with meltwater can burst their boundaries and cause floods.

ABOVE *The south face of Mount Everest. As the global climate warms, concern is mounting over the loss of Everest's glaciers, which are retreating at an alarming rate. Scientists predict the possibility that the Himalayan glaciers could melt away completely in the next 30 years.* UPPER LEFT *The summit of Everest.*

are often moraines, unstable mixtures of rock and ice shoved in a pile by glaciers that have since retreated. Rising water levels strain the banks, causing lakes to overflow suddenly in what are called glacial lake outburst floods.

The flood threat on Everest is here and now, not years in the future. One Nepalese glacial lake flood in 1985 killed nine people, knocked out bridges and tumbled houses and damaged crops and a community hydropower station.

A new threat lies about 40 km (25 miles) from Mount Everest at Tsho Rolpa, Nepal's largest glacial lake. If the surrounding moraines give way, the rush of water could wipe out communities and a power plant and affect life and property as far as 100 km (60 miles) downstream.

Floods are not the only problem. Some 20 million people live in the areas surrounding the Himalayas and depend on glaciers to feed rivers that provide water for drinking and irrigation. The water supply could disappear with the glaciers. And if the glaciers disappear, the Himalayas will no longer reflect the Sanskrit meaning of their name – abode of snow.

The Snows of Kilimanjaro

More than 5,600 km (3,500 miles) and a continent away from Mount Everest, Africa's highest peak, Mount Kilimanjaro, rises dramatically 5,895 m (19,340 ft) from northeastern Tanzania's cultivated plains. Kilimanjaro lies near the equator, and trails leading to its summit wind their way from the rain forest through alpine desert, up to glaciers near the cratered top which date back 12,000 years. Yet a hiker's chances of seeing glacial ice have dwindled. Of the glaciers that existed in 1912, 80 per cent are already gone and the remainder are expected to vanish by 2020.

The highest of the three dormant volcano craters that make up Kilimanjaro still hangs on to 2.6 sq km (1 sq mile) of ice. Within the last century, three large glaciers equalling about four-and-a-half times that expanse covered the area. Today, the three are broken up into smaller pieces, with ice on the eastern side of the summit vanishing the fastest, as the ice to the west is shaded by regular afternoon clouds.

Averaging 30 m (98 ft) thick the remaining ice holds a staggering 6.7 million cubic m (236 million cubic ft) of water. Despite their size, the glaciers feed just two streams on the mountain's southwest side; most of the upper third of the mountain is dry much of the year. Almost none of the water being lost from the ice makes it to the lowlands, since evaporation and sublimation account for most of the glaciers' shrinkage.

RIGHT *Kibo peak, Mount Kilimanjaro, showing one of the three dormant volcano craters on the mountain. Ice in this area shrank dramatically during the last century.*

Several factors have contributed to the plight of Kilimanjaro's glaciers, which have been in retreat since the 1850s. As people in the surrounding lowlands collect wood for fuel and farmers set fires to clear land for crops, they have slowly, inexorably destroyed the mountain's natural humidifier, tree by tree. Forests trap moisture, keeping humidity high, creating cloud cover and reducing glacial sublimation and evaporation. Without the forests, the air is drier, precipitation is minimal – and glaciers are under threat.

The disappearing forests are not the only culprits. Although much of the radiation is reflected when sunlight hits glaciers, some is absorbed, contributing to ablation. Snow can protect the glaciers' icy surfaces from solar radiation and so changes in snowfall can greatly influence the pace of ablation. This means the survival of the world's glaciers is threatened by sunlight and warmth, in addition to the seemingly unrelated issue of deforestation at lower altitudes.

People may not directly suffer from the loss of Mount Kilimanjaro's glaciers, but the white-capped image of this celebrated African peak may soon be relegated to the realm of literature and legend.

ABOVE *Mount Kilimanjaro seen from the south in 1909. At this time, the glaciers had already begun shrinking. Their retreat is estimated to have begun in 1850 and continues today.*

What Will Happen to the Penguins?

To many people, Antarctica represents the world's most pristine environment. Yet populations of seals, whales and commercially harvested fish in the Antarctic have been subjected to overhunting and overfishing for the past 50 years. Climate change on top of these kinds of stresses may spell disaster for many species in the region, including the animal that perhaps best represents Antarctica in the popular imagination: the penguin.

While uncertainty remains about what effect climate change will have on sea ice in the Antarctic, any change is likely to affect these flightless birds. Certainly a reduction in sea ice does not bode well for species that are highly dependent on it, such as Adelie and emperor penguins, whose breeding success appears to be strongly linked to the presence of this ice. Emperor penguins, in fact, site their colonies on winter sea ice, where they breed and hatch chicks even in extreme weather. Sea ice is especially vital during the three or four week moulting period, because open water can be fatal for birds lacking their new coats. Adelie penguins in the Antarctic Peninsula appear to thrive in the years with extensive sea ice, as it brings an abundance of krill – their main food source. On the other hand, chinstrap penguins, which prefer to live in ice-free locations, may flourish in an Antarctica with less sea ice.

However, relationships between populations and climatic factors are not always simple. Although emperor penguins need sea ice during moulting, too much sea ice

LEFT An Adelie penguin in Antarctica shelters its chick.

during their winter breeding season increases the distance they must travel between their colonies and feeding grounds, and may decrease the likelihood that their newly laid eggs can be hatched successfully. Similarly, although Adelie penguins in the Antarctic Peninsula thrive under heavy sea ice conditions, the same conditions in the Ross Sea region appear to lead to a decline in the local Adelie population.

Antarctica's remoteness and the rigours of its climate have not been conducive to long-term scientific observations and much still needs to be learned about how the continent may be affected by a changing climate. It is not yet possible to know whether penguins will find a way to withstand a warming Earth.

BELOW A colony of emperor penguins. The widespread melting of sea ice in Antarctica threatens this species' breeding cycles.

Most permafrost is barely frozen, making it extremely sensitive to temperature changes. In parts of Norway, ground temperatures in recent years have risen four times as fast as they did in the entire last century and permafrost is thawing. Other countries surrounding the Arctic Ocean report similar trends.

An entire landscape can be physically altered when once-solid permafrost becomes soggy. When ice warms quickly, water pools on top, creating slick surfaces; the same is true for permafrost. Landslides can result. In Canada, a direct correlation has been established between the rate of warming and the frequency of landslides. Level land, instead of sliding, sinks when it thaws. In some cases, Arctic land has dropped 4.6 m (15 ft). This type of severe thawing results in huge holes and trenches, called thermokarsts. But even small changes in the firmness of permafrost can threaten the stability of buildings, roads, railroads and pipelines.

Every summer, as the active layer of permafrost thaws, the tundra – the vast, treeless plains of the Arctic and the far North – bursts into life. Insects such as mosquitoes and flies thrive in the wet, soggy environment. Several hundred million birds migrate to the Arctic each summer from as far away as the Southern Hemisphere to take advantage of this abundant food source. They stay there for the season to fatten themselves and to breed, and their success in the Arctic determines the size of their populations worldwide. As the Arctic becomes warmer, however, tundra areas are expected to shrink. Along the northern coast, tundra will likely be eroded by rising seas. At the southern limit, the tree line will encroach on tundra areas. Scientists forecast that some bird species will lose more than half their breeding areas in the Arctic during this century, as tundra area shrinks.

The Antarctic Paradox

The North and South poles are fundamentally different: the Arctic is an ocean surrounded by continents, whereas Antarctica is a continent surrounded by oceans. Climate change affects each of the polar regions differently.

While evidence of warming in the Arctic has been resoundingly clear, interpreting trends in the Antarctic climate is more complex. The latest research has shown that the vast East Antarctic ice sheet has actually been growing in mass for a decade or more – an effect consistent with climate change. This massive ice sheet, which covers an area larger than the continent of Australia, is up to 3.2 km (2 miles) thick. Measurements made between 1992 and 2003 by the European Space Agency's ERS-1 and ERS-2 satellites show that, during that time, the ice sheet has been thickening at a rate of 1.8 cm (0.72 in) a year. This growth is attributed to an increase in annual snowfall

NORTH POLE

SOUTH POLE

ABOVE Two satellite images of the Earth. Watery areas are blue, vegetated areas green and arid areas brown. Snow, ice and clouds are white. TOP the northern hemisphere, with the North Pole at the centre, surrounded by the Arctic ice pack. BELOW the southern hemisphere, with the South Pole, which is on the continent of Antarctica.

in this coldest region of the Earth – a phenomenon that would be expected with warming temperatures, which allow the atmosphere to hold more moisture.

The East Antarctic ice sheet holds enough fresh water to raise sea levels worldwide by as much as 60 m (200 ft). The continued growth of the ice sheet will probably slow the global rise in sea levels by about 0.12 mm (0.005 in) a year. The East Antarctic ice sheet, it appears, is 'the only large body of ice absorbing sea level rise, not contributing to it,' said Curt H. Davis of the University of Missouri, USA, the leader of the team that discovered the thickening of the ice.

At the same time that the East Antarctic ice sheet is growing, the ice appears to be thinning in West Antarctica and other regions of the continent. Temperature records at research stations in the Antarctic Peninsula – the region of Antarctica that extends the furthest north – have shown that at least this region has been warming substantially over the past 50 years. In fact, temperatures on the peninsula have been rising several times faster than the average global rate.

Warming is also eating away at the edges of Antarctica. Scientists who study glaciers and ice sheets have long predicted that warmer temperatures would bring about disintegration of the Antarctic ice shelves – flat sheets of the ice cap that extend into the surrounding sea. In 2002, about 3,250 sq km (1,250 sq miles) of an Antarctic ice shelf – an area considerably larger than the country of Luxembourg – dissolved in just over a month. This was just one of six ice shelves that have drastically diminished since 1945. Living things are also responding to increased temperatures in the Antarctic Peninsula: flowering and non-flowering plants have been thriving under warmer summer temperatures and longer growing seasons.

As the Earth's climate continues to warm, scientists forecast that more ice shelves along the Antarctic Peninsula will break up. Sea ice is likely to diminish as the amounts of carbon dioxide in the atmosphere continue to increase and global temperatures continue to rise.

Despite their differences, the vast areas of snow and ice in Antarctica and the Arctic both contribute to maintaining global climate. Monitoring changes at the poles is crucial to the predicting and understanding of climate change throughout the world.

The Diminishing Greenland Ice Sheet

About 85 per cent of Greenland, the world's largest island, is covered by an ice sheet measuring 2,530 km in length and 1,090 km in width (1,570 miles by 680 miles). Windy and snow-covered for six to nine months of the year, this self-governing Danish territory hardly lives up to its name. Ilulissat Icefjord, an area on the west coast where one of Greenland's glaciers meets the sea, was added to UNESCO's World Heritage List in 2004. At llulissat, the glaciers release enormous ice chunks into the ocean. Warming temperatures could change that, however, and jeopardize the Ilulissat Icefjord.

As the spectacular calving at Ilulissat so sharply demonstrates, the Greenland ice sheet is never static. Snowfall at high elevations adds mass at the same time that the edges of the glacier melt or break off into the ocean. Due to the high pressure at the bottom of the mass, ice may melt. This meltwater acts as a lubricant, accelerating the movement of ice towards the sea and reducing the volume of ice stored in the sheet. This phenomenon has been increasing in response to global warming.

On average, the Greenland ice sheet sheet has a thickness of about 1,500 m (5,000 ft), but its characteristics differ depending on the latitude and the altitude. Above a specific altitude – usually around 2,000 m (6,500 ft) – the amount of ice lost closely approximates the amount gained. At these high altitudes, it does not seem that changes in the ice sheet can be correlated to warming. At lower altitudes, however, thinning now prevails at rates that can exceed 1 m (3 ft) per year in some places.

Greenland and Antarctic Ice Melt

Overall, the Greenland ice sheet is shrinking. Since 1979, according to studies of satellite data led by Konrad Steffen at the University of Colorado, USA, the summer melting that takes place every year around the edges of the Greenland ice sheet has expanded higher and farther up the ice sheet. The part of the ice sheet's surface that melts in the summer increased on average by 16 per cent from 1979 to 2002. That area of expansion is roughly the size of Sweden. In 2002, the area of melt set a record high.

Although a study by the National Aeronautics and Space Administration cites a rise in sea levels of less than 0.13 mm (0.01 in) a year due to melting from Greenland's ice sheet, a long-term trend in this direction could have drastic consequences for Earth's oceans and climate. Sea levels would rise by about 7 m (23 ft) if the entire ice sheet melted, causing widespread flooding and disrupting ocean currents in the North Atlantic.

This outcome is unlikely in the near future, as the ice is predicted to respond to climate change gradually. Scientists believe that between the last ice ages, water draining from Greenland's ice sheet caused a sea-level rise of about 2 m (7 ft). Predictions vary, but if the ice sheet does start to collapse, there is probably nothing humans will be able to do to reverse it. The result would likely be a thousand-year-long disintegration of the ice sheet culminating in its disappearance.

Antarctic melting also remains a source of concern, despite the thickening of the East Antarctica ice sheet, which absorbs some of the rise in sea levels. In West Antarctica, although the ice sheet appears to be thickening at its western extremity, it seems to be thinning rapidly to the north. Another gap in our knowledge is how rivers of ice flow within the ice sheets. Scientists speculate that if

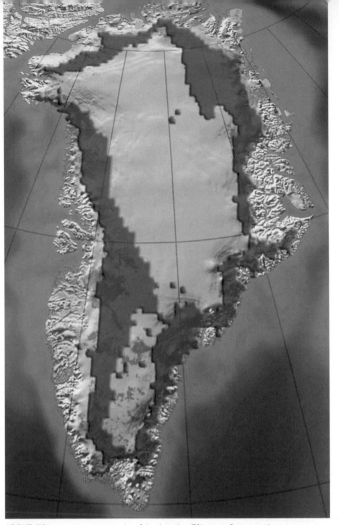

ABOVE *The orange areas in this Arctic Climate Impact Assessment graphic show the extent of seasonal melt of the Greenland ice sheet in 2002. The melt zone in orange is where the summer warmth causes melting around the ice sheet's edges. This area has been expanding inland in recent years.*

West Antarctica's floating ice shelves break off, there would be little to keep the rest of the ice sheet from rushing into the ocean and rapidly raising sea levels by 5 to 6 m (16 to 20 ft).

Most scientists agree that the West Antarctic ice sheet will not disappear soon. Disintegration may start if global mean temperature rises 5.4°F (3°C); complete melting and a subsequent rise in sea levels could take 500 to 500,000 years.

IMPACT ON LIVES

Life cannot exist without water. More than 71 per cent of the planet is covered in water, but only 3 per cent of it is fresh. Water's cycle through the globe is superficially simple: it evaporates from lakes and oceans, forms clouds and precipitates back to the Earth, where it replenishes rivers and reservoirs, ground water and aquifers. The reality is more complex. Moisture can travel for kilometres in the atmosphere; it can disperse thinly and widely or be heavily concentrated; it can stay put, as ice, for millennia.

Less than one per cent of the Earth's freshwater is free-flowing. The rest is frozen, mainly in Greenland and Antarctica. But melting ice around the world does not meaningfully boost the availability of usable fresh water. As global temperatures rise, evaporation increases from tropical waters. Tropical and subtropical ocean evaporation is up an estimated 10 per cent since the 1980s and precipitation does not necessarily fall in its place of origin – the moisture travels much farther north before dropping out of the sky. The result has been fuller Arctic rivers at high latitudes and droughts in the subtropics.

The rate of glacier ablation is particularly distressing in regions that rely on freshwater from mountain sources. Snow and ice reserves in the western United States, for example, could decline by up to 70 per cent during the next 50 years. West Coast agricultural and metropolitan areas, including Seattle, San Francisco and Los Angeles, rely heavily on rivers coming down from the mountains for water and power. In response to a drastic drop in water supplies, substantial restrictions on water use would need to be put in place and enforced.

The effects of reduced snow and ice pack would be felt all over the planet, because much of the world depends on glacial melt for its freshwater supplies. Glaciers in the South American Andes Mountains could disappear by 2015 and 60 per cent of China's glaciers could melt in mere decades. In the world's dry countries, mountain glaciers can account for up to 95 per cent of water in river networks. Even in lowland areas of temperate countries such as Germany, 40 per cent of water comes from mountain glaciers. The problem is not just one of the future: in the village of Dadrapuke in Ghizer, Pakistan, elders reminisce about fertile valleys full of wheat and fruit trees, watered by steady glacial streams. Today's stark reality is one of dried-up streams, failed crops and an exodus of villagers.

Burgeoning populations compound the problem, as do other stresses on water supplies, such as the clearing of forests for agriculture. Millions of people may soon be wondering how to find water to drink and to irrigate their crops.

ABOVE *Icebergs melting in a fjord in southern Greenland. Increased ice melt does not increase the availability of usable freshwater in a significant way.*

ABOVE *A Greenland shark in shallow waters of Arctic Bay, Canada. Many wildlife species here are exhibiting unusual behaviour, perhaps due to warmer temperatures.*

POLAR BEARS FACE EXTINCTION

While many animals would find the Arctic sea ice inhospitable, the polar bear thrives on it. The largest land carnivore on Earth, it is well adapted to the northern cold and to the hunting of seals on pack ice. But climate change is threatening the polar bear's survival by melting the ice on which it depends. The result – growing hunger among bear populations – has jeopardized the survival of one of the great symbols of the Arctic.

The Bear of the Sea

The polar bear spends most of its life hunting on the Arctic sea ice. With both the range and thickness of that ice shrinking significantly over the last forty years, the animal's survival is now in jeopardy. Along with threats posed by pollution, poaching and expansion of humans into previously undisturbed polar bear territory, changes in weather patterns may push the animals to the brink of extinction.

The scientific name for the polar bear is *Ursus maritimus,* or 'bear of the sea' – this is a fitting title for an animal that makes its home along the edges

of the Arctic seas from North America to Greenland to Russia, on the coasts and islands and on ice that covers the waters.

Living mainly by hunting seals on pack ice, polar bears have adapted beautifully to life in their frigid environment. The world's largest land carnivores, they measure up to 2 m (8 ft) and weigh as much as 590 kg (1,300 pounds). Their huge paws function like snowshoes, spreading their weight to prevent them from breaking through the snow and ice. Their dense white fur not only serves as camouflage, but also provides insulation from the cold, as does the thick layer of fat under their skin. These bears are excellent swimmers, who boast marine adaptations such as partially webbed feet. Yet polar bears today face the threat of extinction – principally because of climate change.

The Devastating Effects of Climate Change

Arctic winters are long and severe and summers – despite continuous daylight – are short and wet. But the very qualities that make the Arctic inhospitable to most creatures are what make it a good place to live for polar bears. The danger to such creatures comes not from the severe cold, but rather from the melting of the ice on which they depend.

Polar bears hunt on the pack ice that covers the seas in the autumn, winter and spring. In the summer, when much of the pack ice melts, some bears venture further north onto the ice, but many others stay on land, living off body fat stored during the hunting season. This fasting period can last up to four months, or as long as eight months for pregnant females in certain regions.

LEFT *A polar bear hunts seals on pack ice near the Canadian town of Churchill in Manitoba. Global warming jeopardizes the species' survival by shortening the bear's hunting season.* UPPER LEFT *The polar bear is one of Arctic's great symbols.*

ABOVE *Like this bear, photographed while foraging through a rubbish heap in Churchill, Manitoba, polar bears encroach increasingly on human territory. Dropping birth rates and high infant mortality among polar bears are both linked to diminishing food supplies for the species.*

In recent years, the ice has been melting earlier each spring and freezing later in the autumn, reducing the time during which polar bears can hunt. The result has been growing starvation and malnutrition. The polar bears in greatest jeopardy are those in the southern part of the animals' range, in the Hudson Bay and James Bay regions of Canada. For every lost week of hunting time, the bears there lose about 10 kg (22 pounds), leaving them weaker and less likely to reproduce successfully. Birth rates are reduced and many cubs eventually die of starvation from nursing on mothers with inadequate fat reserves, or from a general lack of food.

An Uncertain Future

If warming continues at its present rate, experts predict there will be no ice in the Hudson Bay by the year 2080. The Arctic Climate Impact Assessment warned that, by 2100, sea ice around the North Pole could all but disappear in summer and that Arctic temperatures would rise by 4°C to 8°C (8°F to 14°F). 'Polar bears are unlikely to survive as a species if there is an almost complete loss of summer sea ice cover,' the report stated.

Polar bears face this dire threat at the same time that other factors threaten their chances of survival. Pollution is damaging their habitat and poachers hunt them. Polar bears are very territorial and as humans encroach on their living space, interactions between human and bear inevitably result. As the bears around Hudson Bay have grown hungrier, they have wandered with increasing frequency into populated areas in search of food.

The polar bear's existence depends on the pack ice of the Arctic waters. If that pack ice continues melting at its present rate, the outlook for the bear of the Arctic seas is indeed poor.

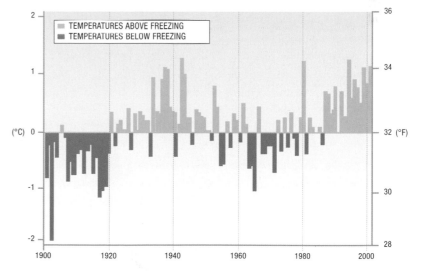

ABOVE *The changes in climate are more drastic in the Arctic than elsewhere. The United Nations Intergovernmental Panel on Climate Change (UNIPCC) has determined that Arctic winters are getting warmer, Arctic springs are coming sooner and the ice is thinner than it used to be, especially in summer.*

Native Cultures

Ten thousand years of history tie peoples such as the Inuit, the Saami and the Chukchi to the frozen Arctic. To survive in such a harsh environment, these groups draw on traditional knowledge and, increasingly, employ more modern techniques to hunt and fish. But since the Arctic is warming twice as fast as the rest of the world, traditional customs are quickly losing ground.

Some groups of Arctic people have herded reindeer since 2000 BCE, using them for food, clothing, shelter, tools and transportation. Yet, despite these animals' marvellous ability to adapt, they cannot withstand climate pressures today. In some cases, climate change will increase precipitation; even if the temperature warms a little, in the Arctic that precipitation will fall as snow. For reindeer that will mean that food sources are concealed under snow and the herd weakened by hunger. Greater warmth would also cause stress. One study in northern Finland found that fewer calves survive after warm breeding seasons. Climate change may jeopardize the livelihood of the Saami people, who depend on income from reindeer products.

In Alaska, almost 70 per cent of rural communities use or hunt wildlife, including salmon, herring, whales, walrus, seals and caribou. Yup'ik and Inupiat people of Alaska have already seen changes in the location, migration patterns and numbers of some wildlife species. Hunting is dangerous on thin sea ice and in increasingly severe weather. Plants and berries are not as abundant and new pest insect species are colonizing the area. Residents are finding food rotting in ice cellars that used to preserve stocks reliably.

Anecdotal accounts from marginalized groups of people are often ignored in discussions of climate

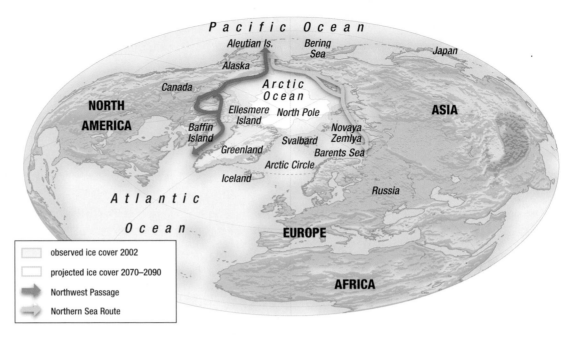

ABOVE *As sea ice melts, shipping in the Arctic may increase. The Northern Sea Route and the Northwest Passage may link ports in northern Russia and ports in eastern Canada with destinations in the Pacific Ocean.*

change, in favour of hard data and long-term studies. But the value of traditional ecological knowledge – environmental knowledge accumulated by generations of observation by native peoples – is gaining recognition and sparking study by the Smithsonian Institution and NASA, among other organizations. The verdict? Warming temperatures hurt the people who make their home in the Arctic.

Effects on Industry and Shipping

The economies of Arctic communities are based on much more than fishing, hunting and herding. Oil, gas, and metal ores contribute to the global interest in the region. They also provide jobs. While warmer temperatures will break up ice in the short term, potentially interfering with shipping and industrial ventures, over time diminished sea ice in the Arctic Ocean is expected to unlock access to valuable resources, including up to 25 per cent of the Earth's remaining oil reserves. Goods will be able to be transported more quickly across the region.

The Northern Sea Route, which opens up seasonally between northern Russia and the Arctic ice, is usually navigable for about a month each summer. Some climate models predict that the shipping season could triple by 2080 and for ships specially equipped to break through heavy patches of ice, the season could be a full five months each year.

For the shipping industry, the potential economic benefits of increased Arctic transportation are significant. Travelling from northern Europe to north eastern Asia through the Arctic passage costs 40 per cent less than taking a circuitous route through the Panama or Suez canals. New opportunities will doubtless develop to fill new needs created by open Arctic seas. Shipbuilders, maintenance companies, icebreakers, charting services and emergency responders could be in high demand.

Opened northern waters will also raise questions concerning access rights, security and environmental protection. New international agreements will need to be forged in order to secure resources, safety of transport, rights to seabed resources and sovereignty of shipping lanes. Environmental crises like oil spills could escalate. Oil spills are always destructive to ecosystems, but their effects can be more catastrophic and long-lasting in the Arctic than in lower latitudes. Because of the cold climate, ecosystems take longer to recover. To date, there is no technology that can clean up oil spills in oceans covered with sea ice. Conflicts are also likely to escalate as

instability that threatens a variety of species, from microscopic plankton to seals to sea birds.

As Earth's climate warms, terrestrial species in central and middle latitudes are shifting their ranges toward alpine regions and the poles. Species on the fringes of those polar and alpine habitats will be the first to move to higher altitudes and latitudes. Those already occupying such regions – living at the figurative ends of the world – do not have the luxury of moving. New parasites and new predators will move in and cause considerable harm. Extinction of

Animals like polar bears and seals face fiercer competition with the disappearance of the sea ice that is their hunting territory. Since ice appears to be lessening at greater rates near Siberia, species there might be at greater risk. Altered snow and permafrost conditions can adversely affect polar and alpine animals' staging areas, migration routes and breeding patterns. Animals cannot feed or find mates if they are sinking into soft ground or spending extra energy trudging through deep snows, which result from temperatures that remain below freezing but rise enough for the air to hold more moisture and release more snow. Across the melting landscape, trees previously supported by firm ground lean eerily sideways. A sudden excess of water can essentially drown whole forests, turning them into grassland and affecting the animals who were once supported by the sylvan vegetation.

ALTERED SNOW AND PERMAFROST CAN AFFECT POLAR AND ALPINE ANIMALS' STAGING AREAS, MIGRATION ROUTES AND BREEDING PATTERNS

commercial fishing operations, shipping outfits, oil companies, tourism developers and indigenous people all vie to protect their interests.

Not everyone agrees that reduced sea ice coverage in the Arctic will benefit industry. Yearly variability might make Arctic passages more unpredictable. Sea ice that builds up behind ice dams during cold years could create even more hazards over longer, warmer periods. Despite growing interest in the commercial possibilities, the benefit of reduced sea ice for Arctic industries remains open to speculation.

Polar and Alpine Ecosystems

Polar and alpine ecosystems have evolved over millions of years and have been shaped by the vicissitudes of natural cycles. Human-induced climate change, however, has introduced a powerful factor of

some species may be the outcome, as many are crowded into smaller and smaller areas.

The adaptation of terrestrial species to cold climates, such as tolerance for short growing seasons, low soil temperatures, minimal sunlight and marginal nutrient levels allow them to carve out an existence in regions of relatively low productivity. These same qualities could work against them if they are forced to compete with newcomers in the neighbourhood. One study in Glacier National Park, in the United States, found that the abundance of four indicator Arctic-alpine plant species decreased by 31 to 65 per cent over a period that was just 0.6°C (1.1°F) warmer than the previous four decades. As taller plants move into regions that once had only short plants, for example, the smaller species are shaded out.

Species living in the Arctic Ocean are also at risk. Melting has increased the layer of freshwater just beneath the ice by one-third, to a depth of 30 m (99 ft), changing the plankton populations on which so many fish, whales and birds depend. Other small organisms, such as marine plants, influence entire food chains. Local variations in sea ice are potentially fatal: narwhal whales can become trapped in new expanses of ice. Birds, fish and baleen whales are vulnerable to changes in ocean currents. Altered currents could fail to transport plankton food sources to some areas.

This picture may seem very gloomy for animal residents of the high Arctic and high mountains. Unfortunately for these species, it appears increasingly as if climate change may squeeze them off the ends of the world.

WATER: TOO MUCH AND NOT ENOUGH

The Dilemma Over Freshwater

Life on Earth depends on water. From rain forests to rice paddies, water binds together every ecosystem. Climate change will dramatically affect the world's supply of freshwater – and the people, plants and animals that rely on it. Precipitation will increase in some places potentially bringing floods and decrease in others, possibly bringing drought. Conflicts between nations may arise over water sources, and plant and animal species may undergo upheavals as some areas of the world suffer from too much water, others from not enough.

CLIMATE AND WATER

Earth's changing climate will affect the world's water supply in many far-reaching ways. It will influence water temperatures, weather systems and the amount of water in streams, rivers and aquifers. Changes in the world's water – how much, where and when it is available – are a matter of universal concern.

Climate change will affect both quantity and quality of water. Algae and bacteria will grow in some areas, and salt content will rise in others. The pace of the water cycle will speed up, producing more rapid evaporation and leading to more intense storms, floods and droughts. Some regions will receive more precipitation and others will be drier. Sometimes, a region may become both wetter and drier than before, experiencing increased flooding in winter and more droughts in summer.

The changes in the quantity and timing of precipitation and available water will profoundly affect humans and other life forms. Ecosystems will change and wildlife behaviour and habitat as well as crop yields will be affected. In addition, climate change will have an impact on water-related economic activity other than agriculture: ski areas, for example, may experience decreased snowfall, squeezing the economies of mountain towns. The length of the Canadian ski season is expected to fall by up to 32 per cent by the 2050s and if resorts cannot increase their snowmaking, the season will be even shorter. According to a study by the United Nations Environmental Programme (UNEP), climate change may cost Switzerland's tourism industry 820 million euros.

Where water is scarce, conflict may arise, either among competing groups of users, such as farmers, manufacturers and fishermen, or among countries. These indirect or secondary effects of climate change on human society may have as great an impact as its direct effects.

Who Will Be Affected?

Water is fundamental to all economic activities, from farming to tourism. People the world over, from Sydney, Australia, to Rio de Janeiro, Brazil, to Ulan Bator, Mongolia, will feel the pressure of climate-induced changes on Earth's water rhythms. But these changes will not affect the world's population equally and they will be experienced in myriad ways, depending upon local or regional contexts. Some regions will undergo great changes in their economies and ways of life; others will face threats to public health. The consequences for some may be slight, but, for others, they may be dire.

Farmers who live in regions of increasing drought may be unable to harvest crops; many will watch as the topsoil that nourishes their plants and retains moisture dries up and blows away. China has seen nearly 9 million hectares (3.6 million acres) of cropland, rangeland and forests turn to desert in this manner during the past five decades alone.

Nomads in sub-Saharan highlands will suffer from infectious diseases such as malaria and dengue fever as warming allows mosquitoes to survive farther up the mountainsides. As greater rainfall leads to larger mosquito populations in East Africa, Rift Valley fever will cause weakness and sometimes death of the cattle that are essential to that region's economy.

From Lima, Peru, to Katmandu, Nepal, people who depend on melting snowpack as a primary source of drinking water may be forced to search for new water sources. When those alternate sources disappear entirely, choices are few: move, invent new technology for water capture, buy water, or steal from someone else's supply.

RIGHT *A couple prepares to pump water into their fields in China's Hainan Province. A drought there in 2004 left short supplies of water for drinking and agriculture and affected 50,000 people. Climate change is already influencing water supplies worldwide, partly by speeding the pace of the water cycle, which leads to more rapid evaporation and, consequently, more storms, floods and droughts.* PREVIOUS PAGE *Clockwise from left to right: flooding caused by El Niño; a girl swims through floodwaters in Bangladesh with empty containers for drinking water; a dead fish stranded on the banks of a drought-stricken pond; parched crop rows in Uzbekistan.*

Disaster for Lima

Cities that depend on snowmelt and glacier melt for their water supply may face major problems because of climate change. Glaciers worldwide are retreating at an accelerating pace. The Andean region of Peru and north and central Chile contain the world's largest collection of tropical glaciers. Because precipitation occurs there mainly in winter, snow accumulating in the mountains provides the main, if not sole, source of water for drinking and irrigation.

Lima, the capital of Peru, is one of many cities in the world where drinking water comes almost entirely from snowmelt. The Quelccaya ice cap, an ancient glacier, provides most of the city's water. Like glaciers around the world, however, Quelccaya, is retreating rapidly. Current shrinkage is at a rate of about 30 m (100 ft) annually, up from about 3 m (10 ft) per year before 1990. The 10 million residents of Lima will be hard-pressed to replace this water source when the glacier's yearly melt no longer meets the city's needs. Demand is expected to outstrip supply within the next few decades.

ABOVE *The Quelccaya ice cap melted 3 m (10 ft) per year during the 1970s and 1980s; in the 1990s, the annual rate was 30 m (100 ft).*

Scientists count on climate models to predict where changes are most likely to occur. They have found, for example, that increased extreme rainfall and greater flooding are expected across Alaska, Canada, northern Europe and northern Asia – regions that already receive heavy precipitation. The predictions are that by contrast, Indonesia may grow drier with global warming. During the 1997–1998 El Niño and again in 2002, climate in Indonesia was, as predicted, much drier than usual. In both 1997 and 2002, smoke and haze from devastating forest fires linked to the drier climate blocked the Sun over Southeast Asia and affected air quality as far away as Europe.

Widespread Problems

People living on coasts may face increased water supply problems as a result of rising sea levels. In Australia, freshwater aquifers already risk contamination from sea water and thus constrain coastal development. Increased temperature in fresh or salt water stimulates harmful algae blooms and can promote bacterial growth, potentially causing health problems.

Generally speaking, people living in poor countries will be worse off than people in richer ones. Richer countries can afford to spend more on infrastructure, such as dams, drainage systems and pumps, to prevent extreme weather events from turning into catastrophes. People in wealthier

ABOVE *Shipwreck Coast, Victoria, Australia. As sea levels rise, freshwater aquifers are being contaminated by salinization, threatening coastal development.*

countries have a better cushion – more food, better health and greater financial resources. For instance, farmers in the United States can often receive some form of government aid during a severe drought year, such as a subsidized loan or disaster relief aid. For a farmer in the African nation of Malawi, however, the failure of a rainy season could mean the difference between life and death for both him and his family.

The effect of changing water resources on the natural world will be widespread. Plant and animal species that already suffer from limited habitat or lack genetic diversity may not be able to adapt and will be at risk of extinction. Animal species that depend on vernal (springtime) pools and stream flow or are sensitive to variations in moisture, for example, will be affected as climate warming speeds evaporation and causes some areas to become drier. Amphibians – frogs, toads, newts – are extremely sensitive to small changes in temperature and moisture, and are expected to have a particularly difficult time adapting to climate change.

Salmon and Humans

The wild salmon is a symbol of the Pacific Northwest region of the United States. Salmon have long been central to the area's Native American culture and are important to the regional economy, both in the United States and Canada. Yet in the Northwest, salmon now share their ancestral rivers with a rising human population and contend with scores of dams that support a large agricultural system. The eight dams on the lower Columbia and Snake rivers provide about 60 per cent of the electricity produced in the Pacific North-west and help ensure the irrigation essential to the region's multibillion-dollar agricultural industry. Yet salmon, also an essential part of the economy, have been brought to the edge of extinction.

Salmon like cold water: it is rich in oxygen, inhibits diseases and lowers fish metabolism, making a long migration easier. Salmon must swim upstream to spawn, generally in autumn and winter.

RIGHT *Map of the Pacific Northwest, USA, highlighting the competition for water resources. Hydroelectric power companies, salmon fisheries and agriculture all draw on water supplies that are increasingly limited. The eight dams on the Columbia and Snake rivers, which generate about 60 per cent of the region's electricity, also threaten declining salmon populations, despite efforts to protect them.*

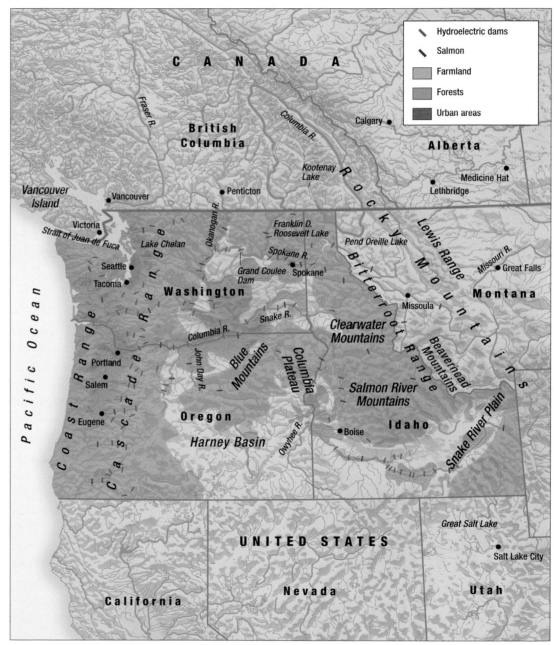

ABOVE *The Lower Granite Dam on the Snake River, Washington, USA. The dam is constructed to allow for the passage of fish, and biologists here regularly collect and mark juvenile and adult salmon to study the effects of the dam on their migration patterns. While mature salmon are usually able to withstand the water pressure created by the dam, young salmon swimming downstream may not survive the rough waters.*

The young fish, called smolts, migrate to the ocean in spring and summer. There are five major salmon species in the Pacific Northwest: chinook, coho, pink, sockeye and chum. Although each species has different migration patterns and needs, all salmon are anadromous: they spend part of their lives in freshwater streams and lakes and part in the salt waters of the ocean. Salmon return to the stream of their birth to spawn, or they die trying.

Before the region's first hydroelectric dam was built in the 1930s, the salmon were already in decline due to overfishing and destruction of habitats by pollution from mining and logging, among other causes. These dams further challenged salmon, preventing the adults' successful migration upstream and the smolts' voyage back to the ocean.

Dams have also affected water levels, temperature, turbulence and nutrient flows on the Columbia River. Climate change and withdrawal of water for use in irrigation have had their own impact on temperature, water levels, water quality and river flow, increasing the stress on salmon, not only in the Pacific Northwest. Europe's wild Atlantic salmon – most of which spawn in freshwater and are also sensitive to increases in water and air temperature – are also showing the effects of climate change. Some scientists predict that Atlantic salmon may soon become locally extinct at the southern edge of their current range, at the same time that they establish new habitats in the north.

Battles over Salmon

For decades, governments, fisheries, farmers, power companies, sportsmen and environmentalists in the US Pacific Northwest have sought to balance competing interests. The conflicts among these parties have been often intense and sometimes violent. Many solutions have been pursued since the 1970s. The federal government of the United States has tried to restore salmon habitats, has modified dams to ease passage for migrating salmon and has regulated salmon harvests. However, the effects of climate change have made this task infinitely more difficult.

By limiting water availability at key times of the year, climate change puts increased pressure on an already fragile web of competing interests. Higher water temperatures and lower water levels will directly affect spring and summer salmon migrations. This will be at the same time of year that agriculture makes its demands on the water supply. Less water will also concentrate the effect of pollution from farm runoff.

Given the many pressures on salmon, it is not clear how they will adapt to a changing climate – or whether they will adapt at all. The effect of climate change on ocean temperatures, flows and nutrients is poorly understood. If climate change forces a choice among salmon, farming and hydropower interests, as some experts predict, hydropower is unlikely to lose.

RIGHT *Salmon trying to jump a waterfall on Kodiak Island, Alaska, USA, as they travel hundreds of kilometres to return to their birthplace for spawning. As conflicts over ever-scarcer water intensify, the Pacific Northwest's salmon could face extinction.*

TOO MUCH AND NOT ENOUGH

As winters warm, it is likely that more precipitation will fall in the form of rain rather than snow. Even if the total amount of precipitation does not change substantially, the consequences of this shift can be serious, as rain and snow behave quite differently. Snow acts as a huge reservoir system, freezing water supplies in place and releasing them slowly during the spring and summer as the snow and ice melt. When winter precipitation falls as rain, it runs downhill, escapes through streams and rivers and is gone before the growing season begins; the snow that has fallen previously also melts faster under the rain. With climate change, peak stream flows will shift toward winter, when they are less useful for agriculture and ecosystems.

Ecosystems have thus been pushed out of balance. In the Pacific Northwest, winter rainstorms have been blamed for scouring young salmon out of their streams before they are ready to migrate into bigger rivers. Similar shifted water patterns have also been observed in parts of Europe and Canada.

In the Himalayas, where average annual temperature has risen each year since the 1970s, snow melting from the mountains used to fill dozens of glacial lakes, which in turn supplied water year round to villages in Nepal, Bhutan and Tibet. In the last few decades, hydroelectric plants have been built on the rivers flowing from these glacial lakes, fueling Himalayan economic development.

But in recent years the snow has been melting too quickly. Overfilled glacial lakes threaten to burst their banks, engulfing the people and economies that depend on them. UNEP has identified 24 lakes in Nepal and 24 lakes in Bhutan that are in immediate danger of bursting because of accelerated melting of Himalayan glaciers.

Until recently, it usually took an outside force to cause overfilled glacial lakes in the Himalayas to flood. One flood was triggered by an earthquake, for example, while another was launched by an avalanche that dumped snow into a lake already full. Scientists are concerned that such disasters could become more frequent as the lakes continue to be dangerously overfilled and as warming continues. Floods, mudslides and avalanches could be extremely harmful to people in downstream villages and the region's entire economy – including infrastructure such as roads, bridges and hydroelectric plants – could be at risk.

International development organizations have earmarked funds to engineer solutions to some of the most dangerous situations. But continued warming makes technological solutions stop-gap at best.

Floods

In many regions, flooding is an annual phenomenon that follows spring thaws and rain. Floods may result from high rainfall or quickly melting snow, or because thawing ice has jammed narrow or shallow parts of a river. As destructive as they can be, floods can have their benefits: they cover land with new sediment, clear waterways and renew the soil. Natural levees, or sediment banks that border large rivers, are the product of flooding. Without a regular flushing by floodwater, deep-river channels can fill with silt. River deltas depend on flooding to bring nutrients that maintain cropland's fertility.

Severe flooding is likely to increase with climate change. In 1996, China's government reported that unusually heavy flood damage cost more than $20,000 million, caused more than 3,000 deaths and nearly 364,000 injuries and destroyed 3.7 million homes, damaging 18 million more. Heavy flooding has continued in the years since then. In 2003, rain fell for 12 days in China's northern province of Shaanxi, resulting in the worst flooding in 40 years.

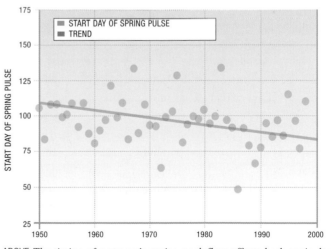

FLOW TIMING TRENDS

ABOVE *The timing of a stream's spring peak flow affects both agriculture and wildlife. Early snowmelt caused the stream in this graph to peak more than two weeks earlier in 2000 than in 1960.*

ABOVE *Hopeful residents await rescue from a rooftop after a 1998 flood caused by Hurricane Mitch in Valle de Sula, Honduras. Climate change will almost certainly cause greater flooding worldwide.*

Climate modellers, by tying expected patterns of rainfall to river models, point to more frequent and severe flooding in the future. The more concentrated extreme weather events that climate change theory is predicting also bring with them a greater likelihood of flooding. As the climate becomes warmer, the atmosphere is able to contain more water. This water may then be released during the passage of a low-pressure system.

Flooding is almost always costly to society, in both property damage and effects on human lives. Damage relief often requires enormous funding. Floods and the landslides that regularly accompany them wreck roads, bridges and buildings and hurt agricultural productivity. By mixing sewage and drinking water and by leaving standing water that becomes a breeding ground for bacteria and mosquitoes, floods can affect public health, causing cholera, hepatitis, diarrhoea and respiratory problems or allergies from moulds. In 2000, flooding in Mozambique led to nearly 2,000 cases of cholera. Floods also destroy crops; in poor countries, they can leave a whole region in the grip of chronic malnutrition. In the years 2003 and 2004, subsistence farmers and agricultural wage labourers in Assam, India, saw their crops and arable land wiped out, and widespread hunger ensued throughout the area.

Landslides

Often triggered by heavy rains, landslides cause thousands of deaths and injuries and billions of dollars of damage worldwide each year. Climate change is likely to exacerbate the problem by increasing the frequency and intensity of rainfall in some areas. At greatest risk of destructive landslides are mountainous regions, especially deforested areas where the land has difficulty absorbing water and plants no longer hold soil in place. Once a slope is weakened, almost any event can touch off a landslide – heavy rain, earthquake, volcanic eruption, wildfires, construction blasting, even thunder. Landslides can wipe out neighbourhoods and entire villages.

Mudslides are fast-moving, water-laden landslides that occur during periods characterized by intense rainfall or rapid snowmelt. Particularly common in mountainous areas with sandy soils, mudslides can range from watery to thick and rocky in consistency. They can sweep away or collapse the houses in their path or fill these structures with mucky, sodden earth.

A mudslide usually starts as a shallow landslide that liquifies and accelerates to a speed of 16 km per hour (10 miles per hour) or more – and in extreme cases, reaches a velocity of up to 56 km per hour (35 miles per hour). More destructive than other types of landslides, mudslides can pick up boulders, trees, houses and cars. As the debris moves into stream beds and rivers, bridges become blocked or even collapse, creating temporary dams that can result in floods in neighbouring areas. Flows of debris from different sources can combine in channels and together cause even greater damage. And while floodwaters eventually recede, mudslides fills streets and houses with a slurry that must be hauled away bucket by bucket.

In Caracas, Venezuela, since the year 2000, more than 20,000 people have been killed and 500,000 have lost their homes in just two mudslides that washed away poor neighbourhoods where uncontrolled development had eliminated trees and disturbed topsoil.

Both landslides and mudslides expose an area to further erosion by wind and water, raising the level of ground water and destabilizing the remaining topsoil. This means that a site where a slide has occurred in the past is at greater risk in the future; and, in fact, landslides generally tend to occur in the same location more than once.

RIGHT *Landslide formation. This diagram shows how a landslide can occur. In this instance, mud and other debris have flowed downhill and through channels, growing in volume with the water. Traverse cracks in the ground weaken the slope and the original ground surface slides as a result of the added pressure. When the debris reaches level ground, it spreads over a broad area, accumulating in thick deposits.*

Heavy rain can trigger a debris flow or other type of landslide, defined as the movement of material down a slope. Loss of vegetation will exacerbate a slide.

The crown of a slope is the nearly undisturbed area just above a landslide; it adjoins the main scarp or cliff that forms when material moves downhill.

The landslide forms a scar in the original ground surface – the uncollapsed slope that existed before the slide took place.

Fractures in the ground called traverse cracks weaken a slope; when they form and widen, the chance of a landslide increases.

As the population rises, development continues on marginal lands. Natural hazards only increase with the deforestation caused by settlements.

The toe of a mudslide is the bottom edge of the material that moves in a landslide; it is usually curved and bulging.

Not Enough Water

The percentage of lands experiencing serious drought more than doubled between the 1970s and the early 2000s, according to an analysis by scientists at the US National Centre for Atmospheric Research, in Colorado. Unusually severe droughts, and long-term declines in rainfall have been measured across the world. In 2005, to cite just one example, the worst drought in 28 years affected one million people in Vietnam and caused crop losses of more than £46 million.

With diminished rainfall, pollution becomes more concentrated in waterways. Aquifers, especially those in naturally arid lands, become increasingly salty. Crops do not grow as well. It has been predicted that the value of California cropland will drop by as much as 15 per cent if current trends continue and that the area consumed by forest fires each year in the western United States will double during the coming decades.

Though climate change simply curtails rainfall in some areas, it can also cause drought in other ways. When the water cycle speeds up, rain falls in ways that are not as useful. Warmer weather leads to faster evaporation. The atmosphere holds more water vapour; moisture does not fall back to Earth as rapidly as it did before. When rain does fall, it comes in shorter, heavier storms; the ground cannot properly absorb and store the runoff.

Where ecosystems depend on water stored as snowpack, climate change can have profound effects even when average annual precipitation does not change substantially. In the Pacific Northwest region of the United States, winter snow acts as a big reservoir of water, stored until the relatively dry summer

Warmth and Drought

Warming itself causes drought in some locations. From the late 1990s to the early 2000s, the southeastern United States experienced a prolonged drought that was linked to rising temperatures; the western United States has suffered a drought of a magnitude not seen since the Dust Bowl of the 1930s. The link between drought and temperature in the West has held true over the past 2,000 years. During the Medieval Warm Period, which lasted from around 900 to 1300, a prolonged drought in an area centred on today's states of Wyoming and Montana was exponentially worse than the current ten-year dry spell. There are geographic links as well. The El Niño–Southern Oscillation, which is tied to warmer sea surface temperatures in the eastern tropical Pacific, causes drought in much of western North America. Warming in the eastern Pacific decreases the east-west temperature difference that drives the trade winds. In turn, a relaxation of the trade winds alters atmospheric circulation around the globe, brining more or less precipitation than average, depending on the location.

ABOVE *A man looks for fresh corn among rows of dead stalks in his garden in South Carolina. This part of the United States has seen several years of prolonged drought, which is only likely to become worse as the climate warms. When the water cycle speeds up, drought is exacerbated by a number of factors, including faster evaporation and erratic or reduced rainfall.*

LEFT *Women near the village of Zinder in the Sahara Desert, Niger, dig pits near a riverbed, in an attempt to find clean drinking water. The practice is intended to filter water through the sediment. Freshwater will be in increasingly short supply as both climate change and population growth continue.*

drop at the same time that temperatures continue to rise. Climate models also predict that several other places in the world, including southern Africa, Central Asia and the Mediterranean region, are at risk of severe droughts in an era of continuing climate change. In the countries of Southeast Asia, the rain-giving monsoon is expected to be weaker than normal in the west; at the same time, climatologists expect rainfall to increase in the east.

Climate change is not the only cause of water problems. Poverty aggravates drought: poor farmers in the nations of southern Africa are unlikely to enjoy the same access to drought-tolerant varieties of crops as farmers in the developed world. Population pressures and harmful land-use practices such as overgrazing create fragile conditions that do not allow much room for error.

The worst humanitarian crises come when drought hits an already dry and overused landscape. Not surprisingly, drought caused by climate change has the greatest impact in places where the population's natural resources are already strained to breaking-point.

months. The snow usually stays frozen on the mountains until spring, when the weather warms up, at which point seeds are ready to sprout and the growing season is ready to begin.

As winters in the region have grown warmer, the winter precipitation has increasingly come as rain, which runs into the streams and out to the ocean directly after falling, despite the needs of plants and wildlife, which are adapted to a more seasonal cycle. Peak stream flows from the US Pacific Northwest to British Columbia in Canada and from Poland to western Russia, have been shifting from spring to winter as mountain-region precipitation has come in the form of rain rather than snow. If there is no snow left to melt during spring and early summer, when both farmers and wildlife need more water, the region will find itself gripped by drought conditions when the hot summer temperatures arrive.

Widespread Droughts

Many parts of the world are likely to experience abnormally dry conditions, at least in some seasons, as continuing warming trends disturb the delicate relationship of life and water.

Australia has become more susceptible to severe droughts because rising global temperatures have strengthened a vortex of winds around the Antarctic; global climate models illustrate that as ocean and atmospheric temperatures become warmer, precipitation in Australia will continue to

DESERTIFICATION

Sand rises to the tiled rooftops of a village in northern China. Dust storms tear through the capital, Beijing and turn the air yellow in Seoul, Korea, hundreds of miles away. During the storms of sand that emanate each spring from northern China, the weather service in Seoul is accustomed to issuing 'yellow dust alerts'. Poor visibility grounds flights and children stay home from school.

China estimates that its deserts, which already occupy 28 per cent of the country, are growing at a rate of over 10,400 sq km (4,000 sq miles) a year. Drifting sand covers hundreds of kilometres of railroads and thousands more kilometres of roads. In northwest China the government warns people to cover wells during the increasingly ferocious black-sand dust storms. The Gobi Desert grew by 52,000 sq km (20,000 sq miles) between 1994 and 1999. China puts the annual cost of desertification at between 16–25 billion yuan RMB.

Yet the problem is worldwide. Although true deserts cover an area that constitutes approximately 7 per cent of the Earth's surface, about 30 per cent of the Earth is 'dry land' – that is, perpetually arid and always at risk of becoming uninhabitable. The United Nations warns that the livelihoods of a billion people in 100 countries are threatened by the spread of deserts. The IPCC estimates that there are already about 30 million 'environmental refugees' who have been displaced from their land and hypothesizes that this number will rise to 150 million by 2050, as huge expanses of land become so dry that they can no longer support agriculture or grazing.

According to the United Nations Convention to Combat Desertification, two-thirds of Africa's arable land is in danger of disappearing, along with one-third of Asia's and one-fifth of South America's. In northern Nigeria, the Sahara overcomes villages as it spreads southward at a rate of 0.64 km each year. In the United States, where water in the arid western states has long been a source of conflict, UNEP estimates that about 40 per cent of the land may eventually become too dry for agriculture or grazing.

Desertification is the transformation of productive land into desert by natural or human degradation. The term was coined in the 1950s, though desertification became notorious in the United States in the 1930s, when poor farming practices, combined with drought, caused a tremendous loss of topsoil that turned the Great Plains into a 'Dust Bowl' and drove farmers to California.

LEFT *A dry bed of the Oued Tensift River in Marrakech, Morocco. The region is at the edge of the Sahara Desert, which is rapidly expanding.* UPPER LEFT *The red cliffs of the Gobi Desert, Mongolia, another of the world's growing deserts, which could eventually spread into areas that are currently habitable.*

Land taken over by sand dunes often cannot be reclaimed, but in many places marginal, arid land can be restored through careful stewardship. Farmers can increase the soil's ability to hold moisture by using tilling methods that maintain vegetation, such as leaving the remains of harvested crops on the soil surface rather than ploughing them under. Land restoration techniques include planting trees and native vegetation, using fences and straw mats to stop sand migration and changing irrigation and water management policies to avoid waste.

Land restoration has scored successes against desertification in dry areas where good science came together with good land policy. In the Turpan depression of western China's Gobi Desert, underground irrigation has reclaimed land for use as vineyards. Direct water management, which allocates water by volume to various users, is one of many agricultural and land use practices that can help ensure that all the people in an area have the water they need.

ABOVE *The Dakhla Oasis in the Western Desert of Egypt. The encroaching sand dunes are spreading to cover this lush agricultural region popular with tourists. Desertification of this kind is worsening as a result of climate change, but it is also causing local climates to become more arid, creating a vicious cycle of dryness.*

Climate change can exacerbate desertification by stimulating conditions that cause drought and increasing evaporation rates. At the same time, desertification itself causes the local climate to become more arid, creating a vicious cycle. Poor grazing and farming techniques add to the problem by causing soil to erode and blow away and by removing the vegetation that once stored soil moisture, making the local climate drier still. Indeed, most experts agree that the greatest cause of desertification is human abuse of the land. Overgrazing, overcultivation and overuse of water – beyond that which nature can replace – all deplete the land.

In many regions, including sub-Saharan Africa, maximizing food production is a life-or-death matter for subsistence farmers; lack of resources and little room for error make it difficult for them to experiment with new, sustainable farming practices. International development agencies are working to break the cycle of poverty and land degradation that is one of the greatest causes of desertification in the world's fragile dry lands.

ABOVE *Farmers planting trees on a hillside in Yanquing County, northwest Beijing, China. This is one of several land restoration techniques employed by Beijing's municipal government to prevent the sandstorms that have devastated the city in recent years. The severity of these storms is blamed on improper land use, high population and rapid urbanization, combined with increasingly dry conditions.*

DRINKING WATER: QUANTITY AND QUALITY

Four out of ten people on earth lack access to safe drinking water and long-term solutions to this very basic problem remain elusive. In Bangladesh, where contaminated surface water was once a leading cause of disease and childhood mortality, international agencies intervened starting in the 1970s, digging thousands of tube wells to tap into the ground water, which seemed cleaner. But in the 1990s, toxic levels of arsenic were discovered in the ground water, the new source of drinking water for at least 40 million people.

People in wealthy countries still take safe drinking water for granted. Governments charge very little for it and people routinely flush expensively treated water down the toilet without a thought. Average water consumption in most of the world is between about 20 and 40 litres (5 and 10 gallons) per person per day; in the United States and Canada, the figure exceeds 135 litres (35.6 gallons) per person per day – three to six times more than the world average.

Climate changes will affect drinking water supplies in both rich and poor countries. Rising sea levels, shortage of snowmelt and dwindling ground water, as well as pollution and increasing demand, will all have an effect on the quantity and quality of drinking water. Rising sea levels will imperil low-lying freshwater aquifers near coasts. Where clean drinking water originates as snowmelt or glacier melt, precipitation falling as rain rather than snow in winter will put pressure on drinking water supplies.

Some ancient aquifers are dwindling as water is 'mined', or pumped out of the ground, faster than it is replenished by rainfall. Aquifer water is used by farms, factories and cities. In the United States, 5.6 million hectares (14 million acres) of Great Plains farmland and ranchland in six western states are currently irrigated by the Ogallala aquifer. Since the aquifer lies under several states, each state has the legal right to control its own use of the water. The Ogallala aquifer has declined in volume steadily since it was first used for major irrigation in the 1940s. In the 1990s, water levels were dropping at a rate of about 50 cm (1.5 ft) per year, and it was estimated that the remaining water would not last

ABOVE *A woman drinks from a groundwater source near Matlab, Bangladesh. Over 60 per cent of wells in the area contain toxic levels of naturally occurring arsenic. An international effort is underway to mitigate the effects of arsenic contamination, but millions of Bangladeshis still lack access to safe water.*

more than another 20 years. Natural springs long considered sacred by the Hopi Indians have run dry.

With their recent experience of severe drought, people in the western United States have begun to realize that cooperation and conservation are vital. Even small steps to save water, such as leaving crops on top of the soil after harvesting instead of ploughing them under, postpone the day when climate change will combine with human use to dry up the aquifer.

Declining Water Quality

A vast amount of drinking water is already lost as a result of pollution, over-irrigation, poor land-use

WATER LEVEL IN OGALLALA AQUIFER

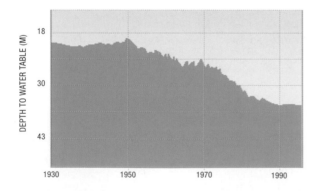

ABOVE *The Ogallala Aquifer in the western United States has been mined faster than it can recharge. This graph from the University of Akron, in Ohio, USA, shows well water levels dropping. Use of aquifer water, especially for irrigation, increased dramatically during the period covered by the graph.*

practices and waste. As the climate warms, more problems are likely to enter this already toxic mix. Warmer air and water temperatures encourage bacterial growth and algae blooms, which can endanger human health. These blooms are occurring more often in reservoirs and water supply systems, lowering water quality as temperatures rise. In Larimer County, Colorado, USA, for example, in 2003, types of algae that produce liver toxins, neurotoxins and derma-toxins were found to be abundant in the water supply.

Even if pollution from farms and industries remains constant, pollution levels in waterways will become more highly concentrated downstream as total water volume diminishes, to the point where water may become undrinkable.

Already, drinking water supplies do not meet the needs of the population of some arid regions, including the River Jordan basin in Israel and Jordan. During a drought in 1998 and 1999, residents of Amman, Jordan, received drinking water only two days each week and Israel cut water supplies for agriculture by 25 per cent. In many African countries, water shortages have long been a fact of life. Women and children spend a considerable amount of time each day carrying water, sometimes from communal wells. In these semi-arid places, water is always scarce and any drop in

ABOVE *Grand Prismatic Springs, Yellowstone National Park, Wyoming, USA. The coloration is a result of algae that thrives in the hot, mineral-rich water. Higher water temperatures worldwide could encourage such algae blooms, further reducing supplies of freshwater.*

rainfall, such as the shortfall expected in southern Africa because of climate change, may mean the difference between survival and death.

Bulk Water Exports

Climate change will not affect all the water in the world equally. While many places worry about drought and scarcity, other places are likely to actually profit from this scarcity. Canada is expected to be one of the climate change water 'winners'. Currently, as much as 20 per cent of the world's fresh water lies in Canada. Although much of it is frozen in ice sheets, at least nine per cent of it is usable, high-quality liquid water and the trend in Canada has been toward increased precipitation. In a much drier future, Canada may become the OPEC of water.

For a few Canadian entrepreneurs in the late 1990s, the tremendous water needs of Europe and

China looked like a profit-making opportunity. A number of companies were formed to begin exporting water in jugs and tanker ships.

At first, the government of Canada was receptive. One company received a government permit to export more than 600 million litres (158 million gallons) of water from Lake Superior to China by 2002. The export proposal of a Newfoundland company was backed by the provincial premier and the mayor of Grand Le Pierre, a town where the unemployment rate had risen to 40 per cent since the collapse of the cod fishing industry.

Bulk water shipments from Canadian lakes and streams to Asia or Europe would indeed have generated profits for the exporters and also jobs for some Canadians. But the public outcry against bulk water exports was swift and strong. One group, the Council of Canadians, mobilized its 100,000 members to oppose the sale of Canada's water, which, they said, belongs to everyone and therefore to no one. The opposition was great enough to induce Canada's government to impose a moratorium on bulk water exports.

The opponents' argument depended, in part, on the IPCC's climate change prediction for the Great Lakes region. Climate models, the IPCC report said, show a high probability of falling lake levels and outflows in the Great Lakes–St. Lawrence system, even as an increased likelihood of rain was predicted for much of the rest of Canada.

Critics of water exports from Canada's Great Lakes maintained that once the door to bulk water exports was opened, Canadian water would become, in legal terms, a commodity. Under international trade agreements, any Canadian company

would be able to take this common resource and it would be difficult to stop or control them, even if dire environmental predictions about declining lake water began to come true.

The Canadian debate over water exports, however, may have sparked a new consciousness about water use and conservation in the Great Lakes region. Lake levels were dropping at the time the

WATER HAS BEEN A SOURCE OF CONFLICT FOR CENTURIES, BUT RECENT DISPUTES HAVE BEEN INCREASINGLY VIOLENT. IN AFGHANISTAN, CHINA, INDIA AND THE MIDDLE EAST, CLASHES OVER WATER HAVE SPARKED RIOTS

water export legislation was debated, but recently the lakes have risen about 46 cm (18 in), to their highest level in years. Meanwhile, the thirst in other countries for water to drink and to use for irrigation shows no sign of abating.

Water Wars

As the population of the Earth continues to grow, the United Nations predicts that the world as a whole will be water-poor by 2025. This could be a recipe for violence. Water has been a source of conflict for centuries, but recent disputes have been increasingly volatile. In Afghanistan, China, India and the Middle East, clashes over water use and allocation have sparked riots. The number of violent incidents relating to water distribution has grown.

In 2000, farmers rioted over how water in northern China's largest lake, Lake Baiyangdian, was

allocated; several people were killed as villagers in Shandong Province clashed with officials. In southern Guangdong Province, officials blew up a water channel to prevent a neighbouring county from diverting water.

In 2002, in Tamil Nadu, India, violence over the allocation of water from the River Cauvery between Karnataka and Tamil Nadu resulted in more than 30 injuries and several arrests.

In 2004, in Somalia near the Ethiopian border, more than 50 people were killed when divisions of a clan fought over who would get access to grazing land and water wells.

The Middle East is often named as a place where water wars could break out. Population growth in that region has led to full use of all available water. The number of countries in the Middle East classified by the United Nations as 'water scarce' has risen from three in 1955 to eleven today. Water levels in the Dead Sea dropped by 10 m (32.81 ft) during the twentieth century.

Population pressure accounts for much of the problem, but it is estimated that climate change will be responsible for about 20 per cent of the increase in water scarcity that is predicted for the Middle East by 2050.

THE GREAT DRYING

Many regions around the world are becoming drier because of climate change. One of the most striking effects of this transformation is an increase in dust storms. Dust storms from the Sahara Desert have increased tenfold in the last 50 years.

Overpopulation and overgrazing in dry areas have contributed to this problem, as have the trucks that have replaced camels for transportation in the desert. Wheeled vehicles rip up the thin layer of lichen that has held the desert surface relatively stable for many centuries. Some scientists have linked Saharan dust carried through the atmosphere to dying coral reefs in the Caribbean Sea as well as to plankton growth in tropical oceans.

Poor land-use practices inevitably contribute to dust storms. Cutting down entire forests for timber or firewood, over-grazing and over-tilling start the sinister process of land degradation. Often these practices are the result of desperate poverty. But poverty is not the only problem: bad land use is caused in equal measure by ignorance, greed and short-sightedness.

At the desert borderlands of northern and western China, poor land-use practices stemming from poverty and from a host of other reason, have combined with a changing climate to make the bad situation in that region even worse.

LEFT *Satellite image of a sandstorm originating from Africa's Sahara Desert (right), the worst ever recorded in the region. It covered the Cape Verde Islands (lower left) in the Atlantic. The Canary Islands are at top. In the photo, arid areas appear brown, water is blue and clouds are white. Plumes of wind-borne dust and sand can be seen reaching far westward.*

The Drying of Lake Chad

At the intersection of four countries in West Africa – Chad, Niger, Nigeria and Cameroon – Lake Chad was one of the three largest bodies of freshwater on the African continent in the middle of the twentieth century. With an area of about 25,000 sq km (9,700 sq miles), it was similar in size to Lake Erie, in the United States. Over a period of fewer than 40 years, the lake shrank to an area of just 1,500 sq km (585 sq miles), about a twentieth of its former size.

Desertification in the Sahel region to the west and climate change resulting in a prolonged regional drought are two reasons for the alarming decline, but there are others. The lake is very shallow – only 7 m (23 ft) at its deepest, with an average depth of 1.5 m (5 ft), and it is especially sensitive to fluctuations in its average depth. Human overuse of lake water for irrigation, much of it in response to the drought, has also taken a toll. Lake Chad exemplifies how humans must figure climate change into their decision-making or risk using up resources permanently.

Rainfall in the Lake Chad region has declined steadily since the 1960s. Yet, beginning in the 1980s, even as the lake was already in decline, water was diverted for massive irrigation projects. During the next decade, irrigation quadrupled from the level of the previous 25 years.

Today, dust from Lake Chad's dry bed is the world's greatest source of atmospheric disturbance from dust storms. A great number of people who lived in villages around the lake have had to move to other places or have become refugees, without a new permanent home. In impoverished West and Central Africa, the influx of refugees has caused considerable social strain; in fact, displaced Lake Chad refugees were one of the stress factors leading to violence in the Darfur region of Sudan. Meanwhile, Lake Chad continues to shrink and is not expected to recuperate soon. Scientists Michael T. Coe and Jonathan A. Foley of the University of Wisconsin at Madison, in the United States, call it 'a ghost of a great lake'.

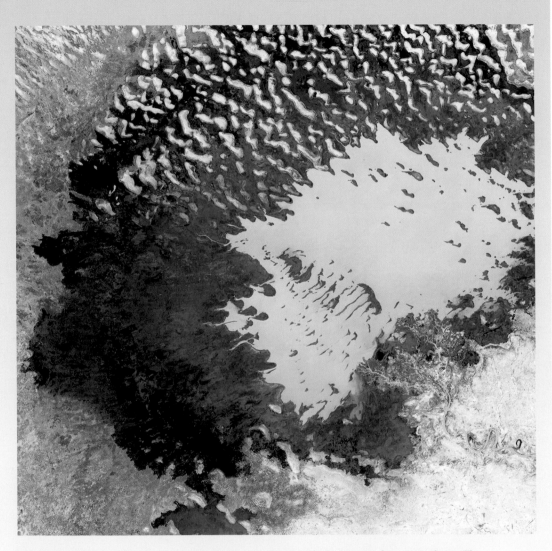

ABOVE *Lake Chad, in Africa, seen in a satellite image, is now only 5 per cent of its original size. The lake is dramatically smaller as a result of a combination of irrigation projects and declining rainfall. Lake Chad provides water to more than 20 million people living in the four countries that surround it.*

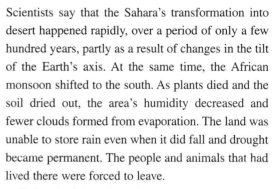

THE EFFECTS OF A SINGLE COUNTRY'S DUST STORMS CAN BE FELT ACROSS THE PLANET. THE FINE PARTICLES CAN REMAIN IN THE ATMOSPHERE FOR WEEKS, CROSSING OCEANS AND TRAVELLING AROUND THE WORLD

In recent years, the Chinese government has started mass tree-planting campaigns. Without mature trees to provide shelter from heavy winds, however, and without sufficient water and good topsoil with vegetation to retain moisture, the newly planted trees have difficulty taking root.

Meanwhile, many of the villages that are downwind of these poorly vegetated areas, in China's Hebei Province and elsewhere, are being swallowed by sand and grit, which sometimes piles up to the roofs of houses.

The effects of just a single country's dust storms can be felt across the planet. The fine particles from these storms can remain in the atmosphere for weeks at a time, crossing oceans and travelling around the world. Landing on water, they deposit iron and other nutrients, causing plankton blooms. In the air, they shade the Earth from the Sun's radiation, affecting climate by absorbing heat in dust clouds high in the atmosphere while leaving the air below cooler than normal.

Dust storms are harmful to the Earth's environment in many other ways. On the ground, for example, dust storms regularly cause topsoil to blow away, which affects the land's ability to hold moisture, support plant life and resist erosion.

The Expanding Sahara

Only 5,000 years ago, the Sahara was covered by shrubs and grasses. Ancient paintings and archaeological finds show that hippopotamuses and alligators lived in the area. Now, the core of the Sahara is a land of giant sand dunes stretching from Egypt to the west coast of the African continent.

ABOVE *The town of El Gedida being overwhelmed by sand from Egypt's Western Desert. About the size of Texas, the Western Desert covers roughly 700,000 sq km (270,000 sq miles) and is sometimes called the Egyptian Sahara.*

Scientists say that the Sahara's transformation into desert happened rapidly, over a period of only a few hundred years, partly as a result of changes in the tilt of the Earth's axis. At the same time, the African monsoon shifted to the south. As plants died and the soil dried out, the area's humidity decreased and fewer clouds formed from evaporation. The land was unable to store rain even when it did fall and drought became permanent. The people and animals that had lived there were forced to leave.

Today the Sahara is the largest desert in the world. It stretches more than 5,630 km (3,500 miles) across North Africa from the Atlantic Ocean to the Red Sea, and more than 1,930 km (1,200 miles) from north to south. Two large rivers, the Nile and the Niger, pass through the desert. Underground aquifers and springs feed the occasional oasis.

South of the desert is a dry-land border region called the Sahel. Still able to sustain life, it is the home of some of the world's poorest people and supports acacia trees, which for centuries were used to produce gum arabic and grasses. The vegetation differs from that of the Sahara and the savannahs and forests farther south. Today, the region faces possible desertification. The runaway change of areas of the Sahel into desert is difficult to reverse. It is a dance between natural and human causes, between drought and poor land-use practices.

SPECIES AT RISK

The Earth is now witnessing the biggest extinction of species since the time of the dinosaurs. Nearly a quarter of all existing species are in danger of becoming extinct within the next fifty years and a major underlying cause is climate change, complicated by competition from invasive species, pollution, poor land management, destruction of wildlife habitats and other human pressure on them and overharvesting of animals, fish and plants for food.

Declining and shifting water supplies due to climate change increase pressures on already stressed wildlife. Native plants that cannot migrate as fast as their water sources are likely to be threatened. Animal species that are very sensitive to moisture, such as toads, or that are dependent on a narrow habitat that is susceptible to climate change, such as bighorn sheep, may not survive.

Hot Spots

Twenty-five biodiversity 'hotspots' covering a mere 1.4 per cent of Earth's surface are the only home for almost half of Earth's plant species and over a third of its bird, mammal, amphibian and reptile species. The effects of climate change in these areas have tremendous implications for global biodiversity. The hotspot in Southwest Australia, for example, is home to a great number of animal and plant species, including nearly 3,000 different plants and 30 reptile species. These species are found nowhere else in the world, and they are now under threat from climate change. Rainfall in this region has dropped by 20 per cent in the past century and may drop another 60 per cent in the next according to research by

ABOVE *In these biodiversity 'hot spots' around the world, a high concentration of species lives in small areas (in red) vulnerable to threats from human development and climate change.* UPPER LEFT *A Rocky Mountain bighorn ram in Alberta, Canada.*

the Commonwealth Scientific and Industrial Research Organization. Species threatened by this rapid drying include Australia's most endangered reptile, the Western swamp turtle and the numbat, a small marsupial anteater. Most of the world has only opossum-like marsupials; losing any Australia's stunning marsupial diversity therefore represents a tremendous loss.

LEFT *The numbat, a marsupial native to Australia, was named the world's most endangered mammal in 1982.*

Predator and prey: bighorn sheep and jaguars

In California, over the past century, desert temperatures rose about 1.8°C (1°F) and annual rainfall decreased about 20 per cent. At the same time, bighorn sheep populations dropped by more than a third. These sheep do not handle change well and research suggests that the drop in population threatening the sheep with extinction was correlated with the changing climate.

In contrast, jaguars are adaptable to change and have been found in rainforests, grasslands and dry areas. Yet they, too, are threatened with extinction. The largest remaining populations of jaguars are in the Amazon rainforests. The increased drought that may come with climate change has the potential to devastate this critical habitat and the jaguars it houses.

Succulent Karoo

Two of the best-studied hotspots, the Cape Floristic Province and the Succulent Karoo in southern Africa, are predicted to lose half their area by 2050, with climate change playing a significant role. Between them, they have over 13,000 plant species; two-thirds of these species, including the king protea, South Africa's national flower, occur nowhere else. Both hotspots are known for their dazzling spring wildflowers; the Succulent Karoo, one of only two completely arid regions to earn the hotspot label, also boasts the world's most diverse array of succulents, including bizarre plants like the halfmens. This tree-like succulent tends to form north-facing clusters; combined with a crown of leaves that suggests a head of hair, it creates the illusion of groups of people gathered together gazing north. The large populations of elephants and black rhinoceroses once hosted by the Karoo have already disappeared.

ABOVE *A jaguar in its native habitat. The largest population of this endangered species lives in the rainforests of Central and South America, an environment that is threatened by climate change and human activity.*

Biodiversity and Water

Changing water supplies threaten biodiversity. Water diverted for agriculture, hydropower, or cities affects animal and plant life, which is dependent on water.

The choices forced by lack of water are illustrated by New Mexico's Gila River, in the United States. The Gila was polluted by mining and its vegetation and soils degraded by livestock. Farmers diverted so much upstream water for irrigation that not enough was left to support farming downstream. Dams reduced the Gila still further. Only in recent decades have court battles by environmental and Native American groups resulted in the funding required to initiate efforts to restore habitats and encourage the revival of native fish, birds, vegetation and wildlife – including jaguars – along the River Gila .

RIGHT *A halfmens, or elephant trunk, an endangered plant species in South Africa's Richtersveld National Park. This region is considered a biodiversity hot spot.*

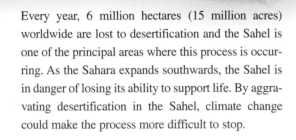

Every year, 6 million hectares (15 million acres) worldwide are lost to desertification and the Sahel is one of the principal areas where this process is occurring. As the Sahara expands southwards, the Sahel is in danger of losing its ability to support life. By aggravating desertification in the Sahel, climate change could make the process more difficult to stop.

The Future of the Sahel

A major drought that extended from the 1960s through the 1980s led many to fear that the Sahara sand dunes would swallow the Sahel. Now, however, scientists believe that the balance between desertification and recovery is under human control, although the economic forces at work have not been positive. Near urban areas, virtually all trees have been cut and used for fuel; wood used to supply about 90 per cent of cooking needs. With wood scarce, dung is burned instead, so little organic matter is put back into the soil for fertilizer. Without trees, soil moisture is lost and erosion becomes difficult to combat. Windstorms carry large quantities of dust into the air.

Yet many people still think that careful land management can begin reversing the desertification of the Sahel. Replanting native trees, grasses and shrubs can begin to stem erosion; teaching nomadic pastoralists, or goat herders, to nurture and not destroy young growth will protect what has been planted. Wind fences and straw grids – a method of covering the base of small dunes with straw, which helps against wind erosion – will also control erosion and begin stabilizing the land.

Cloud Forests

In part because of climate change, the moist, evergreen, tropical forests found at the tops of some mountain ecosystems, known as cloud forests and montane forests, are becoming drier. Named for the clouds that are nearly always present and in constant contact with the vegetation, these regions are an exotic but essential part of the world's water supply system. Though they may seem remote, they are far from irrelevant to people's daily lives. According to a UNEP report, the La Tigra cloud forests in Honduras supply 40 per cent of the water used by the 850,000 residents of the Honduran capital, Tegucigalpa. In Ecuador, Tanzania and Malaysia, cloud forests provide large expanses of land and millions of people with their major source of water during the dry season. Cloud forests have also been documented to influence cloud formation in nearby lowlands, bringing more rain and moisture to lower regions as well.

Cloud forests' altitude, distance from the sea, latitude, rainfall and winds combine to create an ecosystem. The number of species living in cloud forests is greater than might be expected. According to the UNEP report, more than 30 per cent of the 272 mammals, birds and frogs native to Peru live in cloud forests. A cloud forest of 20 sq km (8 sq miles) in western Ecuador has about 90 native species that are endemic, or particular to that location. Wild relatives of food crops such as potatoes, tomatoes, cucumbers and beans grow in cloud forests, preserving the diversity of their gene pools. With so much biodiversity packed into such tight quarters, these forests are good indicators of the health of the water balance.

Cloud forests are already endangered by fires, logging, road-building, invasive plant species, illegal clear-cutting by opium or coca growers, wood harvesting for fuel, gem mining and habitat fragmentation caused by human development.

Added to this already potentially lethal mix are threats posed by climate change.

As warmth increases, evaporation intensifies at low elevations, restricting cloud formation to altitudes sufficiently high and cool for condensation. Without cloud cover and with greater evaporation, the mountain at lower elevations dries out. Animals and vegetation move uphill as the cloud forest shrinks. Toucans, normally resident in lowland forests, have now established breeding populations in the cloud forests of Costa Rica.

Because the cloud forests and the organisms that inhabit them are already at the top and cannot migrate farther, they are threatened by the prospect of being replaced by invading species. The more rapid evaporation and accompanying decreased cloud cover on tropical mountains that climate models once predicted have been observed in recent years. With less immersion in clouds and less contact between vegetation and clouds, less water is captured by the cloud forest. The entire ecosystem is threatened with extinction.

Scientists have only recently begun to appreciate the immense value of cloud forests to lowland water systems. Today, scientists and environmental groups are working to raise awareness of the role they play in the world's water ecology and to protect them from the human and climate pressures conspiring to extinguish these fragile ecosystems.

The Golden Toad

The tiny golden toad *(Bufo periglenes)* is a spectacular orange-yellow resident of a tiny area measuring just 10.4 sq km (4 sq miles) in a single cloud forest on a mountain in Costa Rica. Or at least it was.

Golden toads were very shy and secretive except during their explosive mating season in May, around temporary, vernal pools. The toad, like other amphibians, took its mating cues from the environment – the air temperature and moisture indicated to the toads that the vernal pools were back, ready to receive their eggs. The year 1987 was dry in Costa Rica. It was an El Niño year, which means that unusually warm ocean currents in the Pacific were creating conditions similar to those that could be expected with longer-term climate change. It is speculated that in 1987, the temporary pools dried up before frogs and toads had a chance to get out of their larval stage. Half of all frog and toad species, 25 out of 50, disappeared from the Monteverde cloud forest that year, including

LEFT *A cloud forest in the Bwindi Impenetrable Forest National Park, Uganda. The many plant and animal species here are being threatened by encroaching lowland species even as the boundaries of the cloud forest shrink Almost half the world's population of mountain gorillas lives here.*

ABOVE *The golden toad, which once lived in the Monteverde cloud forest of Costa Rica. Scientists have concluded that climate change is to blame for the disappearance of this species.*

the golden toad. After that, only five golden toads have been seen and none since 1989. The golden toad is now believed to be extinct.

Drought in Australia

The driest inhabited continent on earth, Australia has long experienced cycles of drought and flood. Yet the drought of 2002 and 2003 was an event that was more extreme than usual. Australia's worst drought in 100 years, it sharply reduced the year's winter wheat harvest and curtailed the number of sheep that could be put out to graze to a number lower than that for any year since the 1920s. The country's farm exports dropped nearly 30 per cent and its farm income plummeted nearly 70 per cent. By the time it was declared over, the drought had cost Australia an estimated $A10,000 million (£4,250 million) as well as 70,000 jobs.

The rains that allowed the government to pronounce the drought over brought a scourge of locusts. The following year, drought was back in many cities. Temperatures were over 50 per cent higher than normal levels, causing dozens of deaths. Eastern Australia received two-thirds less rainfall than its average for the month during June 2004; there were dire predictions about the threat of bush fires and continuing problems with farm production.

In Australia's Murray-Darling Basin, where 40 per cent of the nation's food crops are grown, heat and drought caused record evaporation rates, so that the loss of rain was felt even more acutely. During three previous twentieth-century droughts, the monthly evaporation rates were measured at 136, 120 and 131 mm (5.4, 4.8 and 5.2 in); during 2002 the evaporation rate in the Murray-Darling Basin reached 152 mm (6.1 in) in one month.

Many climate watchers blamed the drought on global warming caused by elevated greenhouse gases and predicted that Australia could enter a period of chronic drought and water shortages if warming trends continued. Australia's Commonwealth

climate change, eastern, south-eastern and south-western Australia are likely to experience drier winters and springs, the seasons most directly affecting farm output. Global climate models indicate that because of the way the atmosphere and ocean interact to produce Australia's climate, global warming is likely to increase the already drought-prone continent's average temperature.

A 2003 report by David Karoly of the University of Oklahoma School of Meteorology in the United States concurred that Australia's 2002–2003 drought was not caused by natural climate variations alone. Karoly said that for the first time in Australia's history, during this severe drought, 'The impact of human-induced global warming can be clearly observed.' The Australian Bureau of Meteorology says that rainfall has been decreasing in southwestern Australia for the last 50 years, dropping an overall average of 30 per cent in some places.

Long-Term Drought Damage

In addition to higher temperatures and lower levels of rainfall, climate change poses long-term threats to Australia. Land that had been cleared for

BECAUSE OF HOW ATMOSPHERE AND OCEAN INTERACT TO CREATE AUSTRALIA'S CLIMATE, WARMING IS APT TO INCREASE THE DROUGHT-PRONE CONTINENT'S AVERAGE TEMPERATURE

Scientific and Industrial Research Organization, part of the government's bureau of Environment and Heritage, conducted research showing that with

agriculture, with its vegetation grazed by livestock until there was insufficient check against erosion, became a perfect breeding-ground for dust storms

ABOVE *A dry reservoir in Condobolin, north of Sydney, Australia. The area suffered a protracted drought in 2002, which destroyed crops and threatened the livelihood of farmers. A drought of this magnitude has far-reaching effects, as Australians turn to ever scarcer groundwater sources to alleviate shortages from the dry spell.*

during the 2002–2003 drought. A giant dust storm in Queensland measured about 1,500 km (900 miles) long, 400 km (240 miles) wide and 2,500 km (1,500 miles) high.

As a result of long-term drying trends, ground water use in the Murray-Darling Basin increased by 90 per cent between the 1970s and the 1990s. This had serious ecological consequences. Inland wetlands decreased in size and a greater number of obstacles barring the way of fish migrating up Australia's rivers was recorded.

Rising sea levels associated with warming also threaten aquifers and may limit options for developing new freshwater sources near the shore. Saltwater intrusion in coastal aquifers will worsen if Australians continue to increase the amount of ground water they must pump out of these aquifers, or if the drier climate decreases the amount of fresh water entering and replenishing the aquifers. An Australian government report predicted that within 50 years, the Murray River, which provides drinking water for the city of Adelaide, may become too salty to drink.

A number of factors are contributing to water pollution. Drier conditions, paired with increased evaporation, and reduced stream flows combined with farm runoff and soil erosion also contribute to toxic algae growth and, eventually, eutrophication of the water. This water is so polluted and contains so many salts and dissolved nutrients that it is not fit for use without treatment. According to a 2002 report by the Australian Greenhouse Office, 'Eutrophication is already a major problem in Australia and its incidence may increase with global warming.'

RISING SEAS

Low Lands in Peril

A warming climate is leading to the melting of ice and snow and also to the thermal expansion of the oceans. Taken together these events will raise sea levels, bringing on widespread flooding and loss of life and property. Effects will vary around the globe, but the cumulative impact could be enormous. Beaches will be eroded, islands washed away, coastal wetlands destroyed, farmland ruined by salinization. At greatest risk are low-lying islands and coastal areas prone to flooding.

CLIMATE CHANGE AND SEA-LEVEL RISE

Sea levels are already rising and they will rise even faster as the Earth warms. At first, the effects will be most dramatic in the world's coastal regions. In 1995, an estimated 2,200 million people, or 39 per cent of the world's entire population, lived within 100 km (60 miles) of the coastline. Today, thirteen of the world's twenty largest cities are on a coast. Population densities in coastal regions are three times higher than the global average and they continue to rise.

As people move into coastal regions, more lives and property will be at risk, with or without climate change, although climate change will make the situation worse as sea levels rise faster.

The problems posed by rising sea levels will not go away anytime soon. Recent studies show that even if nations around the world halted the increase in heat-trapping carbon dioxide pumped into the atmosphere, sea levels would still increase dramatically by the end of the twenty-first century.

Higher sea levels mean more destructive storm surges. Coastal ecosystems will be damaged. Some coastal communities will be forced to relocate. More and more islands will disappear under the waves.

Geologists generally accept that sea levels have been rising more or less steadily since the peak of the last ice age, when much of the Earth's water was lodged on the continents as glacial ice. In the last 20,000 years, sea levels have climbed an estimated 100 m (330 ft) and many islands have vanished into the ocean. During the twentieth century, sea levels rose at a rate of 1 to 2 mm (0.04 to 0.08 in) per year. Given the continued rise in temperatures, by 2100 global sea levels are expected to rise by a total of 9 to 88 cm (4 to 35 in) over the levels in 1990.

The Mechanics of Sea-Level Rise

Global warming contributes to rising sea levels in two primary ways. One way, which is not obvious to most people, is thermal expansion. The molecules of a warming ocean move more quickly and pack less tightly, thus increasing ocean volume, but not mass. In other words, water, like most material things, expands when heated. Current estimates are that thermal expansion is responsible for 0.5 mm (0.02 in) of increase in sea levels a year. Despite annual variations, a warming trend is readily apparent in data dating from the middle of the nineteenth century and the Intergovernmental Panel on Climate Change (IPCC) concludes that average sea surface temperature is likely to increase by as much as 3°C (5.4°F) by the end of the century.

The other way warming contributes to rising sea levels is by increasing the mass of the ocean – most obviously by adding water. Since the peak of the last ice age, continental glaciers (ice sheets that cover a large section of a continent) have been melting, their waters running into the ocean. Their edges have also broken off and floated away. The warming effects of greenhouse gases have accelerated this process in recent years. The melting of continental glaciers is likely responsible for an annual rise in sea levels of 0.5 mm (0.02 in). Conversely, the melting of ice already floating in the sea does not contribute to the rise in sea levels, since a piece of ice, when floating, already displaces the same volume of water as its meltwater would displace.

Of the recorded annual increase in sea levels, one part – measuring 1 mm (0.04 in) – cannot be

RIGHT *The Nile Delta and surrounding areas as seen in an image from a spectrometer, which measures light. The dry areas, including the sandy Sahara, are light, and the lush vegetation of the triangular delta is dark. As sea levels rise, the Nile Delta region could be flooded, and residents would suffer costly losses in agriculture and tourism.* PREVIOUS PAGE *Clockwise from upper left: live oak trees are suffocated by sand on Cumberland Island, Georgia, USA, as sea-level rise threatens the barrier island with erosion both on the ocean and the inland side; fields in Bangladesh after a flood in 1988; a fishing boat off the coast of California endures high swells from a fierce storm in 2004; the Malecon, the famous waterfront street in Old Havana, Cuba, which has been named a World Heritage Site by UNESCO. Rising sea levels would further endanger these already decaying historic buildings.*

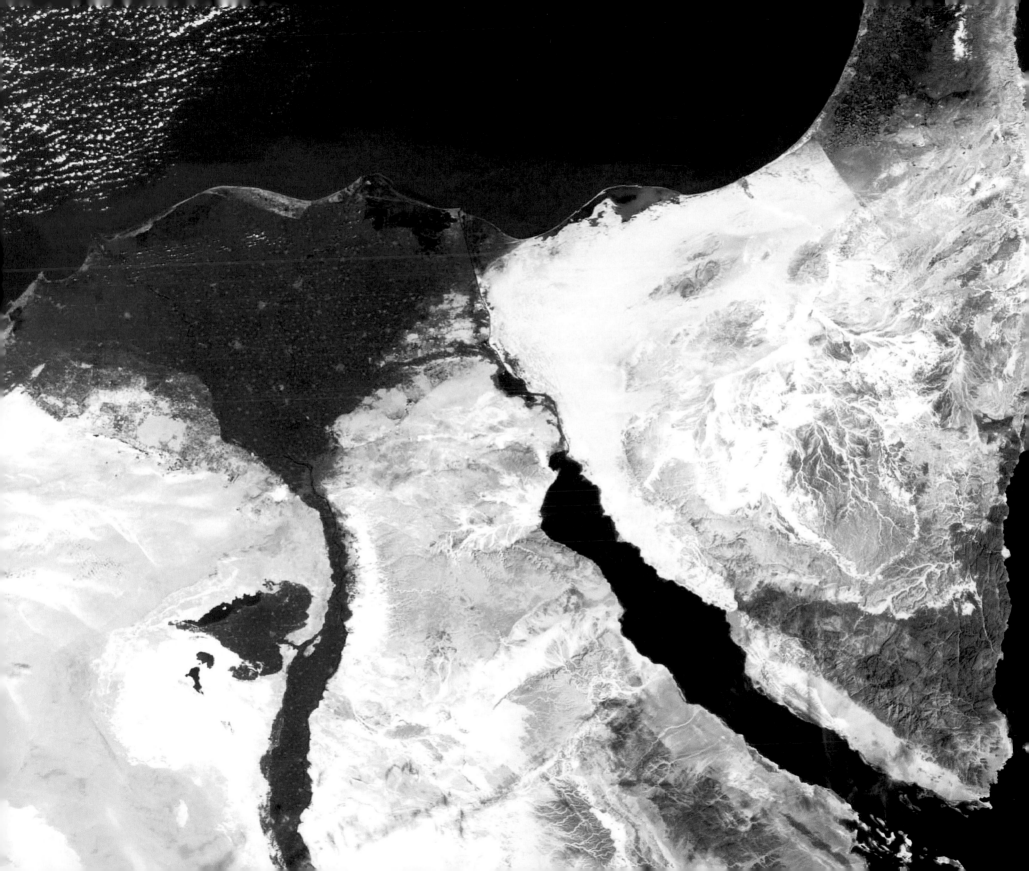

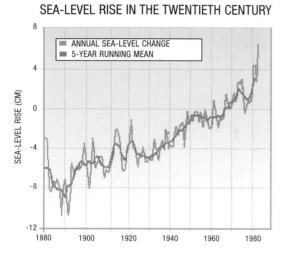

SEA-LEVEL RISE IN THE TWENTIETH CENTURY

ABOVE *Despite fluctuations from year to year, this chart of sea-level change from 1880 to 1980 shows a long-term upward trend. The pattern of rising sea levels is expected to continue as the climate changes.*

attributed to either thermal expansion or melting glaciers. Bruce Douglas, an oceanographer at the Laboratory for Coastal Research at Florida International University in Miami (USA), proposes that the movement, rather than the melting, of ice in Antarctica and Greenland is the most likely explanation of this discrepancy. How does ice increase sea levels without melting? Douglas theorizes that large bodies of ice in Antarctica are cantilevered – that is, partly supported by land, partly floating in the water. As the ocean waters warm, the rate at which large portions of cantilevered ice break from the continent and float free increases. These immense chunks of ice, once partly supported by Antarctica, now contribute their full mass to the ocean. The freed ice is replaced with ice flowing from the continent at a rate of

a few metres per year, which, now partly floating itself, adds mass to the water of the ocean.

Local Variations

For people in coastal communities, of course, the only meaningful measure of sea levels is the relation of the sea surface to the local shore, not its distance from the centre of the Earth. Yet landmasses also move vertically. Along some coastlines, land is actually rising, resulting in a relative drop in sea levels: parts of Alaska, for example, are experiencing a sea-level drop of a few centimetres per decade. In other places, land is subsiding, increasing the rate of sea-level rise by as much as 157 times the global average. In the Fijian islands, some islands are subsiding, some are stable and some mix stable and rising areas.

Geological processes that affect the vertical movement of land include plate tectonics, which can result in either an upward or downward movement. A second phenomenon involves an upward rebound of land that was depressed under glaciers during the last ice age – what geologists call glacial isostatic adjustment. A third process involves settling due to compaction of underlying sediments.

The outer structural, or tectonic, plates of the Earth's crust are in constant motion, 'floating' on top of the lithosphere, the layer of hot, soft rock beneath. When continental plates collide, the result is the uplifting of mountain ranges, such as

the Himalayas. When an oceanic and a continental plate collide, the oceanic plate is pushed beneath the continental plate, which will rise in consequence. When a rift zone opens as a plate tears in two, both hills and valleys are created. These various vertical movements of land can, in relative terms, augment or negate a rise in sea level regionally. For example, local tectonics appear to play a significant role in the apparent differences in rising sea levels of the Mediterranean. The landmasses of Israel and Egypt are rising, and therefore experiencing relatively little rise in sea levels, while the Nile Delta and Gulf of Suez rift region is submerging and therefore suffering from a greater relative rate of sea-level rise.

Rebounding Landmasses

Generally more dramatic is the vertical rebound of landmasses still recovering from the now-removed weight of ancient glaciers. During ice ages, massive glaciers push continents down into the lithosphere. When this weight is removed, the continents rise back slowly, like putty or dense bread dough resuming its shape. This is the force responsible for the rise in the Scandinavian coastline, where sea levels are in relative decline. The glaciers responsible for depressing Scandinavia melted away 10,000 years ago, but the rebound of the land is slow in human terms, though relatively fast on a geological timescale.

Now that the massive and ancient Scandinavian glaciers, as well as the even larger glaciers over the Hudson Bay region are gone, tide gauge data from the nearby coasts show the expected elegant curves depicting the isostatic recovery of the earth's surface.

Sea-Level Measurement Technology

During the seventeenth century and probably much earlier, sea levels were measured with the use of dip-stick-like tide gauges. These consisted of a stick affixed to a pier, with marks by which water level could be measured. Since tide gauges were used for navigational purposes rather than for strict data collection, they were located almost exclusively in harbours. No effort was made to control for the relative movement of the pier, or of the larger landmass.

BETTER INSTRUMENTS
Improvements in tide gauges were made around 1850, when devices called stilling wells were introduced. The stilling well is composed of a pipe with a small hole in the bottom to let water in and a float, which moves a rod attached by wire to a recording device. The small hole

at the bottom of the pipe can let in only a fixed amount of water at a time, making it possible to obtain measurements of sea levels that are relatively independent of wave activity.

In 1990 scientists began to use a method called echolocation to measure sea levels in what National Oceanic and Atmospheric Administration scientists dubbed 'Generation X' water measurement. Like a stilling well tide gauge, echo-sounding uses a pipe with a hole in it, but rather than recording water levels mechanically with a float, echolocation bounces sound waves off the surface of the water electronically to measure sea levels.

MEASUREMENT BY SATELLITE
Generation X devices, however, have been quickly supplanted by what is now the industry standard – satellite altimeter measurement. The TOPEX/Poseidon (T/P) satellite mission was launched in 1992 to record global sea levels with reference to precisely known spacecraft orbits, allowing measurements of unprecedented accuracy, to within 3.1 mm (0.8 in). The satellite bounces radar beams off the surface of the ocean to determine the distance from orbit to sea level, providing a measure of sea level in relation to the centre of the Earth. The successor of the T/P satellite, known as Jason, was launched in 2001 and continues to build on the T/P data by recording global sea levels every ten days.

Data collected by satellite altimeter measurement shows an annual rate of rise in sea levels of approximately 2.8 mm (11 in), almost half again the rate estimated from tide gauge data.

ABOVE *An historic 1931 photo showing portable automatic tide gauge at Buzzard's Bay, Massachusetts, USA. Data was recorded on a roll of paper at the top of the post.*

LEFT *The Jason-1 satellite at Vandenberg Air Force Base in California, USA. The satellite, a joint programme between NASA and the French space agency CNES, was designed to record ocean surface topography and other data. One of its most critical missions is to monitor global climate interactions between the sea and the atmosphere, tracking climate change and weather patterns.*

ABOVE *Ocean City, Maryland, USA, at right, and Assateague Island, at left. The two barrier islands are separated by stone jetties which keep the inlet open for navigation but deplete Assateague of sand. Rising sea levels combined with erosion threaten the island; restoration projects are already underway and geologists are studying the island to determine the effects of sea-level rise on coastal areas.*

Such patterns trace a gradient from dropping sea levels in the areas central to the ancient glaciers – those areas that were most weighed down and are now rising – to increased rises in sea levels in those areas that are now sinking because they were at the periphery of the glaciers, where molten material accumulated beneath the crust, causing the surface to bulge.

In the United States, the shores of Louisiana and Texas are examples of coastlines that are subsiding because of another large-scale geological process – compaction. In those states, the relative annual rise in sea level averages as much as 10 mm (0.39 in), or five times the global average. In Texas, compaction is due

largely to the extraction of fossil fuels from underground. The drained reservoirs simply collapse, lowering the overall height of the Earth's crust. In Louisiana, subsidence is due to natural squeezing of water from sediment as it is deposited and compacted. However, the degree of compaction has been increased by the settling of soil into cavities left by water siphoned from aquifers and by pressure from the weight of the city of New Orleans.

Consequences for Human Habitats

The impact of rising sea levels on any given coast will depend on how well local ecosystems and societies can absorb the changes or adapt along with them. For example, natural buffers like coral reefs and mangrove swamps will reduce the damage that would otherwise be caused by higher storm surges. Steep coastlines – like much of North America's Pacific shore – lose less land for each increment of sea level rise than do gently sloping coasts. Well-

prepared, wealthier societies will be better able to control and minimize the effects of rising waters. However, the consequences of rising sea levels will be experienced to some degree the world over. Erosion of shorelines, increased flooding and storm damage, loss of wetlands and intrusion of sea water into freshwater supplies could result in major economic, social and environmental problems for the most densely populated regions of every continent, particularly in major river deltas.

Agriculture may become impossible, at least for traditional crops, in many parts of the world, as soil is flooded with salt water. Marshes and wetlands may disappear when river sediment growth can no longer keep pace with rising seas. Important fisheries species could be affected, particularly those, like salmon, that depend on healthy shallow-water estuaries as resting places for the young migrating out to sea. Ecosystems could lose the valuable natural filtration services that marshes and wetlands provide.

EFFECTS OF RISING SEAS

In the most extreme cases, many islands could be lost. Some have already disappeared. Sharp's Island in Maryland in the United States was 280 hectares (700 acres) in area when first settled in the late 1600s; it supported a handful of farms and a hotel as late as 1910. Now only its lighthouse is visible above water. The residents of the Pacific island nation of Tuvalu are already preparing for possible evacuation should their islands be inundated. Shorelines on some Pacific islands have retreated

by more than 30 m (100 ft) in the past 70 years. The South Pacific island nation of Fiji is contending with increased coastal erosion, flooding and scarcity of fresh water. For the Micronesian island of Yap, a rise in sea level of 1 m (3 ft) would lead a retreat in the shoreline of 9 to 96 m (30 to 315 ft), according to some calculations. For the same rise, Majuro Atoll in the Marshall Islands would be expected to lose more than 6 per cent of its land surface. For small island nations whose cultural identity is strongly linked to their special geographical setting, coastal erosion on such a scale could be devastating.

Not only low-lying coastal lands will be in danger. Erosion of steeper sandy shores as well as cliff coasts is likely to increase. Rising water levels can mean waves breaking closer to shore and the consequent reshaping of a beach's profile, both of which increase the likelihood of permanent loss of sand, either into offshore depressions or off the edge of the continental shelf.

LEFT *Flooding in western Turkey after a 1999 earthquake and tsunami caused the coastline to fall about 3 m (10 ft). As sea levels rise, coastal developments are in increasing danger from the effects of natural disasters such as these.*

Flooding and Storms

An estimated 10 million people worldwide experienced coastal flooding in 1990. Rising sea levels will increase the number of people at risk from floods and storm surge damage, as waves attack higher on the shore. Studies have forecast that 90–200 million coastal residents will be flooded annually by the 2080s. For the small island nations of the Caribbean, Indian Ocean and Pacific Ocean, climate change is expected to increase the number of people facing high flood risks by 200 times by the 2080s. Large numbers of people will also be at risk in the densely populated delta regions of South and South-east Asia. Worldwide, storm waves are likely to erode significant portions of delta land. In the Nile delta alone, researchers estimate a rise in sea levels of 0.5 m (20 in) would cost £1,150 million in agricultural losses and submerge more than half the region's historical sites and monuments.

ABOVE *View of the destruction on Île-à-Vache, Haiti, after Hurricane Allen struck the island in 1980. Remnants of smashed huts, demolished by high-velocity winds, clutter the shoreline. The hurricane reached Category 5 status, the highest level of force, three times.*

A NATION OF CLIMATE REFUGEES

Worldwide, it is estimated that one million people live on coral islands that are in danger of disappearing because of rising oceans. Low, flat islands built up from coral reefs, these landforms are particularly vulnerable to flooding. As they disappear, the cultures that have developed on them over thousands of years are also likely to vanish, even if the people themselves resettle elsewhere. One such culture, that of Tuvalu, is now in peril.

The South Pacific island nation of Tuvalu, the fourth smallest country in the world, is one example of a community threatened by rising seas. Its population of just over 11,000 citizens resides on nine coral atolls, with a combined area of only 27 sq km (10 sq miles), spread over a vast swath of ocean halfway between Hawaii and Australia.

Tuvaluans have a distinctive history and society. 95 per cent of the islanders are descendants of Polynesians who arrived in Tuvalu from Samoa, Tonga and Uvea within the past 2,000 years. After a period of British rule from the 1890s to 1978, they adopted a parliamentary form of government on gaining independence. Tuvaluans form a tight community, with 100 per cent literacy and virtually no violent crime. Their villages are mostly built around a church and a meeting-house; their houses have open sides and thatched roofs. Most Tuvaluans wear lightweight, brightly coloured clothing made of cotton. Historically, their subsistence has depended on seafood, coconuts, pigs, rice, breadfruit, bananas and a taro-like root called pulaka.

Until recently, the only garbage produced on the atolls of Tuvalu was fish bones and coconut husks.

In recent years, foreign revenue from Tuvaluans working overseas, foreign-aid organizations and creative money-making schemes generated by the government – including producing stamps for the international philatelic trade – have brought relative prosperity and its inevitable changes. But all Tuvalu's history and recent good fortune may become insignificant in the face of one hard fact: no part of Tuvalu is higher than 4 m (13 ft) above sea level. As American science writer and filmmaker Julia Whitty has observed, 'No one here has ever lived a moment without hearing the thunder of surf.' Sea levels rose 20–30 cm (0.4–1.2 in) in the last century and may rise 100 cm (39 in) in the next.

RIGHT *Fishermen on a boat in the waters of Tuvalu. The low-lying island nation is in danger of being inundated as sea levels rise.*
ABOVE *A woman dressed in traditional Tuvaluan dancing costume.*
UPPER LEFT *The Tuvalu island of Nukufetau.*

Inexorably, Tuvalu faces increased flooding and higher rates of erosion, which constantly eats away at the edges of the islands. At the same time, salt-water intrusion contaminates drinking water and hampers attempts at agriculture. The Tuvaluans have a stark choice to make: stay put and die, or emigrate and watch their culture disappear. So far, they have chosen to stay on Tuvalu.

Nevertheless, the unusually severe storms of the last decade have forced Tuvaluans to reexamine their options. As government secretary Panapasi Nelesone put it, 'We don't know when the islands will be covered [by the sea] but we need to start working on this now. …We cannot just float on the water hoping that the sea will go down again.' Tuvalu's government has entreated Australia and New Zealand to accept refugees, should the rising seas force them to evacuate. It was proposed that Tuvaluans be granted one of the uninhabited islands at the end of Australia's Great Barrier Reef. At least with an island of their own, Tuvaluans might stand a chance of maintaining their cultural identity.

ABOVE *View of encroaching seawater on the island of Funafuti, Tuvalu. During World War II, pits were dug here to extract coral. While sea walls are sometimes used to protect islands at risk, the atoll of Tuvalu is also being flooded from the inside out, as water seeps up through the porous coral rock.*

In 2002, Tuvalu made plans to file lawsuits against the governments of the United States and Australia in the International Court of Justice in the Hague. The Tuvaluans argued that the carbon dioxide emissions from the United States and Australia – the largest per capita emitters of greenhouse gases, respectively – brought about an unfair restraint of trade on Tuvalu. By causing raising sea levels those emissions are destroying the Tuvaluan economy and way of life.

LEFT *High tide in Tuvalu. In recent years, this small island nation has become more prosperous through tourism and other business development. However, the atoll of seven coral islands may find itself under water within the next century and the ancient culture of the Tuvaluans could be lost.*

In the El Niño years of 1982 and 1983, heavy surf stripped large amounts of sand from the beaches in Malibu, Santa Monica and Venice, in California, and deposited it into offshore canyons.

Coastal Wetlands

Rising sea levels also endanger coastal wetlands, the marshes and swamps that are home to countless species adapted to frequent and variable inundation. These wetlands play an essential role in the ecological and economic health of coastal communities. They serve as nursery habitats for fish and shellfish, collecting nutrients and organic material from rivers while filtering pesticides and other pollutants from the water. They stabilize shorelines by preventing erosion and protecting inland areas from the impact of flooding.

Wetlands depend on a balance of sediment – as some arrives from rivers, some is lost to erosion. Too much sediment from rivers causes wetlands to become dry land; too much erosion and they are swallowed by the sea. If sea levels rise slowly enough and the adjoining land is undeveloped, wetlands can adjust; tidal marshes and mangrove stands, for example, can simply migrate up the shore along with the rising seas. But if sea levels rise too quickly or coastal development prevents the wetland ecosystem from migrating, wetlands will disappear. Rapidly rising sea levels also can drown wetland vegetation; although adapted for life in salt water, most wetland plants can tolerate only a certain amount of flooding.

Scientists estimate that sea levels rising due to climate change could lead to the loss of as much as 22 per cent of the world's coastal wetlands. Mangroves and tidal marshes, along with coral

ABOVE *The Aukurun wetlands, Queensland, Australia, showing magpie geese and pandanus trees. Wetlands such as these, which are natural buffers between land and sea, are endangered by rising sea levels and human development.*

reefs, dunes and beaches, now provide significant natural protection against coastal erosion. But as more of these natural safeguards disappear, increasingly severe coastal erosion can be expected.

The damage caused by coastal erosion, along with the infrastructure required to prevent or cope with such damage, can be overwhelmingly costly, especially for developing countries. One study estimated that a rise in sea levels of 0.5 m (1.7 ft) on the coast of Montevideo, Uruguay, would cause £13 million in damage if shoreline protections such as seawalls and dikes were not reinforced. In Poland, the cost of land losses if sea levels keep rising has been estimated at 10 billion Polish zioty; protecting the coast along the Baltic Sea would cost 2 billion. Another study concluded that Venezuela could not afford the required shoreline protection or the damages resulting from the same rise in sea level.

Natural Coastal Protectors

Coral reefs are the most massive coastal protection structures in the world. They cause waves to break offshore, sheltering tropical shorelines from heavy seas and reducing erosion caused by large waves. The tropical cyclone that hit Orissa, India, in 1999, provided one illustration of the value of coral reefs in protecting coastal communities from powerful storms. In Orissa, sections of the coast that had no protection from coral reefs lost considerably more beach in the 1999 storm.

RIGHT *Unprotected by reefs or man-made barriers, highly populated waterfronts such as the one shown in this rendering of a hypothetical Thai resort, are vulnerable to storm surges, which grow more dangerous as sea levels rise. Damage to a tourist hotel may cost more to repair than losses sustained by a fishing village. The cost in lives lost along crowded coastlines cannot be measured.*

Under normal conditions coral reefs can grow fast enough to keep up with the expected rate of sea-level rise. Increasing carbon dioxide levels in the water, however, may slow the growth of reefs, as may coral bleaching, which is expected to increase as oceans warm. Coral reefs worldwide are also under tremendous stress from overfishing, marine pollution and destructive forms of tourism. As with many threatened natural systems, the danger to coral reefs becomes life-threatening when a new stress, such as the warming of oceans, is piled upon existing strains. Some scientists estimate that more than half the world's coral reefs may be lost over the next 30 to 50 years.

The 1999 Orissa tropical cyclone also helped demonstrate the value of mangrove swamps in mitigating storm damage. The dense tangle of mangrove growth serves as a coastal buffer by reducing current, wind and wave action, while the web of roots holds sediments in place. Scientists like Vivek Kulkarni of the Soonabai Pirojsha Godrej Marine Ecology Centre in Mumbai, India, point out that mangroves are more effective in protecting coastlines from erosion than the more expensive concrete barriers relied upon by many urban centres. The cyclone that hit Orissa washed away entire villages, killing 10,000 people, but the settlements around Bhitarkanika, protected by the second-largest mangrove swamp in the world, were mostly spared.

LEFT *A child snorkelling on a coral reef wall in the Red Sea, Egypt. Unregulated tourism can be destructive to delicate coral reefs, which are threatened by many other factors, including marine pollution and overfishing. Warming oceans add a new dimension of stress to this already fragile ecosystem.*

More Salt, Less Freshwater

Another potentially devastating consequence of rising sea levels is the contamination of agricultural land and drinking water by seawater. Seawater can swamp land and surface waters during floods or seep into the soil at the shoreline. In many locations, overuse and depletion of ground water has already encouraged more seawater to flow into the soil. In Lebanon, the salinity in some wells near Beirut is more than 60 times levels noted earlier.

Once groundwater becomes saline, it can take decades for the salt to be flushed out through the hydrological cycle. Ecosystems that cannot adapt to the increase in salinity are threatened. For example, 70 per cent of the original pine forests in the Florida Keys in the United States have died off as ground water has become increasingly salty with rising sea levels. Irrigation with saline groundwater can contaminate agricultural land and reduce production. One example is the Batinah coastal plain of Oman, which had to be abandoned for agricultural use.

Unfortunately, marginally saline ground water may be all that is available for irrigation, despite its long-term dangers to the land. This is expected to become the case in the Yangtze delta in China, particularly around the southern city of Shanghai, if ground water in the region becomes significantly contaminated by seawater.

The aftermath of the giant tsunami of December 26, 2004, provides a case study of salt contamination as a result of surface flooding. Salt water from the great wave poisoned the soils of many agricultural fields, making them inhospitable to crops. It may take as long as a decade for rains to wash the salt from the soil and make it productive again. Thousands of wells in Indonesia, Thailand, Sri Lanka, the Maldives, the Seychelles, Yemen and Somalia were tainted with salt water, as well as other pollutants. To restore the wells, salt water will need to be pumped out and the remaining freshwater treated for contaminants. If the aquifer itself was contaminated by the inundation, then there is nothing to do but wait until rainwater can restore a floating surface layer of fresh water. In these cases, recovery may take decades.

The Tigers of the Sundarbans

The Sundarban Islands, which sprawl in the delta of the mighty Ganges and Brahamaputra rivers on the border of India and Bangladesh, are the site of the most extensive mangrove swamps in the world. But their future looks particularly bleak. Satellite imagery data suggest that, for the last two decades,

RIGHT *Mangroves on the southern coast of Kenya, near Mombasa. With their strong root system and their ability to thrive in brackish water, mangroves are responsible for anchoring sediment and helping to prevent erosion. The depletion of mangrove forests to make way for agriculture is implicated in the widespread destruction caused by the giant 2004 tsunami.*

sea levels have been rising in the Sundarbans at an annual rate of 3.14 cm (1.2 in), or about 157 times the average global annual rate of rise in sea levels. Several of the islands, such as Lohachara and Suparibhanga, have already disappeared, leaving thousands homeless; other islands, including Ghoramara, which has 5,000 inhabitants, are rapidly disappearing.

A grim future for the Sundarbans spells a grim future for Bengal tigers. Four national parks, three in Bangladesh and one in India, have become a final refuge for the 700 or so remaining Bengal tigers in the Sundarbans. The tigers are well-adapted to their watery environment and are capable swimmers, but they are still suffering the consequences of efforts in the early twentieth century to convert the formerly expansive grasslands of the Sundarbans to rice fields.

A century later, not only is there no rice production in the area, but there are also no sambar, gaur, buffalo or swamp deer, all formerly prey of the Bengal tiger. In desperation, some of the remaining Bengals have turned to a new food source – that provided by fishermen. Of the 350,000 people who make a living from the Sundarbans, an estimated 34 are eaten by tigers annually. The tigers, in turn, are prey to poachers: a Bengal's pelt will fetch around 190,000 Bangladeshi Taka, a small fortune

LEFT *A Royal Bengal tiger photographed in a zoo in Dhaka, Bangladesh. The Bangladeshi population of Bengal tigers is imperiled as its native habitat, the Sunderban Islands of India and Bangladesh, are inundated by seawater. Through the destruction of habitats for agricultural use, this tiger species was deprived of its natural prey and has turned to preying on humans.*

in the local economy. Paws, bones and other body parts can be sold separately as ingredients in Chinese medicine. Corrupt park officials have been known to assist in poaching efforts.

Threats from poachers, however, pale in comparison to the threat from the sea, which is predicted to swallow an estimated 15 per cent of the Sundarbans' habitable land by 2080. When the Sundarbans are gone, there will be nowhere left for the islands' Bengal tigers.

Population and Storm Impact

With over 2,200 million of the world's people living within 100 km (60 miles) of a coast, average population density along the coastline is around three times higher than the global average population density. Further, people live closer to the actual shoreline than they have in the past, largely owing to the influence of tourism. Demand continues to grow for housing, hotels, restaurants and shops on beachfront property that historically was not considered suitable for development.

Oceanographers Robert Nicholls of the Flood Hazard Research Centre at Middlesex University, UK, and Stephen Leatherman of Florida International University, USA, estimate that 100 million people live at elevations of 1 m (3.3 ft) or less above sea level. Nicholls and geophysicist Christopher Small of Columbia University, USA, estimate that 450 million people live 10 m (33 ft) or less above sea level. Populations along coastlines are expected to grow at a rate that outpaces growth in inshore areas.

As coastal flooding continues to increase and beaches around the world are lost to erosion, people

ABOVE *Most of Bangladesh lies less than 15 m (50 ft) above sea level, making it particularly vulnerable to flooding. This map shows how a sea-level increase of just under 1.5 m (5 ft) would alter the country's coastline. In the next century, 15 to 17 per cent of Bangladesh is expected to be lost to flooding or salinization.*

in coastal communities will be forced to migrate inland. Many coastal cities, such as Ocean City, Maryland, in the United States, combat the losses in beach with expensive sand replenishment programs. By 2001, Ocean City had already spent more than $82 million on this endeavour. Many countries with far fewer resources than the United States will have to face the problem of relocating people displaced by rising tides. The cost of this relocation would come at a time when those nations would also be suffering economically from a decline in tourism. In the worst situations, countries such as New Zealand and Australia, will need to take on the challenge of absorbing entire island populations displaced by rising waters.

CHANGING ECOSYSTEMS

Cascading Effects in the Kingdoms of Life

All over the world, climate change is transforming the interactions of plants and animals. Birds' eggs are hatching when there is not enough food for the emerging chicks to eat. Whitebark pines are falling victim to infestations related to global warming, endangering the grizzly bears that eat the seeds. From the sea-dwelling starfish to the guinea pig-like pika, which lives in alpine meadows, every species is part of a larger group of living creatures that are interdependent. Now the environments in which these groups live are changing and the consequences will reverberate for many years to come.

A MATTER OF TIMING

In a biological community, species interactions vary immensely. Species compete with each other for food, space and resources, facilitate each other's survival by providing shelter, support or nutrition, and kill or parasitize each other. Climate plays a major role in shaping interactions between the members of every biological community. Foreseeing how climate change will affect ecosystems is difficult, however. Such predictions cannot be based simply on the effect of climate variables on a single species.

One of the clearest indicators that climate change is already affecting biological communities is the widespread change in phenology, or the time at which plants or animals perform seasonal tasks such as migration or blossoming. In the United States, ecologists Camille Parmesan of the University of Texas in Austin and Gary Yohe of Connecticut's Wesleyan University examined phenological data for 677 plant and animal species from studies covering periods of between 16 and 132 years (the studies had an average length of 45 years). They found that 62 per cent of those species have been performing spring routines such as nesting, breeding, migrating, flowering or leafing an average of 2.3 days earlier per decade. For some of the species, starting earlier had strong positive effects. Birds that nest earlier, for instance, may be able to rear more than one brood per season. On the other hand, if species shift their timing at different rates, critical interactions may be

disrupted. For instance, what if the food sources the nesting birds rely upon to feed their young are not available earlier? Hidden in the world of plants and animals is a great deal of synchronicity. Organisms that rely on others for food, pollination, or other resources and services – that is to say, most plants and animals – depend on the others being available when and where they need them. This synchronicity is being lost as a result of climate changes.

Migration

Timing is critically important when it comes to migration. For many animals, finding the right combination of food, shelter and company means moving to different places at different times of the year. Many species of birds, insects, fish and even some mammals migrate seasonally, usually for reasons related to food or breeding. For instance, Arctic and sub-Arctic wetlands provide a rich food

source for birds in summer, but are brutally cold in winter. And although snow-covered in winter, alpine meadows provide an abundance of flowers whose nectar feeds butterflies and other pollinators in the spring. Some species simply move up or down in elevation as seasons change. Others fly or swim thousands of kilometres to find the right place to spend the winter or lay their eggs. The uneven effects of climate change around the globe may mean a hard adjustment for animals that call different places home at different times of year.

For migratory birds, certain conditions affecting the area where they spend the winter may spark an earlier or later move to their breeding-grounds – whose conditions may or may not be hospitable for the birds' breeding activities. If the situation is reversed, so that the breeding grounds change while the wintering grounds do not, the same problem occurs. This mismatch is likely to be most severe for species that migrate long distances.

RIGHT *Climate change affects migration and breeding patterns of many bird species, particularly those that travel long distances each season, like these geese.* PREVIOUS PAGE *Clockwise from upper left: A forest of lodgepole pines after a wildfire in Yellowstone National Park in the western USA; an endangered Quino checkerspot butterfly, whose range is being squeezed in the south by a warming climate and in the north by Los Angeles sprawl;* Ulva *(formerly known as* Enteromorpha*), an intertidal estuarine algae, which can tolerate large changes in temperature and salinity; the Montana grizzly bear, now endangered.*

Shorter-distance migrants and residents may find it easier to adjust. Some European butterflies that migrate short distances customarily change how far they travel depending on temperature. In warm years, more migrants appear in the United Kingdom than in cold years. The ongoing warming trend seems to have caused some traditionally migratory butterfly populations to give up migrating altogether, becoming permanent residents of the United Kingdom.

Changes in migration timing may not affect all members of a species equally. In some migratory bird species, males arrive on the breeding-grounds well ahead of the females to establish mating territories. For at least one of these species, the European barn swallow, recent climate warming corresponds with earlier migration of males, while female migration and nesting time has remained unchanged. Evolutionary biologist Anders Moller at the Pierre and Marie Curie University in Paris suggests that males are able to arrive earlier because the cost of doing so is reduced when food is plentiful earlier, as happens during periods of climate warming.

Current understanding of what determines migration timing in bird populations is incomplete. Migrating birds are believed to have an internal calendar that determines the start of migration, with day length usually serving as a trigger. But the success of individuals migrating at different times, and thus the migratory trend for the population as a whole, is determined by shifting factors such as temperature and food availability. Some species migrate earlier now than decades ago and some later; others have not changed the timing of their migration much at all. What makes the difference is still unknown.

Earlier Egg-Laying

To determine whether birds used climatic cues to decide when to lay their eggs, Humphrey Crick of the British Trust for Ornithology and Timothy Sparks of the Institute for Terrestrial Ecology in Huntingdon, UK, analyzed records of egg-laying dates for 36 bird species in the United Kingdom over a period of 57 years. For 86 per cent of these species, temperature and rainfall levels could be used to predict when birds would lay their eggs. Many species showed long-term trends toward earlier laying that could be linked to warming trends at their breeding-grounds. Crick and Sparks used models to show that by 2080, some of those species would lay eggs, on average, 18 days earlier than at present. A related study, led by Christian Both of the Netherlands Institute of Ecology in Heteren, recorded the egg-laying dates in 25 populations of flycatchers around Europe between 1990 and 2002. For 20 of those populations, breeding-grounds had warmed during the study and egg-laying occurred earlier; for a few populations, breeding-grounds had cooled and egg-laying had been delayed.

For some bird species, early nesting increases reproductive success: not only is there a possibility that the parents can raise more than one brood during the season, but also, the chicks may have longer to mature before winter. If, on the other hand, the cycle of seasonal foods used by parents, such as insects or seeds, do not change at the same rate, the hatchlings may have no food.

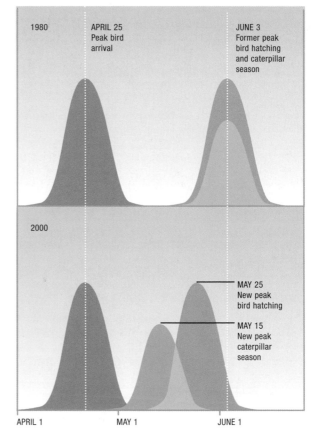

FLYCATCHER MIGRATION AND FEEDING PATTERNS

ABOVE *Flycatcher hatches used to be timed perfectly to the peak season of their main food source, caterpillars. But as this graph shows, the two seasons have shifted out of sync, because of the influence of warming temperatures.*

The pied flycatcher faces this very problem. This striking songbird spends its winters in western Africa, then migrates to Europe to breed, building nests in holes in trees or in nest-boxes. In Spain, Juan José Sanz and his colleagues at Madrid's National Museum of Natural Science found that the timing of the egg-laying for at least two of the southernmost

ABOVE *A blue tit, a European songbird threatened by a mismatch of timing between egg-laying and caterpillar hatching, which could leave it without food for its young. So far, this species has proved adaptable to the change brought on by warming temperatures and has been able to adjust its nesting-time from year to year.*

The United Kingdom has a different situation. Predatory newts are entering the breeding-ponds of frogs earlier and the frogs' breeding times are generally unchanged. As a result, newts consume frog embryos and larvae at greater rates than previously.

Mismatch in a Watery World

Terrestrial animals are not the only ones whose reproductive period coincides with a plenitude of food. The tiny, floating animals that are known collectively as zooplankton, which live in lakes, oceans and rivers, also produce their young when food is in bountiful supply. The timing for some of these zooplankton has remained the same, despite climate change; for other zooplankton, as for some terrestrial animals, it has been altered.

For instance, in the three Swedish lakes, Vänern, Vättern and Mälaren, the spring bloom of phytoplankton, a primary food source for zooplankton, occurred about one month earlier during the 1990s than it had during the 1970s and 1980s. The zooplankton that feed on these photosynthesizing organisms are also peaking early.

A more extensive survey of 66 types of planktonic organisms over a 45-year period in the central North Sea found different issues. While a few types of algae peaked earlier, the overall phytoplankton peak did not change; in contrast, most larval zooplankton peaked 27 days earlier than at the beginning of the 45 years. Since zooplankton feed on phytoplankton, including algae, this timing difference means starvation or delayed development for them and suggests that food limitation could become more problematic as climate change progresses.

pied flycatcher populations had not changed over an 18-year period of warming, although the oaks on their breeding-ground were leafing out significantly earlier. Earlier leafing-out means that the caterpillars on which pied flycatchers feed their young appear earlier. During the course of this same 18-year period, the growth of baby flycatchers in the nest declined, as did their survival after leaving the nest. Sanz and his colleagues attribute this decline to the mismatch of the timing between the period of egg-laying and the abundance of caterpillars to feed on.

Another European songbird, the blue tit, is also experiencing a caterpillar/egg-laying mismatch. During some 23 years of warming in the Netherlands, the period of peak caterpillar populations advanced significantly, while at the same time egg-laying by tits did not. Blue tits, however, appear to be fairly adaptable. They tend to adjust their nesting date from one year to the next in response to how well they matched the food peak the previous year; they also use a variety of food items to feed their young.

ALTERED INTERACTIONS

In addition to its effects on timing, climate change will influence interactions among species, and between species and their environment. If an animal's main food source goes locally extinct, that animal species will be in trouble even if the changing climate causes it no direct distress. Likewise, if a new species starts moving into the territory of another species and competes for resources, survival then becomes more difficult for the original species. This is the case for the whitebark pine – the lodgepole pine encroaches on the whitebark's territory.

Climate change is bringing transformations to the landscape of the world's ecosystems. American beech trees, shown in the map on the right, may shift almost entirely out of their current range. Under these conditions, how will ecosystems and species interactions change?

Hundreds of species, both terrestrial and aquatic, have shifted their ranges toward the poles over the past several decades and many mountain-dwelling species have expanded their range to higher altitudes. This shift is consistent with the climate envelope model of species ranges, which states that a species's range is determined primarily by temperature and precipitation patterns. The problem, however, is that not all species can change their range as fast as the climate around them changes.

RIGHT *Populations of several European butterfly species, including two whose shifting ranges in Sweden and Finland are shown on the right, have expanded northwards; in at least one case, they have become extinct at the southern end of their historical range. These shifts are probably due to climate-related temperature change.*

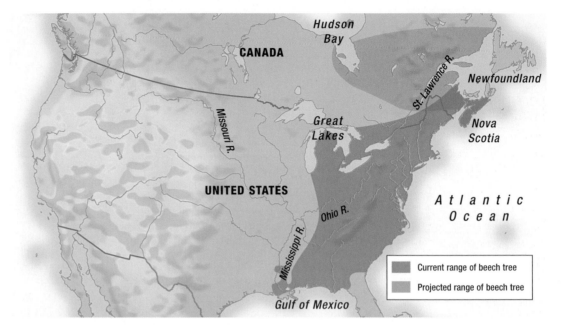

ABOVE *According to the US Environmental Protection Agency, a warmer climate will probably cause deciduous forests to shift northward, replacing coniferous forests in many areas. American beech trees, now common throughout the eastern United States, may move out of their current range almost entirely.*

Current range of beech tree

Projected range of beech tree

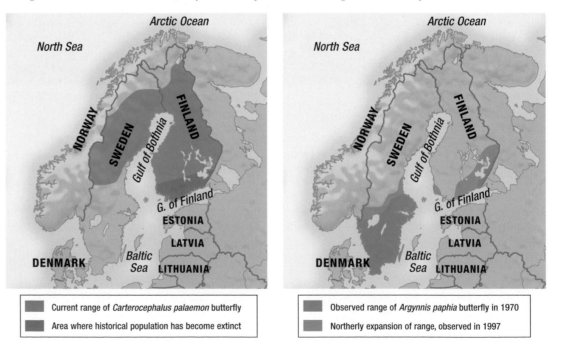

Current range of *Carterocephalus palaemon* butterfly

Area where historical population has become extinct

Observed range of *Argynnis paphia* butterfly in 1970

Northerly expansion of range, observed in 1997

For some species, the problem is what ecologists call limited dispersal ability: if adults cannot move (as is the case with most plants) and there is only limited dispersal of young, the species will not be able to expand its range very rapidly. For other species, migration might be blocked by cities or roads, or by once appropriate habitat converted by humans or other forces to less appropriate habitat. And if an organism already dwells near the poles or the top of the mountain, there may be nowhere else to go.

And what if an organism's neighbour, on which it depends, responds to climate change differently? Different species shift at different rates and over different distances. Some species' ranges have not changed, while some butterfly species have shifted 200 km (120 miles) farther north and some marine copepod species have shifted polewards by as much as 1,000 km (620 miles). As climate changes, associations of plants and animals that are now thought of as normal may disappear. As an ecosystem changes, not all the plants and animals in it are likely to survive.

Cascading Effects

Grizzly bears are a symbol of untamed nature. They are the world's second-largest land carnivore – only polar bears are larger – and some members of the Kodiak subspecies may actually equal polar bears in size. Although grizzlies hunt moose, bison and musk ox, the survival of some grizzly populations may come down to what climate change does to a single species of tree and a pigeon-size bird

RIGHT *Although grizzlies in Yellowstone National Park, USA, are endangered, Alaskan grizzlies like this mother and her cubs are expanding their range north into regions previously considered polar bear country.*

called the Clark's nutcracker. This case exemplifies how climate change can have cascading effects on species relationships.

The Clark's nutcracker lives throughout the mountain regions of the American West. It feeds almost exclusively on large pine seeds, in particular those of the whitebark pine. These birds use their powerful beaks to pry seeds out of the cones and stash them for later use. Clark's nutcrackers prefer recently disturbed or open areas for their seed caches, burying seeds at the base of trees, under rocks, or in thick moss beds. A single bird may stash between 50,000 and 120,000 seeds in up to 8,000 separate caches. Although Clark's nutcrackers have excellent memories and can find food caches even under deep snow, some portion of the cached seeds inevitably are left to germinate.

While it might seem that seed-eaters make survival more difficult for whitebark pines by consuming their seeds, the trees depend on Clark's nutcrackers.

LEFT *A Clark's nutcracker with a pine nut in Utah's Bryce Canyon National Park. The nutcracker inhabits pine forests of the western United States. It uses its long beak to extract the nuts from pine cones. The survival of some populations of grizzly bears may depend on this bird and the seeds of the whitebark pine tree that it stores.*

The cones of the whitebark pine do not naturally open to release their seeds like the cones of many other conifers. Only decomposition or the Clark's nutcracker can release the seeds and decomposition takes too much time for it to be the only method.

Whitebark seeds make excellent food, because they are bigger and more nutritious than the seeds of other conifers. They have a high fat content essential to mammals such as grizzly bears, which need to build up energy stores to make it through a winter hibernation. Also, the period of peak whitebark seed availability coincides with grizzly hyperphagia, the period in which the bears' physiology shifts in such a way that they think about nothing but eating enough to get them through the winter.

Given the choice, bears will choose whitebark seeds over any other food when the seeds are plentiful.

In at least three different ways, climate change is threatening whitebark pines and therefore the animals that depend on them – including Clark's nutcrackers and grizzlies. The first is by creating conditions that encourage the spread of pests and diseases, in particular the fungus white pine blister rust. Both the rust and the destructive bark beetle that spreads it are normally limited in their range and severity by temperature and moisture conditions. When warmer temperatures create conditions favourable for the disease, whitebark pine forests can suffer up to 90 per cent mortality.

A second problem is that as temperatures rise, sub-alpine species like lodgepole pine begin to shift their range farther up the mountains. In a process called competitive displacement, these lower-altitude trees outcompete alpine trees for food, space and other resources. Whitebark pine can shift its own range upslope, but ecologists are concerned that as warming continues, the pines will be essentially pushed off the tops of the mountains.

Yet another climate-related challenge for these pines is the expected increase in large forest fires.

LEFT In the United States, near Prudhoe Bay, Alaska, a research team examines shrubs. As global temperatures warm, arctic species are being displaced by subarctic species, such as these shrubs. This same process is threatening trees like the whitebark pine, which is losing space to the lodgepole pine.

Although whitebark pines can handle small fires, the large conflagrations that result from a combination of increased drought, heat and disease could weaken them significantly and hasten their replacement by sub-alpine species. For instance, during the summer of 1988, major fires in Yellowstone National Park on the border of Idaho, Montana and Wyoming reduced whitebark pine stands by 26 per cent.

The future of whitebark pines is in jeopardy. It may take their disappearance for us to realize how important they were to the Clark's nutcracker and some endangered grizzly bear populations.

Mediterranean Spread

Worldwide, climates with mild, rainy winters and dry, warm-to-hot summers are called Mediterranean. These climates usually occur at latitudes from 30 to 50 degrees north and 30 to 40 degrees south of the equator. In summer in these climates, high-pressure belts keep rainstorms from forming; in winter, rain-bearing low-pressure systems enter and give the areas most of their annual rainfall. In the United States, coastal central and southern California are the perfect examples of this climate, receiving 350 to 450 mm (14–18 in) of rain annually from November to February, but less than 100 mm (4 in) the rest of the year. Other examples include the Cape area of South Africa, the central coast of Chile and portions of coastal Western Australia.

As climate change takes hold, more of Earth's temperate regions – the British Isles, for instance – are expected to begin to experience Mediterranean conditions. Winters will be milder and wetter, summers warmer and drier. This will encourage the spread of new kinds of vegetation, particularly the evergreens known collectively as the Mediterranean forest, wood and scrub biome. Although this biome is characterized by different plants in different places, and called by different names – maquis in the Mediterranean, chaparral in North America, fynbos in South Africa, mallee in Australia and matorral in Chile – the dense growth and fire-prone nature of these plant communities makes them recognizable worldwide. Already, beech forests and heather heathlands in some cold-temperate ecosystems in Europe are being replaced by holm oaks and other plants characteristic of a Mediterranean climate.

Agriculture will be affected by Mediterranean spread as well. The winter rains of a Mediterranean climate permit crops to grow during a limited period, but not the same crops currently grown in temperate regions. Grape-growing and wine production are well suited to a Mediterranean climate, and expansion of such climates may allow wine production to spread in more northerly parts of Europe. In the south and midlands of Britain, corn, peaches and nectarines may become popular crops. Warmer seas will attract fish that until now lived in more southern waters, such as the unusual numbers of mullet, anchovies and shark recently reported in northern European waters by local anglers.

Perennial plants in regions shifting toward a Mediterranean climate will be replaced by those hardy enough to withstand summer droughts. In existing Mediterranean climates, many perennial plants have thick leaves that retain moisture. Insect pests may change, too. British farmers may see growing populations of destructive Colorado beetles. Cockroaches, fleas, mites, ticks and scorpions are all likely to spread in Mediterranean climates.

Barnacles and Algae: the Big Squeeze

The rocky intertidal zone is a world of extremes. Covering the zone between mean high tide and mean low tide, it can experience temperature shifts of 20°C (40°F) within 12 hours; during the same time period, it goes from being completely submerged to being completely exposed. The distribution of many plants and animals in this world is determined by the interplay of what are called biotic (biological) and abiotic (non-biological) forces. Generally, abiotic factors, such as heat and desiccation, or drying, limit how high in the intertidal zone organisms can live, while biotic factors – competition and predation – determine their lower limit. Climate change will increase the extremes of heat and desiccation, potentially making it impossible for species to survive as high in the intertidal as they do now. If the competitors that keep those species from living lower in the intertidal area do not shift their range downward at the same time, the high intertidal species may well become locally extinct.

In the United States, Chris Harley of the Bodega Marine Laboratory in Bodega Bay, California, has shown this sort of relationship between a species of algae and the herbivores that eat it. On north-facing, wave-exposed rocks, the algae can live high enough that its herbivores cannot reach it. At similar tidal heights on the hotter south-facing shores, however, the algae bleaches and dies. Unfortunately for the algae, the herbivores live at the same height on these shores and the algae, squeezed between certain death above and below, cannot survive on south-facing shores.

In a different scenario, cold- and warm-water barnacles (*Semibalanus* and *Chthamalus*, respectively) coexist on Europe's Atlantic coast. *Semibalanus* can usually outcompete *Chthamalus* if temperatures are cool, restricting *Chthamalus* to very high in the intertidal and to generally warmer locales. When local climate heats up, *Chthamalus* gains a competitive advantage and expands both lower in the intertidal and to more northerly locations.

ABOVE *Barnacles on the rocky shore of San Juan Island in the northwestern United States. As on Atlantic coasts, these barnacles on the Pacific shoreline compete for space.* Balanus glandula *(upper right) thrives in cooler locations and is less tolerant of warm temperatures than the smaller* Chthamalus dalli *(lower left).*

LAST DAYS OF THE PIKA

The American pika, a relative of the rabbit, is finely adapted to the cold climate of high mountains. Now climate change is threatening pikas. Warmer weather is killing them by overheating their bodies and altering the vegetation on which they depend for food. Without the pika, eagles and weasels lose one source of food. And the environment will no longer be reshaped by the piles of grasses the pika make as they thriftily prepare for winter.

ABOVE *Wildflowers in southwestern Colorado's San Juan Mountains. In this part of the western United States, there has been limited change to the pika's habitat so far, but local populations of the animal have already become extinct in other areas.* UPPER LEFT *A juvenile North American pika.*

The American pika *(Ochotona princeps),* which measures about 18 cm (7 in) long, lives in mountain crevices of the western United States, and subsists on a diet of flowering plants. Once abundant in the Great Basin region of the United States between the Sierra Nevada and Rocky Mountains, the pika has disappeared from more than one-fourth of sites surveyed since the mid-1990s, according to a 2002 study conducted for the US Geological Survey (USGS). Although cattle grazing and proximity to roads may have had some effect on the pika, its alarming rate of disappearance is believed to be due primarily to a changing, warming ecosystem. Rising temperatures may make this mammal one of the growing number of casualties of human-induced climate change.

The USGS study found that populations of pikas had disappeared at 7 out of 25 sites where they had been documented in California, Oregon and Nevada. A follow-up field study months later showed extinctions at two more sites, meaning that 36 per cent of pika populations in the region had gone extinct in less than a decade. Head USGS researcher and ecologist Eric Beever reports, 'Population by population, we're witnessing some of the first contemporary examples of global warming apparently contributing to the local extinction of an American mammal at sites across an entire ecoregion.'

Long a common sight on the jagged slopes of the High Sierra and Rocky Mountains, the pika, with its small size and squeaky whistle, has endeared itself to many generations of mountaineers. The pika was voted one of North America's three cutest animals

ABOVE *A North American pika carries a mouthful of grass. The animals cut and dry their food to store it for the winter, in a process called haying. If the pikas disappear completely, other species will suffer, including those who rely on them for food.*

in an Internet survey by the World Wildlife Fund (WWF). Pikas are often photographed with their mouths full of wildflowers. Besides its whistle, the pika makes a variety of vocalizations by rubbing the sweat glands in its cheeks on rocks. Beever says the pika's call makes it easy to locate and study. 'One of the nice things about pikas,' Beever says, 'is that if they are there, you'll know it.'

A Tragic Disappearing Act

Finely adapted to its mountain life, the pika typically dwells at altitudes above 2,100 m (7,000 ft), and an Asian species has been found as high as 5,800 m (19,000 ft) on Mount Everest. With a normal body temperature of 40°C (104°F) and a thick, furry pelt that keeps it from shedding heat, the pika can withstand the cold mountain climate. The females give birth to two to four offspring in winter, one month before the snows melt. But these alpine animals are not adapted to warm weather. They can die in minutes from overheating if their body temperature rises a few degrees.

According to Beever, the warmer and drier conditions of recent decades have been a major factor in the rapid disappearance of the pika. Earlier studies found that some Canadian Yukon pika populations suffered 80 per cent mortality after extremely warm winters. Without insulating snow blankets over their rock tunnels, animals can freeze when temperatures fall between warm spells. And if rain occurs rather than snow, ice forms on the pikas' food supply and makes it inedible. 'There are several contributing factors, but climate seems a very strong one,' Beever observed. The comprehensive survey found that on average, the sites that lost pikas were drier, warmer and at lower latitudes than sites where the animals remain.

Ecosystem Engineers in a Time of Change

Warming temperatures can not only shrink the habitat available to pikas, but can also reduce the population of the pika by altering the ecosystem on which the animal depends. A warmer climate can reduce the vegetation they rely on for their diet, providing them with less to eat.

The small mammals, weighing between 170 and 396 g (6 and 14 oz) subsist principally on wild flowers, which they collect and dry in the sun to use for provisions in winter, during a process known as haying. The haying work is done over the peak of summer, during which time the pikas create a storage system of hay piles in areas of talus (rocky rubble). The haying helps pikas meet their food needs year-round. These little 'ecosystem engineers', as researchers call them, help shape their environment through their haying activity. Their disappearance would affect not only the local vegetation, but also the animals that feed on pikas, such as eagles and weasels.

As global temperatures rise, alpine animals like the pika are expected to seek refuge at higher latitudes and altitudes. Where the area of high mountains they live in is very large, pikas may be able to move northwards. But they are so sensitive to temperature change that they cannot survive any warm weather at all along their migration route. In California, Nevada and Utah, pikas are marooned on isolated mountaintops, separated by valleys whose warm weather would likely be fatal to them.

Brooks Yeager, vice president of the Global Threats Program at the WWF, sees a lesson in the loss of the American pika: 'Extinction of a species, even on a local scale, is a red flag that cannot be ignored.' The loss of local populations is especially troubling, notes ecologist Eric Beever, because most local extinctions are driven by habitat loss. In this case, habitat changes have been limited, and it is warming temperatures brought on by climate change that are to blame.

Starfish and Mussels

It was once assumed that climate change would alter ecosystems in a gradual way, over a long period of time. But a study of starfish and the mussels on which they feed suggests that transformation can happen much more rapidly if a keystone species in a biological community is temperature-sensitive. Slight warming or cooling can spark cascading changes and affect every organism in the community.

On North America's Pacific coast, the ochre sea star, a species of starfish, is a common predator. Feeding on mussels, barnacles and other invertebrates, it is a keystone predator in that its presence or absence determines how diverse the ecosystem is. In its absence, the mussel, its preferred food, tends to outcompete and eliminate other species.

A number of laboratory and field studies that were conducted by the US ecologist Eric Sanford at the Bodega Marine Laboratory at the University of California, showed that the feeding rate of starfish plummeted if water temperature dropped 3°C (6°F) below the normal temperature for summer months. If this were to happen, the population of California mussels would explode and crowd out barnacles, algae and other organisms. Ecosystem diversity would be lost.

Conversely, when the water temperature rose by 3°C (6°F), starfish went on a feeding binge, sometimes wiping out entire mussel populations.

RIGHT Ochre sea stars off the coast of British Columbia, Canada. Warmer temperatures induce these sea stars to eat far more of their favoured food, mussels. Fewer mussels mean more diversity overall, although species that prefer dense mussel beds may decline in number.

Although the presence of too many mussels reduces diversity, too few mussels is not beneficial either. Mussel communities form reefs that are homes to crabs, sea cucumbers and worms. As the mussel population declines, the reefs fall apart and the ecosystem changes. Whether the water becomes colder or warmer, the changes affect the whole ecosystem.

The temperature ranges Sanford studied reflect ranges seen on the Oregon coast during normal and upwelling conditions. During upwelling periods, strong winds move warmer surface water offshore and allow cold, deep water to surface, lowering water temperatures by about 3°C (6°F). Global atmospheric change may influence the wind patterns that determine the frequency and intensity of upwelling, thereby determining how often sea stars feed and how many mussels survive through the summer.

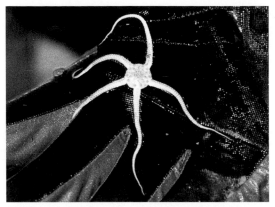

ABOVE *A brittle star. Such variables as salinity, temperature, minerals and plant life directly affect the survival of species that live on the sea floor.*

Trees in a Changing Climate

The most obvious human impacts on forests are direct, such as clear-cutting; or the planting of large areas with a single species of tree; or the introduction of new species to be used for timber. Yet humanity can influence forests just as profoundly in an indirect way, through climate change.

In Spain's mountainous regions, beech and heather are being replaced by holm oak with the expansion of the Mediterranean climate. In Norway, the range of beech and Norway spruce is predicted to stretch northwards, potentially displacing dwarf birch, which may in turn move into previously tree-less tundra. For plants, the rapid rise in carbon dioxide that is driving global warming is at least as important as the accompanying changes in temperature and precipitation. Like a fertilizer, the extra carbon dioxide stimulates photosynthesis and plant growth.

Increased carbon dioxide is also believed to be the cause of a change in the rainforests of the Amazon basin, also known as Amazonia. Here, scientists from the United States and Brazil have found dramatic changes in research plots they have monitored for decades. Large, faster-growing types of trees are becoming more dominant and dense; there has been an accompanying decline in smaller, slower-growing trees like those that make up the understory, the part of the forest between treetops and the ground cover. Lianas, woody vines usually characteristic of disturbed forests, are increasing.

Because there is no evidence in any of the plots being studied of current or past disturbance from logging, fires, hunting or major windstorms, the changes could only be from some pervasive factor, the most likely being rising carbon dioxide levels.

ABOVE *The Monteverde Cloud Forest in Costa Rica. High-elevation cloud forests may face different challenges from the rainforests of the Amazon basin. Most of the moisture in these forests comes from clouds, not rain and a combination of climate change and clear-cutting are reducing the clouds that nourish these forests.*

All over Amazonia, both growth rates and mortality rates in forests appear to be increasing – the biological turnover in these forests is much higher than it was just a few decades ago. This change is significant on a global level, because tropical rainforests are thought to be one of Earth's major carbon sinks – regions that seem to remove large amounts of carbon dioxide from the air before it can add to the greenhouse effect. Since young trees take up carbon more quickly than old trees, rapid growth up carbon more quickly than old trees, rapid growth rates could lead to increased carbon uptake, perhaps slowing global warming.

But the increased mortality of older trees is simultaneously releasing the carbon stored in them right back to the atmosphere. Although young trees take up carbon more quickly than older trees, they can never store as much carbon in total as majestic old-growth trees. Changes in the dynamics of tropical rainforests have the potential to shift the global carbon budget radically.

PARASITES, PESTS AND DISEASE

Parasites, pests and diseases are everywhere, living in or on other organisms and getting food and shelter at their hosts' expense – these hosts range from sea urchins to musk oxen. Some have little effect on their hosts; others are fatal. And like so many other biological interactions, the relationships among these creatures can be strongly influenced by a changing climate. The ecologist Drew Harvell of Cornell University, USA, who led an investigation of climate-related pests and pathogens, put it this way: 'What is most surprising is the fact that climate-sensitive outbreaks are happening with so many different types of pathogens – viruses, bacteria, fungi and parasites – as well as in such a wide range of hosts, including corals, oysters, terrestrial plants and birds.'

For warm-blooded animals such as humans, external temperature does not strongly affect early developmental rates; the parents are usually able to maintain constant, optimal temperature and moisture levels for developing embryos. Many pests and pathogens, however, cannot generate their own heat and are temperature-sensitive. For them, higher temperatures often mean speedier development, higher transmission rates and more generations per year. In cold environments, warmer winters can dramatically increase the survival of pests and pathogens.

Musk Oxen and Invertebrates

In 1988, a new parasite was discovered in northern Canada's herd of musk oxen. Like most parasites, this one has a complex life cycle. As a microscopic roundworm, it lays its eggs in the musk ox's lungs; after hatching, first-stage larvae find their way out of their first host and into their second, a slug. The worms go through three larval stages in the slug, then move back to a warm-blooded host. When an infected slug or contaminated grass is eaten by a musk ox, the larvae transform into adults and find their way back to the lungs to begin the cycle again. In a normal climate, larval development is slow. The parasite takes two years to go through the three larval stages and winter mortality of both larvae and their slug hosts is high. But in warm years, more common in the past several decades, the parasites can complete their entire life cycle in a single summer and winter mortality is much lower.

These conditions can lead to staggering infection rates, sometimes up to 100 per cent. Highly infested musk ox herds are now half as big as they were in the late 1980s. Hunters report that these herds often include several animals that are bleeding from the nose and that lag behind the others, presumably because of the lung parasites. In Norway, outbreaks of related reindeer parasites occur when temperatures rise only 1.5°C (3°F) above normal.

In Denmark's Wadden Sea, warming climate is exacerbating the effects of parasitic flatworms on snails and tiny crustaceans known as amphipods. In warm years, when flatworm populations swell,

ABOVE *A group of musk oxen in Nunavet, Canada. Some populations of musk oxen have declined by 50 per cent since the 1980s. Scientists believe this is due to warmer temperatures, which are hospitable for roundworms, parasites that sometimes infect entire herds of musk oxen.*

thriving amphipod communities can simply vanish in just five weeks, while snail populations plummet by 40 per cent. Loss of amphipods means loss of their sediment-stabilizing burrows in the sand, more erosion and a major shift in the surviving invertebrate community. The Wadden Sea's tide flats are critical feeding grounds for hundreds of species of migratory birds, so changes in these invertebrate communities could have worldwide repercussions.

RIGHT *Sea urchins in New Zealand. In nearby Tasmania, urchin populations have soared, perhaps due to climate change and overfishing of an urchin predator, the rock lobster. This leads to areas being denuded of kelp, which would normally prevent erosion and support vibrant marine life.*

Urchins, Kelp and Amoebas

The importance of invertebrates goes well beyond their value as bird food. A classic in ecological textbooks is the so-called otter-urchin-kelp triangle. Sea otters eat sea urchins, and sea urchins eat kelp. In the absence of otters, urchin populations can soar, leading to the creation of urchin barrens, large areas completely denuded of kelp. Healthy kelp forests provide critical habitat for juvenile fishes and invertebrates, and the transformation from kelp forest to urchin barren can have devastating ecological and economic results. In Tasmania, for instance, sea urchins caused an estimated 70 per cent reduction in productivity for the abalone and rock lobster fisheries on affected reefs. Kelp beds are also important in reducing erosion, minimizing storm damage and moderating land-based pollution.

Off the eastern coast of Canada, there is now a different triangle at work: the urchin-kelp-amoeba triangle. The green urchins native to the region are host to parasitic amoebae and during outbreaks, the amoeba can almost eradicate the host population.

The range of these parasites is influenced by currents and especially temperature – both their development and transmission rise exponentially with temperature. Thus, both the size and location of parasitic outbreaks have the potential to be strongly influenced by climate change. Indeed, long periods of high ocean temperatures off Nova Scotia in 1995 led to almost complete urchin mortality.

FAR RIGHT Tracks of elm bark beetles on the trunk of a tree infected with Dutch elm disease. The beetles, which carry the disease, are increasing in number as warmer temperatures allow them to thrive more or less unchecked by winter mortality. RIGHT A bark beetle.

Although ocean temperatures are expected to rise from Nova Scotia southwards along the Atlantic coast of North America, there is a cooling trend from New Brunswick north; thus the effect of climate change on the urchin-kelp-amoeba triangle will vary by region.

On the east coast of North America, disease is currently the primary mechanism to shift the ecosystem from urchin barrens back to kelp beds. From an ecological perspective, this makes it seem that increased disease outbreaks would be beneficial to the ecosystem. The problem is, however, that warming oceans may lead to more frequent and intense outbreaks of the parasite than in the past, with unknown ecological consequences. Also, Maine and Nova Scotia have extensive and profitable sea urchin fisheries; increases in disease outbreaks could make this industry less reliable.

Bark Beetles on the Attack

Bark beetle outbreaks are growing in range, severity and cost worldwide, largely due to changing climate. These dark, cylindrical beetles are a major source of injury, mortality and reduced timber value in trees.

Each bark beetle species has a preference for a particular tree species. Jeffrey pine bark beetles, western pine bark beetles, southern pine bark beetles, mountain pine bark beetles, eastern ash bark beetles, native elm bark beetles, smaller European elm bark beetles and hickory bark beetles are a few of the species all named for the tree on which they feed. Adult bark beetles burrow between a tree's bark and wood to mate and lay eggs. Grubs and adults feed on a tree's live tissue and mature grubs burrow out through the bark and fly off to find a new tree.

Birth and Death Inside a Fig

The relationship between figs and fig wasps is highly intimate. Each species of fig depends on one or two species of wasp for reproduction while fig wasps cannot reproduce without their particular species of fig.

Figs begin as inside-out bunches of flowers called syconia. Fig wasps start life as eggs inside a fig. When the eggs hatch, male and female wasps mate; males then die, while females leave the fig, covering themselves with pollen in the process. The females instantly seek other syconia into which to deposit their eggs, pollinating the flowers inside the new syconium as they crawl in. Because females die within a day of birth, wasp populations rapidly die off if there is not a constant supply of fresh fig flowers. A severe drought in northwestern Borneo in 1997–1998 led to several flowerless months for most fig species. The result was that all species of pollinator wasps went locally extinct within two and a half months and most had not reappeared within six months of the drought. If climate change makes such severe droughts more common, it could mean the end of many species of fig wasps and the figs that cannot reproduce without them.

Figs and fig wasps would not be the only organisms affected. Extinctions would likely have cascading effects on vertebrates such as birds, bats and monkeys that depend on figs as an essential food. That, in turn, could harm tree species that depend on the services of those animals to disperse their seeds.

ABOVE *Fig wasps inside a strangler fig. The strangler fig survives through a symbiotic relationship with the fig wasp. The wasp pollinates the plant and uses it to reproduce. Climate change events such as droughts can interfere with the short lifespan of these wasps, which in turn threatens interdependent fig species.*

Bark beetles kill trees in one of two ways. The most direct route is for beetles to girdle the tree, that is, to eat away the live tissue around the entire circumference of the tree's trunk. This prevents the roots from obtaining any of the sugars that have been produced in the leaves by photosynthesis and it keeps the leaves from getting water and nutrients from the roots. A girdled tree will die within one or two years.

In addition to this direct damage, bark beetles carry various strains of fungus with them and these organisms further damage the trees. One fungus carried by bark beetles was the cause of Dutch elm disease, which killed more than half of the elms in the northern United States during the course of a few decades and wiped out more than 90 per cent of mature Dutch elms during epidemics in France in 1919 and 1972. Another common hitchhiker is the blue-stain fungus. This fungus can block the flow of nutrients, food and water between the roots and crown of the tree, mimicking the effects of girdling.

In most regions of the world where bark beetles are native, their populations are held in check by predators, weather, host resistance or lack of sufficient food supply. Periodically, conditions for the beetles improve. A warm winter or a drop in the population of beetle predators will often lead to a beetle epidemic or outbreak. However, conditions usually return to normal after a few years, causing beetle populations to drop and giving tree populations a chance to recover. Tree ring records in some areas indicate that this cycle of outbreaks and recovery goes back at least 400 years. The story is different in the wake of our changing climate.

Threatened Oysters

Around the world, the shellfish industry is suffering from the effects of climate change, in the form of both diseases and invasive species. Beginning in 1990, the east coast of the United States experienced a series of warm winters. Subsequently, outbreaks of an oyster disease known as Dermo were recorded as far north as Massachusetts. The disease is caused by a parasite, a microscopic protozoan called *Perkinsus,* which is fairly common in southern waters but was rarely seen north of Chesapeake Bay because of its inability to tolerate cold. As northern waters grew warmer during the 1990s, the parasite spread as far as Maine. Many scientists and aquaculturists are now concerned that the warmer ocean temperatures brought on by climate change may make Dermo outbreaks a permanent problem well beyond their historic range.

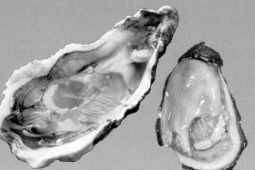

ABOVE *A healthy oyster, left, and an oyster with the disease Dermo, right. Warm winters and the related ocean warming have let the disease spread northwards along the eastern seaboard of the United States from Chesapeake Bay.*

Europe has seen similar problems. During the 1970s and 1980s, another protozoan parasite that favours warm waters, *Bonamia ostreae,* spread throughout Europe's traditional rearing areas of the European flat oyster, or *Ostrea edulis.* Causing an illness called hemocyte disease, the parasite has contributed to a serious and lasting drop in the oyster population. In France alone, oyster production fell almost 90 per cent in 1995.

Climate change can also threaten native species of oysters by increasing the spread of invasive oysters. In Britain, the Japanese oyster, which prefers warm waters, is threatening to spread invasively because climate change has been raising the temperature of the water, allowing the oyster to breed widely. The Japanese oyster has already spread to Australia and Argentina, displacing other oyster species of greater commercial value.

Loss of oysters means more than just the loss of jobs and income to commercial harvesters. Oysters play a key role in keeping estuarine waters clean and clear, and they create extensive reefs that provide critical habitat to other organisms. Overharvesting, pollution and introduced diseases have led to significant declines in many oyster populations. By increasing outbreaks of oyster diseases and the spread of invasive species, climate change may make recovery difficult.

ABOVE *Oystering in Brittany, France. The parasite* Bonamia ostrea *affects north and central Europe's oyster farms; some fear that warming may allow it to move north, as* Perkinsus *did in North America.*

Climate Change and Bark Beetles

In the United States, conditions have not become cool enough in south central Alaska since 1987 to stop bark beetle outbreaks; in Wyoming, the winter mortality of bark beetle larvae has dropped from 80 per cent, typical during cold winters, to 10 per cent during recent winters, which have become increasingly warmer. In the Canadian province of British Columbia, bark beetles have infested more than 8.9 million hectares (22 million acres) of lodgepole pine – enough area to supply timber for 3.3 million homes, the equivalent of the entire United States housing market for two years. Around the world, bark beetle attacks are increasing in range and severity.

One reason for this increase is that bark beetles prefer stressed and weakened trees. The drought and temperature stress that has been imposed on many trees by a changing climate makes them more attractive targets for bark beetles. At the same time, the warming climate has dramatically increased the rate at which bark beetles grow and reproduce; in many forests, the time it takes for bark beetles to mature has dropped from two years to one. This rapid development, coupled with lower winter mortality, means that, come spring, the numbers of beetles ready to attack trees has greatly increased. The beetles do not generally attack healthy trees when their populations are low, but that is not the case when populations are soaring. The bark beetles must then attack healthy trees in order to survive.

The economic stakes are high. In just two decades, British Columbia alone lost trees worth $4.8 billion to the mountain pine beetle and the situation is worsening. Ecologist Alan Carroll of the Canadian Forest Service notes that there has been

'a significant increase in the number of infestations in areas that were historically climatically unsuitable for the [mountain pine] beetle.' If climate change progresses as predicted, the range of beetles will continue to expand to higher latitudes and altitudes. An increase in average global temperature of just 2.5°C (5°F) would allow pine bark beetles to move northwards by seven degrees of latitude, about 780 km (480 miles).

Parasites and Invaders

It is hard to predict how the interactions among parasites and diseases, their hosts and climate will play out in the future. Scientists have studied the interactions in the short term. Over the long term, however, what happens depends on the course of climate change, the organisms' evolutionary responses and the state of the biotic communities

wetter conditions would worsen outbreaks, warmer, drier conditions would cut their number and severity. It may also be that as species' ranges shift toward cooler locales, the organisms will encounter new parasites and new diseases that will limit their ability to thrive. These types of problems affected the introduction of wheat to Brazil from the Old World and the growing of coffee in Asia from Africa. The future of parasites in a world of climate change is also related to the future of invasive species, species with a tendency to expand their range quickly. Invasive non-native species are one of the top threats to global biodiversity, and climate change aggravates the situation.

Climate change often stresses native species so that it is easier for non-native species such as pests and parasites to move in. The spread of bark beetle species and the diseases they carry exemplifies this

ABOVE *A swarm of locusts. This species has been known to decimate crops in a short span of time. Climate change makes it easier for invasive species to take hold, as native species can be stressed by loss of habitat or change of overall temperature.*

CLIMATE CHANGE STRESSES NATIVE SPECIES, ALLOWING NON-NATIVE PESTS AND PARASITES TO MOVE IN. SUCH INVASIVE SPECIES MAY HAVE AN EDGE IN THIS UNCERTAIN WORLD

where these interactions take place. If a new species arrives that a parasite prefers to its traditional host, that host may end up flourishing under the changed conditions. If alternative hosts die off or migrate out of the range suitable for the parasite, remaining host species may be harder hit.

The interactive effects of temperature and moisture changes are also unclear. For at least some parasites and pests, it seems that while warmer,

phenomenon. Such invasive species may share characteristics that give them an edge in this uncertain world. Their generation times tend to be short, and their ability to grow rapidly lets them spread into adjacent areas quickly as conditions become suitable.

Successful invasive species are also more likely to tolerate a broader range of climates than non-invasive species and may have attributes that predispose them to successful survival in a warmer

world. The invasive fire ant, for instance, which spread to Australia in 2001, has habits that help it thrive that native ants do not. The fire ant remains active all day, while populations of native ants stop working during afternoon heat.

In the past, most of scientists' concern about invasive species has focused on species carried over long distances by humans, species such as the zebra mussel or the green crab, which have ended up in an environment to which they would never have been able to travel on their own. As climate changes, however, new species may often be able to arrive under their own power from warmer locales.

By bringing together new competitors, predators, pathogens and hosts, climate change has the potential to dramatically reshape the form and function of the biological communities and ecosystems that are now natural to the planet.

CITIES UNDER SIEGE

From the Urban Frying-Pan

Dense with concrete and bristling with skyscrapers clad in steel and glass, the world's cities seem to live apart from nature in an artificial realm of their own. But, in fact, cities are especially vulnerable to climate change because of their concentrated populations and high temperatures, and because many are situated on the coastline near rising seas. The urban areas of the globe – from Alexandria to Melbourne, from Paris to Quito, from New York to Shanghai – can expect much turmoil as the global temperature rises.

CITIES AND CLIMATE CHANGE

Climate change and its symptoms – including higher temperatures, water shortages, increased flooding and rising sea levels – present problems that are not new to urban populations. But as the global climate continues to warm, cities will be especially vulnerable, most notably in developing nations. Urban populations are growing fast in these countries and they have fewer of the resources needed to adapt or respond to the impact of climate change.

With their high population density and concentration of valuable real estate, cities have always been particularly prone to disruption by the forces of climate. For example, traffic stalls because of heavy snowfall; a hurricane stops a day of trading and business. Modern cities have added new dangers of their own. Vast expanses of concrete, stone, brick and asphalt absorb summer sun and exacerbate heatwaves. High temperatures, coupled with air pollution from industrial emissions and vehicle exhaust, increase the incidence of respiratory illnesses in city dwellers. Pavement and underground drainage systems prevent rainwater from seeping into the ground and contribute to flooding.

Urban activities themselves – including industry, vehicle traffic and the heating, cooling and lighting of buildings – produce a large percentage of the greenhouse gases that are causing climate change. Climate and cities have a long relationship that is about to become more complicated.

Cities as Part of the Land

Despite their artificial nature, cities still exist within a larger, regional climate system. Natural climate change, sometimes severe, is part of their heritage, if not of their immediate history.

Present-day Los Angeles, for example, is known for its warm, dry climate. But about 10,000 years ago, the Los Angeles basin was cold and wet. When it underwent severe warming, cool-weather mammals such as the sabre-toothed cats and woolly mammoths that populated the area were driven northwards or went extinct. The magnificent redwoods died. Drastic temperature change altered an entire ecosystem. Given this regional history, the stability of the future climate of Los Angeles should not be taken for granted.

In ancient times, cities learned about climatic instability the hard way. Near the end of the third millennium BCE, during the time of ancient Egypt's Old Kingdom, a flourishing city of the Akkadian empire rose at what is now the site of Tell Leilan, Syria. But around 2200 BCE, this community began to suffer from the effects of a drought that would last three centuries. Ash from a volcanic eruption probably made conditions worse by blocking sunlight and leading to several summer-less years.

The fields that supported Tell Leilan became a dust bowl; irrigation canals dried up. Between 14,000 and 28,000 people fled the city, leaving it virtually abandoned for hundreds of years.

Modern technology notwithstanding,, today's city dwellers are ultimately as dependent on the vicissitudes of climate as the people of antiquity. The difference is that today nearly half of the world's population lives in cities.

The United Nations forecasts that by 2050, some 65 per cent of the world's population will be living in cities – the same percentage living in rural areas only one hundred years earlier. Cities have become the world's centres of culture, industry and economy. Across the globe, they produce an average 60 per cent of a country's economic output. City populations are also growing to sizes that are

RIGHT *Thick smog in Hong Kong in 2004, when air pollution reached a record high. Cities like Hong Kong, with its high population density and corresponding levels of greenhouse gas emissions, exert profound effects on both regional and global climates.* PREVIOUS PAGE *Clockwise from upper left: traffic clogs a highway in the United States; smog hangs over Mexico City; houses perch above the highly polluted Tiete River in São Paulo state, Brazil; children in Bandar Seri Begawan, Brunei, wear gas masks on smoggy days.*

WORLD'S LARGEST CITIES (IN MILLIONS)

	2001	2015		2001	2015
TOKYO	26.5	27.2	SHANGHAI	12.8	13.6
SAO PAULO	18.3	21.2	BUENOS AIRES	12.1	13.2
MEXICO CITY	18.3	20.4	JAKARTA	11.4	17.3
NEW YORK	16.8	17.9	OSAKA	11.0	11.0
MUMBAI	16.5	22.6	BEIJING	10.8	11.7
LOS ANGELES	13.3	14.5	RIO DE JANEIRO	10.8	11.5
CALCUTTA	13.3	16.7	KARACHI	10.4	16.2
DHAKA	13.2	22.8	MANILA	10.1	12.6
DELHI	13.0	20.9			

ABOVE *These United Nations Population Division figures project the growth of the world's largest urban agglomerations – cities and their outlying surburbs. Many of the most rapidly growing cities are in the developing nations of Asia, Africa and Latin America, where poverty makes people more vulnerable to climate change. (Census years vary from country to country, and figures past 2001 are not available for all cities listed here.)*

unprecedented. The United Nations Population Division forecasts that by 2015, 33 cities – more than half of them in coastal zones – will have populations that exceed 8 million.

While modern cities still cannot be considered separate from the larger regional climate systems, they exert profound effects on both regional weather and global climate. Cities are concentrated islands that absorb heat. They are also generators of the heat and the carbon dioxide emissions that have been driving global climate change.

Cities and Coasts

As colonization and commerce grew, at first in ancient times and then again from the fifteenth century onwards, the coast offered natural sites for cities. Many of the largest cities in Africa, Asia and Latin America are port cities, sharing a colonial past

dependent on maritime commerce. Yet the very symbols of commercial success may now be at the mercy of the waters on which they so much depended. Global sea levels are rising at an annual rate of between 1 to 2 mm (0.04–0.08 in), and that rate is expected to double or even quadruple in response to climate change. The IPCC (Intergovernmental Panel on Climate Change) projects that by the year 2100, sea levels will have risen by a total of somewhere between 9 to 88 cm (3.5 to 35 in) above 1990 levels. Even a moderate rise in sea levels of 40 cm (16 in) – the midpoint of the projections for the end of the century – would have serious consequences for low-lying coastal cities such as Alexandria, Egypt; Tianjin, China; Jakarta, Indonesia; and Bangkok, Thailand.

The problems of shoreline erosion, increased flooding and storm damage, and intrusion of seawater into freshwater supplies would all be amplified in densely populated urban regions. In China, rising sea levels are threatening the coastal region that is home to 40 per cent of the nation's population and more than half its national wealth. Nearly half of China's largest cities – among them Hong Kong, Guangzhou (Canton), Shantou, Hangzhou, Shanghai, Qingdao and Tianjin – are on the coast, each with a population of over 1 million. In Shanghai alone, a .91 m (3 ft) rise in sea level

RIGHT *Residents of the Brasilia Teimosa shantytown in Recife, northern Brazil. Houses here are poorly constructed and sit precariously close to the water, making them especially vulnerable to rising sea levels.*

would flood up to a third of the city, displacing as many as 6 million of its inhabitants.

Coping with rising seas will be expensive for coastal cities. At the minimum, improvements to flood control systems and storm drains will be needed. In Bangkok, the cost of pumping alone, required because of rising sea levels, is expected to be an additional $20 million a year.

Rich and Poor

In developing countries, it is often the poorest people who inhabit the lowest and most hazardous areas of cities. Cities usually develop on the most suitable areas first; new growth spreads to the more flood-prone areas. In Egypt, the Old City of Alexandria was built on land 12 m (40 ft) above sea level, while the port area and newer suburbs were constructed on lowlands that have been kept dry only with the aid of coastal protection. Dakar, the capital

ABOVE *A woman collects bricks from the site of her family's house in Banda Aceh, Indonesia, after the the December 2004 tsunami. Disasters such as this give a* *foretaste of the potentially catastrophic consequences of rising sea levels for urban areas in southeast Asia, which are expanding into lower-lying areas.*

city of Senegal, was first settled on the relatively high and non-erosive Cap Vert peninsula. Settlement, however, has spread into lower-lying lands. New immigrants and the poor often have no choice but to live in such undesirable and hazardous areas. In Recife, Brazil, the favelas, or shantytowns, are expanding into mangrove swamps, situated at the normal high-water level or often below it.

Rising sea levels will also be a problem for many coastal cities in the developed world. Boston, Miami, and New York City in the United States and Amsterdam and Rotterdam in the Netherlands, are all less than 15 m (50 ft) above sea level. The Japanese metropolitan areas of Tokyo, Osaka and Nagoya, which together account for more than half the country's industrial production, are on coastal land that is below the mean high-water level. Osaka is sometimes called the 'Venice of Japan' for its canals and rivers. A rise in sea level of .91 m (3 ft) would put 4.1 million people and 102 billion yen (£0.5 billion) worth of assets at risk of coastal flooding. It has been estimated that it would cost Japan about 8.9 billion yen (£46,000 million) to adapt existing coastal protection measures to avoid or reduce flood damages.

Cities Underwater

Filmmakers love to drown New York City. The popular films *AI: Artificial Intelligence* and *The Day After Tomorrow* both depict an inundated Manhattan. But how real is the threat? Both the US Environmental Protection Agency and the National Assessment of the Potential Consequences of Climate Variability and Change conclude that massive flooding is unlikely in New York City during the next several decades. But the Statue of Liberty need not be submerged for life in this city to be severely affected. A coastal storm in 1992 resulted in terrible

flooding and very severe winds. Had storm waves been higher by only 0.3 to 0.61 m (1–2 ft), subway tunnels and rail lines would have been inundated. That did happen several years later, in autumn of 2004, when storms led to subway flooding that virtually shut down the greater metropolitan area, causing huge economic losses.

The National Assessment of the Potential Consequences of Climate Variability and Change estimated that, by the 2080s, sea levels in New York City may rise between 24 and 108 cm (9 and 43 in). The result would be frequent flooding of subways, airports and low-lying coastal areas – but not a drowning of the city's skyscrapers.

The Greenland ice sheet has been melting rapidly over the past decade, and scientists estimate that it may disappear completely if the average annual temperature in Greenland increases by more than 3°C (5.4°F). By 2100, greenhouse gases in the atmosphere may be sufficient to increase the temperature over Greenland by more than that amount and set in motion the gradual disintegration of the ice sheet over the course of the next thousand years. That scenario would lead to a global sea-level rise of 7 m (23 ft), which would indeed flood the hypothetical New York of the future – but only if no steps had been taken to protect the city.

LEFT Artist's conception of New York City in extreme flood conditions. While the melting of the Greenland ice sheet could raise global sea levels more than 7 m (23 ft), scientists have concluded that it is extremely unlikely that our generation would live to see New York City inundated. However, life in New York could be severely affected by even a small rise in sea levels; most subway tunnels and rail lines in New York City and its surrounding airports are 3 m (10 feet) or less above sea level.

And, of course, measures might be taken to save New York City, although the costs would be exceedingly high. Pumping systems could be redesigned

ALREADY NEW ORLEANS IS RINGED ALMOST COMPLETELY BY LEVEES AND WALLS. BUT BY 2100, IT COULD BE 2.5 TO 4 M (8 TO 13 FT) BELOW SEA LEVEL – OR MORE

to keep the subways dry. Lower-lying coastal roadways could be elevated and seawalls built around the city to protect other areas. Airport runways could also be elevated. However, none of these measures would address the underlying problem.

Delta Blues

The city of New Orleans is on the Mississippi River delta, and most of the town lies below sea level – some parts by as much as 3 m (10 ft) The city has been sinking steadily, as the clay-rich sediments beneath it have become increasingly compacted. Like the Italians of Venice, the people of New Orleans have grown accustomed to this situation. They have even developed appropriate burial customs. The dead are entombed in above-ground cemeteries to prevent them from floating upward when the water table rises. The city is ringed almost completely by levees and walls designed to withstand the storm surge of a Category 3 hurricane. However, global climate change and rising sea levels are expected to plunge the city even further below the waterline: by 2100, it is expected to be at least 2.5 to 4 m (8 to 13 ft) below average sea level.

Louisiana's coast, like other coastal areas around the world, has lost many of its natural defences against storms. About one third of its barrier islands have disappeared since 1880, while its coastal wetlands are slipping away at the rate of several hectares per hour. In anticipation of continued rising sea levels, the US Geological Service has recommended that New Orleans levees be raised by 0.3 m (1 ft), to enable the city to withstand Category 4 hurricanes. This would cost at least $100 million. Tony Waltham, senior lecturer of the Civil Engineering Division of Nottingham Trent University, UK, points out, 'Much of New Orleans is now so low that it would not be built today.'

Venice, Italy

More than any other city in the world, Venice has helped focus attention on the peril of rising seas. Because of the city's unique position – sitting on mudflats in a tidal lagoon – flooding is nothing new to its residents. Venice has lived with the threat of storms and periodically high waters since its founding in 452 CE. The entire city is built on piles driven into the mud below and like New Orleans it is sinking because of the natural compaction of that mud. Between 1920 and 1950, there was an overextraction of groundwater to supply an industrial complex.

Art to the Rescue in Venice

For years, Venice has fought the double problems of rising sea levels and the sinking of the soil that supports the city. Now, an artist who lived 300 years ago is helping the researchers who are studying these chronic problems. The works of eighteenth-century painter Giovanni Canal, better known as Canaletto (1697–1768), capture Venice and its life on the water in exacting detail, down to the lines of green scum at buildings' foundations, traces of algae that represent a record of the tides. Canaletto's images are considered accurate because they were made with the use of a camera obscura – a device that projected an image of the artist's subject onto paper or canvas, allowing him to trace its outlines. Using this method, noted by Italian Renaissance artist Leonardo da Vinci more than 200 years earlier, Canaletto produced images so accurate that they depicted the dark crusts of smoke pollution on buildings that had not been washed by rainfall. Canaletto's patrons were mainly English collectors during his lifetime; today his work is of special interest to Venetians concerned with the fate of their city.

Scientists in Venice are using Canaletto's views to fill in gaps in their knowledge of tides, which were not measured with instruments until 1872. Dario Camuffo, a climate change specialist associated with the Italian National Research Council, took new photographs from the original viewpoint of each Canaletto painting and superimposed them on the painting to see how they differed, using computer graphics as needed to correct optical distortions or perspective effects. The result – Canaletto's paintings of Venice reveal considerable alteration in sea levels since the eighteenth century. For example, a 1767 painting of Saint Mark's Square shows algae 34 cm (13.6 in) lower than today. Such tide marks, scientists and public officials believe, can help them research future sustainable water levels for the city.

Scientists have already started analyzing the paintings to chart water levels and map tide dynamics. They hope to use the images to predict future flooding and establish an optimum flood level that could be allowed when building the flood barriers at the openings to the Venetian lagoon. According to Camuffo, 'Today we can measure in the picture what the tide level was some three centuries ago.' He believes the paintings hold much promise, noting that, 'You can do a project with information from the past because the past is the key to interpreting the future.'

ABOVE *A Canaletto painting of Santa Maria della Salute in Venice, Italy (top) as seen from the Grand Canal. A modern photograph (bottom) of the same view makes comparison of water levels possible. Canaletto's paintings are considered to be highly accurate and scientists can use these comparisons to chart sea-level rise in the imperiled historic city.*

This accelerated the sinking, increasing the frequency of what the Venetians call *acqua alta,* or high water (at high tides) in the plazas and alleyways of their city.

The Venetians have learned to live with the *acqua alta.* Raised boardwalks are installed for pedestrians, furniture is temporarily moved to high ground, and lower shelves are cleared in shops. *Acqua alta* normally occurs between September and April and episodes of high water do not last long – typically only one to two hours. However, St Mark's Square, in the centre of the city, now floods ten times more per year than it did in 1900, or about 100 times a year. The average water level in Venice is 23 cm (9 in) higher than it was a century ago. In addition, the water – once brackish – has now become saltier. Seawater is intruding increasingly into the shallow, ecologically rich lagoon around the city. Many scientists believe that Venice's situation is gradually becoming untenable.

In May 2003, the Italian state began an ambitious, controversial project to save the city from the rising waters. Project Moses, as it is called, involves constructing 78 movable, underwater dams, each weighing 300 tonnes, to close the lagoon's entrances to the sea when the worst flooding is forecast. Named for the biblical leader who parted the Red Sea, Project Moses is expected to cost £4,500 million and take eight years to complete.

Amid this drama, some experts contend, by contrast, that climate change will actually benefit Venice. In 2002, scientists at the University of East Anglia, UK, reported that global warming is weakening Mediterranean storms, reducing the chances that the city's 118 islands will be flooded. Nevertheless, scientists meeting at the 2003 Venice in Peril Fund

ABOVE *A street in Dhaka, Bangladesh, after a flood in 2004. The city is flooded nearly every monsoon season, in part because it lies near to sea level. Increased snowmelt and higher river levels caused by climate change are likely to increase the danger to residents of this city of more than 12 million.*

Conference in Cambridge, UK, declared that if the rise in sea level accelerates, Venice will 'almost certainly' be uninhabitable by 2100.

Dhaka, Bangladesh

Dhaka attracts little attention from international news media except at the time of its catastrophic floods during the monsoon season. This city of more than 12 million inhabitants lies between 3 and 7 m (10 and 23 ft) above sea level on the Bengal delta, at the confluence of the Ganges, Brahmaputra and Meghna rivers. These Bengali rivers discharge over 1 billion cubic m (35 billion cubic ft) of water every year, second in volume only to the Amazon. Climate

change is likely to worsen annual monsoon flooding here. Warmer weather leading to increased Himalayan snowmelt is expected to increase river levels and flooding risks. A rise in sea level of .91 m (3 ft) would displace about 60 per cent of the population, potentially swamping this already crowded city with 'climate refugees'. During the 2004 monsoon, 40 per cent of the city was under water as rivers overflowed during heavy rains and mountain snowmelts. The city's drainage system was overloaded, and 500,000 cubic m (18 million cubic ft) of raw sewage escaped into the water daily. Authorities estimated that waterborne diseases caused by the floods afflicted more than 100,000 people.

VARIETIES OF URBAN IMPACT

Beyond the dangers of coastal flooding, there are many ways that climate change will affect cities, beginning with too much or too little fresh water. Floods are very likely to result from the increased precipitation and the melting of glaciers; droughts will occur when there is too little rainfall. In addition, the same city might see both increased flooding and increased drought – too much water during some periods of the year and then not enough water during the rest of the year.

Although flooding from increased rainfall and snowmelt in the mountains is a risk to cities like Dhaka, melting snow and eroding glaciers will bring the opposite problem to other cities in Asia as well as those in South America. In India, cities such as New Delhi and Calcutta depend on runoff from the Gangtori glacier in the Himalayan region to feed the River Ganges. The Gangtori, together with other Himalayas glaciers, has been receding and thinning during the past 30 years, and runoff is diminishing. The situation is similar in South America's northern Andes. Peru's Yanamarey glacier lost a quarter of its area during the last 50 years; Ecuador's Antizana glacier shrank seven to eight times faster during the 1990s than during previous decades and Bolivia's Chacaltaya glacier is expected to disappear by 2010.

In Bolivia, the 1.4 million people of La Paz and El Alto depend on the glaciers that surround their cities, not only for water, but also for 75 per cent of their electricity (hydropower). In Peru, Lima residents also depend on glacial runoff for their water.

Glacial runoff is not the only water supply that will change in a warming climate. The expert panel of the IPCC concludes that climate change is likely to bring more rain to the tropics, and the mid and high latitudes, and less rain to the subtropical belts. And more of the rain is likely to come in heavy downpours, meaning that areas receiving more precipitation could have a higher risk of flooding and areas with less precipitation could experience longer periods of drought. In Richmond, British Columbia, a city in Canada's middle latitudes, the average annual precipitation has been rising between two and four per cent per decade since 1929. The sea level along the coast has also risen from 4 to 12 cm (1.6 to 4.9 in) during the last 100 years. In this city, where 165,000 people live on the floodplain of the River Fraser, at an average of about 1 m (3 ft) above sea level, the impact of changes in precipitation can be enormous.

In China's cities, a water shortage, rather than a surplus, is the problem. Already, nearly half of China's 500 major cities – including the nation's capital, Beijing, and the capital of Shanxi Province, Taiyun – lack sufficient water, and in the future, many more are expected to reach a level of chronic water scarcity.

GLOBAL POPULATION DENSITY ESTIMATES FOR 2015

PERSONS PER SQ KM
- 0
- 0–4
- 5–24
- 25–249
- 250–999
- 1000+
- no data

LEFT *This map shows estimates for global population density in 2015. Darker areas, such as in India, eastern China, the Philippines and central Europe, indicate a greater number of persons per square kilometre. Climate is likely to have a concentrated impact on cities in these regions.*

More than one hundred Chinese cities have acute problems with their water supplies; fewer than five per cent of them have water that is safe to drink. Research conducted by China's Ministry of Water Resources shows that by 2030, increased water shortages due to climate change could lead to losses of up to $500 million in the Beijing-Tianjin-Tangshan area during serious drought years.

Water, Hydropower and Health

Climate change is likely to affect the production of electricity from hydropower. As flows in glacial rivers diminish and the probability of drought and

Both too much freshwater and not enough can increase health risks. Flooding often results in contamination of drinking water, while drought can concentrate pollution and encourage unhygienic practices because of water shortages. Heavy rainfall events in Europe and the United States have been linked to outbreaks of cryptosporidiosis, giardia and other infections. Diaorrhea affected 14,000 people after the 1988 flood in Dhaka, Bangladesh and was the most common cause of death for those under 45 years of age. Cholera outbreaks in Peru have been traced to the growth of algae in warming waters and high incidences of flooding, which

The Far-Flung Connections of Climate

Cities are linked to one another through climate, just as they are connected with distant regions of the world through economics and culture. As a result, events in the air or water of one part of the world may bring simultaneous change for the air or water supplies of another.

For instance, the violent tropical storms known as typhoons typically originate in Pacific waters near the equator but travel great distances westwards to strike Asian communities. In 1983, one such storm, Typhoon Ellen, first took shape as a tropical depression near the Marshall Islands on August 29. It then moved toward China, striking Hong Kong on September 9. Ten people were killed and twelve reported missing in the ferocious winds and torrential downpours.

Elsewhere in the Pacific, an eastwards-moving weather system, the Pineapple Express, periodically connects Hawaii with Seattle, more than 4,200 km (2,600 miles) away. This system takes shape as part of the the Madden–Julian Oscillation, an equatorial rainfall pattern that gives moisture to the Express. The Pineapple Express brings warm, moist air from Hawaii to the west coast of the United States and Canada, often resulting in torrential rains. In Seattle, a city already known for its high precipitation, Pineapple Express storms increase rainfall, bringing 2.5 to 5.0 cm (1 to 2 in) per day and increasing flooding.

Bergen, Norway's second largest city, is also wet; it is Europe's wettest city, with an average annual precipitation of 200 cm (79 in), and has four times the annual rainfall of London. This condition stems from its geography: It is in the path of 80 per cent of the low-pressure systems that originate over the North Sea and the North Atlantic.

Such far-flung connections provide more than enough reason to pay attention to the changing of climates all over the world.

> MORE THAN ONE HUNDRED CHINESE CITIES HAVE ACUTE PROBLEMS WITH THEIR WATER SUPPLIES; FEWER THAN FIVE PER CENT OF THEM HAVE WATER THAT IS SAFE TO DRINK

subsequent reduced river-flow increases, the electricity production is likely to drop.

In Ghana, the 1982–1983 droughts necessitated four years of government electricity rationing in the capital, Accra. In other areas – as in the United States, in the city of Seattle – years with below-average snowfall can lead to contentious summertime battles over water use, with the need for hydropower pitched against agricultural and residential requirements. Elsewhere, rainfall concentrated in heavy downpours would mandate more conservative water storage strategies to prevent flood damage. All factors point to less flexible hydroelectric capacity in the future.

carried sewage (and cholera germs) into clean waters. Cholera transmission is also associated with increases in surface water temperatures.

Heat in the City

Scientists forecast that the mean global temperature by 2100 may increase by anywhere from 1.5°C (2.5°F) to as much as 5°C (9°F). As a result of the heat-island effect (see pages 86–89), cities will experience even larger temperature increases. The 2003 heat wave in Europe led to much higher death rates in Paris than in surrounding rural areas. In Caracas, Venezuela and San José, Costa Rica, temperatures already remain high throughout the year.

CITIES IN THE 21ST CENTURY

How climate change will affect cities by the middle of the century will vary greatly from one region to the next. Effective planning will require local models and predictions and so far these predictions present a sobering prospect: flooding in cities in West Africa and on islands in the Indian Ocean, water shortages in Melbourne and Quito and oppressive heat in Paris and Washington, DC.

Banjul, Gambia

Banjul, the capital of the western African nation of Gambia, sits at the mouth of the River Gambia at an average of less than 1 m (3 ft) above sea level. Unless protective measures are taken, if sea levels rise to double that distance during the next 50 to 60 years, as climate models predict, the entire city will disappear. About 50,000 people will be displaced and an estimated $217 million of property lost.

Gambia already averages 300 per sq km (115 people per sq mile), so the refugees from Banjul may cross into neighbouring Senegal, increasing pressure on that nation, which will probably be coping with its own flooding problems.

BELOW *Concrete blocks prevent beach erosion on the island of Male, the capital of Maldives in the Indian Ocean. Because the island's highest point is only 2.4 m (8 ft), it is at great risk from rising sea levels.* UPPER LEFT *In the United States, the city of Washington, DC – site of the Jefferson Memorial, shown here – is also threatened by rising sea levels and higher temperatures.*

Male, Maldives

Male, the capital of Asia's smallest independent country, Maldives, lies on Male Island in the Indian Ocean. The maximum elevation of this island, which has a population of 75,000, is only 2.4 m (8 ft). During the December 2004 tsunami, waves swept over as much as half the island, pouring down streets and spilling against the president's office.

In anticipation of the expected continued rise in sea levels, Male has built a concrete seawall 3 m (10 ft) high around the island. The government is also building an artificial island called Hulhumale, which is already higher than the average atoll in this nation of islands. Hulhumale would become a refuge for Maldivians if the rising tide submerges Male or other islands.

Melbourne, Australia

Between 1997 and 2002, Australia's second largest city, with a current population of 3.6 million, suffered its worst drought in recorded history – six long years. Record low rainfall combined with record high temperatures and high evaporation. Higher temperatures and even more severe droughts are expected.

Facing drought conditions, Melbourne residents have cut water use by 22 per cent since the 1990s, largely by identifying efficiencies and using water-saving devices. But other changes may also be needed. The city has considered a plan to increase available drinking water by relying exclusively on recycled sewage in Latrobe Valley power stations. Some in the state of Victoria are also calling for a change in domestic power use habits, given that Victoria relies on brown coal reserves that produce heavy greenhouse gas emissions, contributing to the climate change affecting Melbourne's water supply.

ABOVE *Children in Paris cooling off during the summer heatwave in 2003, which killed thousands across Europe. If current warming continues, such lethal heat may become common.*

Paris, France

In a mere eight days during the scorching summer of 2003, some 100 people died from heat-related causes around Paris alone. With the mercury regularly reaching 40°C (104°F) and above, morgues filled to overflowing, forcing authorities to store the dead in refrigerated trucks and tents. This European heatwave, which killed thousands, may have been a sample of what may become regular.

'By the end of the century, a summer with temperatures such as those we had in 2003 will be considered cool,' said Michel Deque, a scientist at the French meteorological agency Meteo France. It is expected that in 2100, a Parisian summer will be, on average, about 6°C (10.8°F) hotter than now. The French capital will have to learn to live with heat that is often lethal.

Quito, Ecuador

Quito, the capital of Ecuador, sits in a valley in South America's Andes Mountains, within view of Cotopaxi, a snow-capped, cone-shaped volcano that towers 5,897 m (19,347 ft) over the surrounding countryside. But from 1976 to 1997, Cotopaxi lost nearly a third of its ice cover. This disappearance is part of a more widespread loss of glacial ice and snow occurring region-wide.

Quito, which lies virtually on the equator, depends on snow-covered mountains for 80 per cent of its drinking water supply. By mid-century, the diminishing glaciers will result in water shortages. Tourism may suffer as well. Ecuador's Avenue of the Volcanoes, a breathtaking strip of ice-capped mountains that is a favourite destination for travellers to the country, may lose some of its allure.

Washington DC, USA

In mid-century, America's capital will feel like the Deep South. By the 2030s, the number of days per year that top 32.2°C (90°F) will nearly double, from 36 to 60. By the 2050s, the mercury will reach 37.8°C (100°F) an estimated twelve times a year, a twelvefold increase over the present.

The higher temperatures are likely to bring increased illness and mortality from heat stress and mosquito-borne diseases. With increased warmth, the ranges of malaria, dengue fever and eastern equine encephalitis are expected to expand. The rising sea levels may increase the risk of flooding of the River Potomac, threatening low-lying monuments such as the Jefferson Memorial and the entire National Mall. Because cherry tree roots are sensitive to standing for long periods in high water, flooding may prevent the appearance of Washington's landmark cherry blossoms.

ABOVE *Quito, the capital of Ecuador, could face severe water shortages if the snow cover of the surrounding mountains is lost due to global warming.*

ABOVE *A haze of smog hangs over Santiago, Chile. Air pollution here is chronic and the city's residents are likely to suffer from associated health problems as temperatures rise.*

Research shows that mortality rates from cardiovascular and respiratory illnesses are likely to increase in these cities. In some very large cities, such as Atlanta in the United States and Shanghai in China, scientists project up to several thousand more heat-related deaths annually by 2050. In general, people who are more than 65 years old would prove the most susceptible to heat. Urban populations in developing countries are particularly vulnerable because their living conditions are crowded and their homes poorly ventilated. At the same time, milder winter temperatures are also likely because of climate change. In some places, including Santiago, Chile, where the climate tends to be relatively cold, the death rate associated with cold temperatures is likely to be reduced and may offset the increase in deaths that is associated with hot temperatures.

Prolonged heat will also aggravate air pollution. Residents of cities with chronic severe pollution, including Mexico City and Santiago, are likely to suffer most. Research shows that a combination of high temperatures and air pollution can lead to increases in respiratory complaints, as well as increased risk of heatstroke, stroke, and brain damage. In São Paulo, Brazil, the mortality of elderly people has increased following peaks of air pollution. In Tokyo, the occurrence of heatstroke in males age 65 years and older appears related to the daily mean temperature, relative humidity, and atmospheric concentration of nitrogen dioxide.

For those who can afford it, air-conditioning is one way to reduce the stress of heat on the human body. Estimates predict that air-conditioning requirements in mid-Atlantic cities in the United States, such as Baltimore, Maryland and Knoxville, Tennessee, will increase by as much as 20 per cent. Electricity generation and transmission systems will be strained in cities where air-conditioning already accounts for the majority of total electricity use, such as Riyadh, the capital of Saudi Arabia and Hong Kong. Energy consumed by the use of air-conditioning will also contribute to the heat-island effects of cities, as well as the greenhouse gas emissions that are driving climate change. It is ironic that, in order to get cool, humans make the world even hotter.

Cities and Disease

Other health problems related to a changing climate may be especially serious for urban residents. Many disease-causing microbes become virulent only in areas of high population density, so newly arrived diseases may pose special risks for urban areas. Cities also serve as transportation hubs, mixing people and diseases from all over the world. Some particularly dangerous infectious diseases, such as dengue and cholera, thrive in urban environments.

The ecology and the transmission of infectious diseases are very complex. The breeding success of insects that transmit diseases may depend on the availability of vegetation or water, while temperature can influence their survival rates. On the other hand, various rodent-borne diseases depend on environmental factors that determine rodent population size and behaviour. Despite the complexity of disease causation, climate often plays a role and urban conditions frequently provide the medium for disease to move in and quickly become epidemic. For example, in 1996, the city of Salvador, Brazil, underwent severe floods. Two weeks later, it suffered the peak in an epidemic of Weil's disease (leptospirosis), which is borne by rodents and spreads more easily under flood conditions.

How diseases are transmitted to humans depends on complex human factors, including population density, individual behaviour and housing location. Here again, urban conditions can spur the spread of disease. The slums and shantytowns that are part of many of the world's large cities provide a breeding ground for disease organisms and disease hosts, including rats, mice, cockroaches and flies. Such areas are likely to suffer disproportionately as climate changes.

How the range and transmission of infectious diseases will change in cities under global warming is still an active area of research. Numerical modelling studies show that, by 2080, there could be a two- to four-per cent increase in the number of people living in the potential transmission zone of malaria. Reduced rainfall around the Sahel, the Sahara's semidesert fringe, may decrease transmission of malaria in this region, which includes N'Djamena, the capital of Chad.

In Rio Grande in southern Argentina, if temperature and rainfall increase, conditions may become favourable for the transmission of malaria. In the cities of the developed countries of Australia, Europe and North America, existing malaria control programmes and public health infrastructure have effectively minimized the risk of importing the disease and can continue to do so. In urban areas of the former Soviet Union, such as Baku, Azerbaijan and Rostov, Russia, lack of funds for maintaining a public health infrastructure may increase the risk of reintroducing the disease.

Other mosquito-transmitted diseases, such as dengue fever, eastern equine encephalitis and West Nile virus have recently seen a resurgence in the United States and have even appeared in urbanized northßeastern states such as Massachusetts, New Jersey and New York. These diseases are spread by mosquitoes that gather near wooded areas alongside swamps

LEFT *Micrograph of the West Nile virus. Scientists are studying the effects of climate change on the range and transmission of infectious diseases like this.*

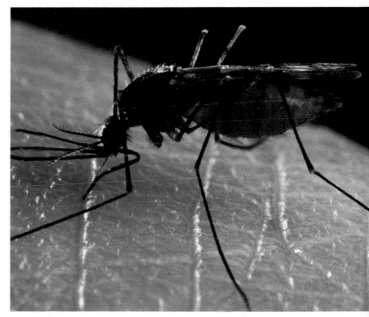

ABOVE *The mosquito, seen here drawing blood, is a growing threat. Some diseases, spread by mosquitoes, have seen a resurgence in the United States, partly because of more favourable conditions, as when receding floodwaters leave swampy areas.*

and marshes. Floods can temporarily lower mosquito populations, but once floodwaters recede, the swampy environment may become even more favourable for the growth of the mosquito population. Numerical modelling studies project that a rise in global temperature of 2°C (3.6°F) could lead to an increase in the range and transmission season of dengue. However, as is the case with malaria, the high living standards and high level of public health protection in developed countries make it unlikely epidemics will take hold here. But for many of the world's cities, vulnerability to dengue is another reason to be concerned about climate change.

EXTREME WEATHER

As Climate Changes, Weather Grows Fiercer

Extreme weather has always been with us, but its character may change as the global climate changes. Extreme weather lies outside a region's climate pattern and is therefore, by definition, infrequent or rare. It can include heatwaves, droughts, heavy rainstorms, hurricanes, tornadoes and ice storms. Not all extreme weather events end in disaster, but they are nearly always potentially destructive and are likely to become more so in the future.

PRESENT TRENDS

Extreme weather events capture public attention. Broadcast images of death and destruction that follow a natural disaster, though sometimes sensationalized, nevertheless reflect real human suffering and often staggering economic costs. In 2004 alone, weather-related events around the world took over 10,000 lives and cost the insurance industry over £20,000 million.

How will this change in a warmer world? Will extreme weather events be more frequent or less frequent? Will they be more intense or less intense? Current scientific understanding of extreme weather and climate change is based on observations of how extreme weather has been changing in the recent past. From there, scientists can begin to extrapolate how extreme weather is expected to change as Earth continues to warm.

Compared to what we know in general about changes in climate over time, our knowledge of the climate extremes of the past is meagre. The biggest problem is a lack of dependable weather data going back before the middle of the twentieth century. Since extreme weather events are relatively rare, identification of a sufficient number of these events to draw meaningful conclusions requires long data records. For many countries, consistent records of temperature, precipitation, moisture, wind or air pressure have been available only since the Second World War. Until the advent of satellites, storms that did not touch land in populated areas often went unrecorded. Localized and short-lived events such as thunderstorms and tornadoes often are recorded only if they occur close to weather stations or populated areas. Even when data records are available, they may not be of comparable quality; in addition, definitions of extreme events of the past have sometimes used different criteria. Despite these difficulties, some trends are discernible from existing records.

Weather and Storms

The storms that have watered and shaken societies throughout history arise from the natural workings of weather. They begin with the everyday feature of weather maps: lows and highs. In the Northern Hemisphere a low, also called a depression or cyclone, is a centre of low pressure around which winds circle inwards in an anti-clockwise direction; a high, or anticyclone, is a centre of high pressure around which winds circle outwards in a clockwise direction. In the Southern Hemisphere, air circulates clockwise around a low, but counter-clockwise around a high. Air blows from an area where the pressure is higher to one where it is lower – across a pressure gradient. The direction of the winds is a result of the combination of pressure-gradient force and the Coriolis effect, the apparent deflection of currents of air or water owing to Earth's rotation (see page 24).

A low is associated with clouds, storms and precipitation. Air moving into the low forms a rising column. As moisture-laden air rises, it cools, reducing the amount of water vapour it can hold. The result is cloud formation and the falling of rain or snow. By contrast, a high is associated with fair, dry weather, because air flows out of the anticyclone near the surface, with more air descending from above. Descending air is usually compressed and warmed.

Because the high, or anticyclone, is associated with fair weather, it gets less attention than the low, or cyclone, which is associated with stormy weather. In fact, the word cyclone is commonly used for several kinds of storms: a type of storm occurring in the Indian Ocean; a tornado; or any storm with spiralling, swirling winds. In the western Pacific or Indian Ocean, a cyclone is also commonly called a typhoon.

RIGHT *A woman and her child return from market across a monsoon-swollen stream, a tributary of the Narmada River near the village of Walpur, Madhya Pradesh, India.* PREVIOUS PAGE *Clockwise from upper left: lightning storm over Lake Tanganyika, Tanzania; tornado over the American plains; a motorist trapped by waves on Cape Hatteras, North Carolina, USA, during Hurricane Isabel in 2003; men carry bags of rice through the flooded streets of Dhaka, Bangladesh, after a heavy monsoon rain.*

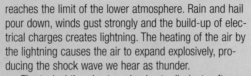

Thunderstorms

There is a reason why thunderstorms happen more often in summer than in winter: they require warm air. As the Sun heats the land and the land heats the air above it, warm air rises. If the air is moist, a strong updraft of rising warm air can develop, often as a result of the interaction of two air masses, one warm, one cold. The rising air condenses into a growing cumulus cloud.

As the cloud grows, the water droplets in it become large and heavy and begin to fall as rain. As they fall, drier air around the cloud is sucked in, and a powerful downdraft develops to accompany the updraft in what is known as a cell. The thunderstorm has now reached its mature stage. The cloud has become a towering cumulonimbus, a behemoth that flattens at the top into an anvil shape as it

reaches the limit of the lower atmosphere. Rain and hail pour down, winds gust strongly and the build-up of electrical charges creates lightning. The heating of the air by the lightning causes the air to expand explosively, producing the shock wave we hear as thunder.

The typical thunderstorm begins to dissipate after about 30 minutes. It is estimated that about 40,000 thunderstorms occur around the world everyday. At any given moment, 1,800 thunderstorms are in progress.

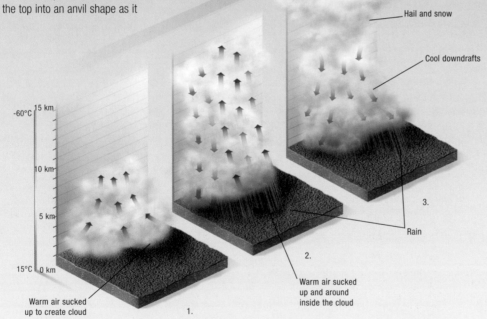

Hail and snow

Cool downdrafts

Rain

-60°C | 15 km

10 km

5 km

15°C | 0 km

Warm air sucked up to create cloud

1.

2.

3.

Warm air sucked up and around inside the cloud

ABOVE *Thunderstorms form in distinct stages. From left to right: 1. Warm, moist air rises in an updraft and condenses into a cumulus cloud. 2. The mature stage of the thunderstorm is marked by heavy precipitation, warm updrafts and cool downdrafts in part of the storm. 3. When the warm updrafts disappear completely, precipitation becomes light and the cloud begins to evaporate.*

Tornadoes, or twisters, are often associated with a cyclone in the middle latitudes. A tornado is a column of air that rotates at high speed under a large thundercloud or developing thundercloud. Usually it forms a funnel-shaped cloud that touches the ground. With wind speeds that can exceed 480 km per hour (300 miles/h), tornadoes can cause tremendous damage to people and property. When a tornado occurs over water, it is called a waterspout.

Fronts and air masses are also important in understanding storms. Air masses account for much of the everyday local weather. An air mass is a large body of air, usually at least 1,600 km (1,000 miles) across and several kilometres thick, characterized by similar temperature and moisture throughout. As it moves, it carries with it the temperature and moisture conditions of its region of origin. When a mass of cold, dry air from northern Canada moves south, the southern region will experience cold, dry weather. When a mass of warm, humid air from the Caribbean moves north, it brings with it warm, humid weather. A front is the boundary where two air masses of different origins and characteristics meet. Changes in the weather – say, from cold to warm – usually occur along fronts. A cold front forms when a cold air mass pushes into a warmer one, while a warm front, which often brings days of rain, takes shape when a warm air mass pushes into a colder one. The winds in a low-pressure system often bring contrasting air masses together to form fronts, which can lead to stormy weather.

Hot and Cold, Wet and Dry

In much of the world, the baseline for what is considered extreme weather is changing – that is, 'normal' weather is getting warmer. Studies in Asia, the South Pacific, Australia, New Zealand, Europe and North America all show the trend of an increase in the number of hot days and warm nights and a decrease of cool days and cold nights over the course of the twentieth century. The number of days below freezing has significantly decreased in the central United States, Australia, New Zealand and Europe. For example, one study shows that the start of the frost-free season in the north-eastern United States now occurs 11 days earlier than in the 1950s. Some locations in Europe have as many as 50 fewer days of frost per year than in the 1910s. In China, studies show that the number of hot days has been increasing and the number of extremely cold days has been decreasing, while studies in Russia show that the number of extremely hot days has increased significantly over the past 50 years.

Recent research shows that global land precipitation has increased by about two per cent since the beginning of the twentieth century – but also significant is how it has increased. While the change in quantities of rain, snow, sleet and hail is not uniform across the globe, middle and high latitudes in the Northern Hemisphere have been getting 7 to 12 per cent more precipitation and a larger proportion of it has fallen in heavy episodes. Over most of the United States and Europe, precipitation has fallen in longer spells and the total acreage of severely wet areas has increased.

In contrast, the northern subtropics and the equatorial regions have experienced fewer heavy rainstorms and have become drier. The extent of severely dry areas over Africa's Sahel region, eastern Asia and southern Africa is increasing. Increased numbers of droughts have also been recorded in Hungary and China.

Hurricanes and Cyclones

The number of storms has increased in the middle latitudes of the Northern Hemisphere during the latter half of the twentieth century and decreased in the Southern Hemisphere. There has been an increase in severe gale days around the North Sea, includ-ing the United Kingdom. The number of strong depressions in the Great Lakes region of North America has also increased significantly. Conversely, cyclone activity in the Southern Hemisphere increased up to 1972 but decreased during the 1990s.

Still, there is little scientific evidence for a long-term global trend in the change in intensity and

RIGHT *Giant hailstones are lumps of ice that form in cumulonimbus clouds. This kind of extreme weather can cause widespread and costly damage.* ABOVE LEFT *Lightning storm, New Mexico, USA.*

frequency of tropical storms – hurricanes, cyclones and typhoons – and severe local weather, including tornadoes, thunder days, lightning and hail.

The Cost: High and Rising

In a crowded world, extreme weather can be ruinous, both in terms of human health and property damage. The cost of an extreme weather event is affected by many factors, foremost among them population density and the wealth of the population. An extreme cold spell in northern Canada or Alaska would probably cost less, in purely economic terms, than a moderate-strength hurricane passing over Miami, Florida. Human losses, however, are not as easy to compare or quantify. In any case, public attention, and hence records, are often focused on the most costly events.

Heatwaves, though sometimes considered a mere nuisance, can be extremely costly, as Europe learned during the summer of 2003, when an extended and severe heatwave took an estimated 30,000 lives and cost 16,000 million euros (£10,000 million) in damage. About 600,000 hectares (1.5 million acres) of forests were burned and important economic losses were sustained in the agricultural and forestry sectors. During a heatwave in the United Kingdom in 1995, heat-related road repairs and train delays cost the transport sector an extra £16 million. High temperatures caused roads to rut. Trains had to travel more slowly to reduce risks of tracks buckling in the hot weather.

Droughts and floods are also growing increasingly costly, and not only in heavily populated areas. In 1999, a drought in the United States affected 25 per cent of the country's harvested cropland and led to farm net income losses of approximately $1.35 billion. Extremely dry conditions, coupled with high winds and temperatures, can also lead to severe fires. Temperatures above 38°C (100°F) and winds of 65 km per hour (40 miles per hour) led to more than 200 bush fires in the state of Victoria, Australia, in January 1997. More than 40 houses were destroyed and total losses were estimated at over $A40 million (£16.9 million).

On the other end of the wet-dry spectrum, in 2002, heavy rainfall from storms crossing central Europe during early August triggered sequential flood waves along two major river systems in Austria, the Czech Republic and Germany. The subsequent severe flooding took 100 lives and caused economic damage twice the island's annual gross domestic product. More than 38,000 people were left unemployed.

The greatest destruction from a coastal storm is frequently caused by the flooding associated with the storm surge, the rapid rise in sea level caused by winds that blow ocean water ashore. In Nigeria, storm surges and associated floods have decimated coastal agricultural areas and dislodged oil producing and export handling facilities. In 2002, a storm surge caused heavy flooding in almost all the streets and buildings on the Victoria Island beach in Lagos.

It is not only tropical cyclones and hurricanes that can cause damage. Although coastal storms in the United States account for about 70 per cent of annual disaster losses, tornadoes also cost the

IN A CROWDED WORLD, EXTREME WEATHER CAN BE RUINOUS, BOTH IN TERMS OF HUMAN HEALTH AND PROPERTY DAMAGE

in excess of 15,000 million euros (£10,400 million). In the Czech Republic alone, 100 towns and villages were flooded completely. In Germany, 180,000 buildings and homes were damaged.

Hurricane Ivan in 2004 was one of the strongest and most destructive storms ever, maintaining a strength of Category 4 to 5 on the Saffir-Simpson scale (see page 217) for more than five days. The storm caused serious damage in the Caribbean and to offshore drilling platforms in the Gulf of Mexico. On the island of Grenada alone, the hurricane led to 28 deaths, and close to 90 per cent of the houses were damaged or destroyed. Damage costs were country $1.1 billion annually in damages. Hailstorms over the southern plains and lower Mississippi River valley in April 2003 caused damages of over $1.6 billion. In eastern Canada, an ice storm in January 1998 stranded nearly three million people without heat or electricity and caused over $1.3 billion in damage.

RIGHT *Palm trees blowing in monsoon winds, Trivandrum, Kerala, India. The monsoon occurs annually in southern Asia, delivering heavy rains. Variations in strength and timing of these rains can devastate crops and livestock. Southwestern monsoons are expected to become stronger as global temperatures increase.*

HURRICANES

As global temperatures rise, hurricanes will probably become stronger. The peak intensity of these powerful storms, also called typhoons and cyclones, may increase 5 to 10 per cent, with precipitation rates increasing by 20 to 30 per cent. If the frequency of these storms remains the same during the twenty-first century as in the past, warming may bring more Category 5 storms – hurricanes with winds of more than 250 km per hour (156 miles per hour).

Hurricanes, the world's most destructive storms, are intense tropical weather systems with powerful circulations and sustained winds of 119 km per hour (74 miles per hour) or more. The word hurricane is derived from the name of the Mayan god of wind and storm, *Hurakan*. In the western North Pacific, it is called a typhoon, from the Arabic *tufan*, which appears in the writings of Aristotle and in the Koran. In the Indian Ocean and some other regions, these storms are also called cyclones, from the Greek *kyklon*, which means 'moving in a circle like the coil of a snake'. They are feared around the world.

ABOVE *A hurricane lashes the coast near Corpus Christi, Texas, in the United States.* UPPER LEFT *A satellite image of Hurricane Andrew over the Bahamas in 1992, when winds registered 250 kms per hour (156 miles per hour).*

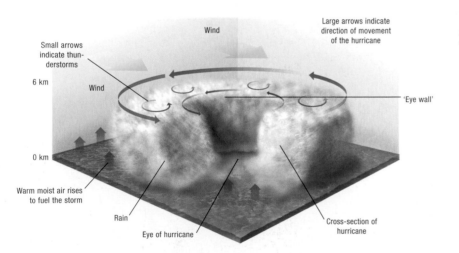

Small arrows indicate thunderstorms

Wind

Large arrows indicate direction of movement of the hurricane

6 km

Wind

'Eye wall'

0 km

Warm moist air rises to fuel the storm

Rain

Eye of hurricane

Cross-section of hurricane

ABOVE *Hurricane formation. A hurricane consists of many individual thunderstorms spiralling around the eye of the hurricane, at the centre. The ring of thunderstorms immediately surrounding the eye is the eye wall.*

Warm temperatures above 27°C (81°F) are more likely to produce hurricanes. For every added increment of heat, the chance of a storm multiplies. At high temperatures, water evaporates from the ocean surface and rises, condensing as it rises to form water droplets. The condensation of the water evaporated from the ocean's surface releases heat, which fuels the hurricane's winds. The centre, or eye, of a hurricane is relatively calm, with sinking air, light winds and few clouds. The most violent winds and rain occur in the eyewall, the ring of thunderstorms immediately surrounding the eye. Most of the air is propelled outwards at the top of the eyewall, increasing the air's upward movement. The winds swirl anticlockwise in the Northern Hemisphere and clockwise in the Southern Hemisphere.

A hurricane's severity is rated according to the speed of its winds. The Saffir-Simpson Hurricane Intensity Scale was created in 1969 by Herbert Saffir, an engineer and Bob Simpson, director of the US National Hurricane Centre, in a report for the World Meteorological Organization. It includes five categories of potential damage from a hurricane. They range from the mildest, Category 1, with wind speeds of 119–153 km per hour (74–95 miles per hour), with no real damage to buildings and minor flooding, to the severest, Category 5, with winds over 250 km per hour (156 miles per hour), severe structural damage to buildings and flooding up to 4.6 m (15 ft) above sea level.

Many attempts have been made to tame hurricanes, though they have not proved successful. Various government agencies abandoned such efforts decades ago and now focus on tracking and forecasting severe storms. However, some scientists are still trying. One idea, proposed in 1976 by hurricane scientist William M. Gray of Colorado State University, in the United States, is to release black soot into the air by burning petroleum on ships near a hurricane. The colour black absorbs heat from the Sun that would then create updrafts to break up the hurricane's normal wind patterns. Researcher Hugh Willoughby of the International Hurricane Research Centre in Florida, USA, has come up with another idea – placing a large foil mirror in space to reflect sunlight to heat the ocean in precisely the correct spot to divert a hurricane. Scientists are taking such ideas seriously, because they believe that if they could weaken hurricane winds by even 10 per cent, it could prevent millions of pounds' worth of damage. Others argue that changing building and land use practices is a more effective and affordable approach to reducing vulnerability to hurricanes.

ABOVE RIGHT *Doppler radar image of Hurricane Isabel, which came ashore on September 18, 2003, in North Carolina, on the US east coast. The colours relate to rainfall, from blue and black (the lightest rain) through green and yellow to orange and red (the heaviest). The heaviest areas represent a downpour of over 50 mm (2 in) per hour. A Category 2 storm when it struck, Isabel took 50 lives and caused more than $2.7 billion in damage.* BELOW *A hurricane creates storm waves at sea.*

RALEIGH

COLUMBIA

FAYETTEVILLE

CAPE HATTERAS

WILMINGTON

MYRTLE BEACH

EXTREME WEATHER

How will extreme weather events change as the world's temperature continues to increase under greenhouse warming? For one, in a warmer world, more water will evaporate from Earth's surface and will be released from plants (transpired) into the atmosphere. Because the atmosphere is warmer, it can also hold more moisture. On the other hand, hot air rises over the warmer land and sea surfaces and the movement makes the atmosphere unstable. A moist and unstable atmosphere has the potential to deliver more precipitation in the form of heavy rain and snow storms. Consequently, as Earth continues to warm, heavy precipitation events are likely to become more frequent.

Beyond this simple scenario, computer climate models can tell us more about changes in extreme weather as they include other components of the climate system in their calculations. A number of techniques have been developed to improve the reliability of climate models at local scales and over short time periods. One involves the development of high resolution, or more data-intense, regional climate models. Another uses statistical methods to extend results from global climate models to regional scales. Much of what climate model studies predict about weather and climate extremes in the future is in accord with our understanding of how the climate system works. Indeed, several

LEFT *A tornado moves across the central plains of the United States. With wind speeds of up to 480 km per hour (300 miles/h), tornadoes can cause extensive damage. As Earth's climate changes, destructive storms such as tornadoes are likely to become more frequent.*

changes in weather and climate extremes predicted by climate models have already been observed in many different regions of the world.

Hotter, Colder, Wetter, Drier

Not surprisingly, a warmer average temperature is likely to increase the probability of days with extremely warm temperatures and decrease the probability of days that are extremely cold. Night-time temperatures are also expected to rise. The greatest increases in extreme temperature are expected in central and southeast North America, Central Asia and Southeast Asia and tropical and northern Africa.

An increase in the level of greenhouse gases in the atmosphere is also likely to bring heavier rain or snow storms almost everywhere, although changes should vary by region. Extreme precipitation could occur twice as often in North America. Asia's southwest monsoon is expected to become stronger, while in the South Asian monsoon region, precipitation is likely to decrease in the west and increase in the east.

Paradoxically, an increase in average global precipitation could also be accompanied by more frequent droughts in many parts of the world. In areas that receive more precipitation, evaporation from soil and transpiration from plants may remove a significant portion of the heavier rainfall. Warmer winters mean less snow, reducing a water source critical for dry summer months. In addition, as more of a region's rain comes from heavy storms, less precipitation arrives at other times, leading to more dry days. In southern Europe, the probability of a 30-day dry spell is expected to increase two to five times with a doubling of carbon dioxide in the atmosphere.

Monsoons

From ancient times, an important part of the climate of southern Asia has been the monsoon. This is a seasonal wind that blows from the south-west over the Indian Ocean from April to October, then reverses course, blowing from the north-east between November and March. The south-western monsoon, which is essentially a giant sea breeze delivering moist air from the ocean, brings heavy rains to southern and south-eastern Asia, including India, Bangladesh, Myanmar and Thailand. Farmers in that region have always depended on the regular rains of the monsoon for their harvests, and through the years have often seen crops or livestock devastated by monsoons whose strength or timing vary too greatly from the norm.

The monsoon's cause is the difference in the heating and cooling of air over land and ocean. In summer, land heats up much more quickly than the surface of the sea. The heated air over the land rises; replacing it is a southwesterly wind bearing moist, warm air from the Indian Ocean. As the air rises over the land, water vapour in the air condenses and forms clouds and rain. In winter, the land cools off much more quickly than the sea. The cool air over the land sinks and travels out to sea as a northeasterly, dry wind.

The monsoon of southern Asia is the best known of the world's monsoons, but the phenomenon also exists in the southwestern United States, eastern Asia, parts of Africa and northern Australia.

ABOVE *Two women huddle beneath woven straw umbrellas on a rainy monsoon-season day near Pokhara, Nepal. In a warmer, wetter world, the south-western monsoons in Asia are expected to become more intense.*

Wetter, Wilder

As the atmosphere becomes warmer and moister, condensation of the extra water vapour provides more energy to sustain powerful storms. Studies show that the number of strong winter storms is expected to increase in both hemispheres. On the other hand, the number of weaker storms is likely to drop as the high latitudes warm more than low latitudes and the smaller temperature difference between the latitudes reduces the amount of energy available to drive storms.

The warmest 10 years of the past 100 have all occurred since 1990, the same period during which greenhouse gas emissions reached a record-breaking high. Climate change of this degree influences ocean

South Asia and Southeast Asia, where there is a concentration of highly populated delta land.

Existing climate models are not yet sophisticated enough to make projections about extreme weather phenomena – thunderstorms, tornadoes, hail and lightning – that affect relatively small areas.

Extreme Weather, Extreme Costs

How the human and economic costs of extreme weather may change in the future is of concern not only to the insurance industry, but to all levels of government and society.

The costs of extreme weather events have risen rapidly in recent decades. For the most part, this rise can be explained by the new reality: there are now

As Earth continues to warm there will likely be stronger storms and more heat waves, floods and droughts. The cost of extreme weather will probably continue to escalate. A part of this cost will be carried by governments and the global insurance and financial industries. But a large part will also be carried by the poor who cannot afford to purchase insurance. In the 1990s, only 20 per cent of the damages caused by extreme weather events was covered by insurance. If extreme weather increases in frequency and intensity, as seems likely, and as the world's population continues to grow, governments will need more than ever to make good city planning and land management decisions to reduce vulnerability to extreme weather.

AS THE ATMOSPHERE BECOMES WARMER AND MOISTER, CONDENSATION OF THE EXTRA WATER VAPOUR PROVIDES MORE ENERGY TO SUSTAIN POWERFUL STORMS

conditions and is likely to increase the severity of hurricanes and cyclones. But even if hurricanes do not increase in frequency or intensity, the risk of storm surges and accompanying flood damage will increase along with rising sea levels. By the 2080s, because of the rise in sea levels and increases in coastal populations, it is estimated that the number of people directly affected by storm surge flooding in a typical year will be more than five times higher than at present. The areas most vulnerable to flooding include the southern Mediterranean, much of the African coastal lowlands and most particularly,

more people on the planet, people are generally wealthier and more people are living in vulnerable areas. The values of insured properties, especially in industrialized countries, have been rising hand in hand with extreme weather costs. Exposed areas such as coastal regions and river and lake floodplains have been increasingly developed commercially and have seen their population densities grow. As a result, if the same extreme weather events that took place in the 1940s and 1950s were to occur today, societal impacts and economic losses would be substantially greater.

Temperature Extremes

Increased heatwaves will bring with them more heat stroke, heat exhaustion and deaths. The very old, the very young and the frail will be most susceptible. Heatwaves have a much bigger health impact in cities both because of the heat island effect that causes an urban area to be hotter than the surrounding region, and higher levels of air pollution, which are worsened by heatwaves. One study in the United States estimates that by 2050 the annual excess summertime mortality attributable to climate change, will increase severalfold, to between 500 and 1,000 deaths for New York and 100 to 250 deaths for Detroit, Michigan.

In a strange calculus, some scientists debate whether increases in summer deaths will be balanced by decreases in winter deaths under climate change. One study in the United Kingdom estimates a decrease in annual cold-related deaths of

20,000, or 25 per cent, by the 2050s. However, another study in the United States estimates that increases in heat-related deaths will be greater than decreases in deaths related to cold by a factor of three. In general, it is likely that the net impact will vary among regions and populations.

The Cost of Storms

A warmer climate is likely to bring an increase in rainstorms and droughts. A US research team estimated that increased rainstorms may lead to a doubling in the country's corn production losses during the next thirty years. This would translate into additional damages totalling an estimated £1,700 million per year nationwide.

Heavy precipitation events bring a higher probability of floods. In Switzerland, studies indicate that frequent intense precipitation in autumn, winter and spring will become more likely in the future. In turn, this will lead to a higher probability of floods in the winter. The Swiss government estimated that additional regional floods would cause 80 250 million euros (£57–170 million) per year in damages.

Rising sea levels, compounded by increases in storm intensity, will raise the costs of storm damage. In Australia, studies estimate that, by 2050, sea levels may rise 10–30 cm (4 to 12 in) and tropical cyclone intensity around Cairns in

RIGHT *Buildings flattened by a violent storm in Jelonka, Poland. Insurance claims related to weather are on the rise, and worldwide economic losses from extreme weather are on the increase. Some large insurance companies could even be forced into bankruptcy.*

northern Queensland could increase by up to 20 per cent. This translates to a flooding area about twice that historically affected. In Denmark, research shows that an average rise in sea levels of 50 cm (20 in) will increase the frequency of severe storm surges on the west coast by five times. An even greater increase can be expected if northwesterly storms become more frequent in the future.

The Insurer's Perspective

The costs of extreme weather events have skyrocketed in recent decades. Worldwide, annual economic losses from extreme events have increased ten times, from £2,300 million per year in the 1950s to £23,000 million per year in the 1990s. The insured portion of these losses rose from a negligible level to £5,300 million annually during the same period. Weather-related claims have already forced some large insurance and reinsurance companies into bankruptcy. Some, such as Lloyd's of London,

have faced solvency crises at least in part because of losses from weather disasters. As this trend continues, insurers will have to raise premiums, limit their liability or even back out of existing markets that have become too risky, as many hurricane insurers have done in the Caribbean.

As many insurance companies now grapple with the challenges presented by climate change, some have gone a step further, seeing climate change as an opportunity to diversify business. For example, Swiss Re, the world's second largest reinsurance company, now provides insurance to buyers and sellers of credits of carbon dioxide. Also known as emissions trading, buying and selling of carbon credits is one solution devised to reduce overall emissions through a flexible and market-oriented mechanism. Companies that keep their emissions below permitted levels are awarded emission credits, which they can then sell to companies that fail to do so. This creates a financial reward for lowering emissions.

OUR FUTURE: BIOME BY BIOME

Earth's Shifting Mosaic

The world's landscapes are impressive in their diversity and scale. From the tundra to the temperate grasslands and the lush tropical forests, each ecological zone, or biome, is occupied by distinctive forms of life – oak trees and squirrels in temperate deciduous forests; cacti and snakes in deserts; palms and primates in tropical forests. Exactly how climate change will affect each individual region cannot be predicted. It is inevitable, however, that climate change will alter Earth's biomes.

BIOMES AND CLIMATE CHANGE

Our world can be divided into what scientists call biomes – that is, large regions distinguished from one another by their climate and dominant vegetation and animals. Systems of biome classification differ. Major land biomes, however, are usually categorized as alpine, tundra, evergreen coniferous forest (or taiga), temperate deciduous forest, temperate rainforest, temperate grassland, desert, chaparral, tropical rainforest and tropical savannah and woodland.

Although every biome has a distinctive climate and includes specific kinds of plant and animal communities, each biome may be found in many parts of the world. For example, the plains of North America, the steppes of Asia, the pampas of South America and the veldt of southern Africa are all grassland biomes. Grasses are always the dominant type of vegetation there, although the species and height of grasses may differ depending on the location. The dominant animals on grasslands are grazers, although these vary from continent to continent – bison in North America and zebras, antelopes and gazelles in Africa.

Rather than being fixed, the distribution of biomes across the globe is shifting. It is likely to shift even more as a result of climate change. As temperature and precipitation change, some grasslands may become desert and others forest. As climates become warmer, biomes characterized by cooler temperatures, such as coniferous forests, may expand into previously cooler biomes, such as

tundra regions. Biomes that are already at the coolest limits of the globe, such as tundra and alpine regions, will be forced to shrink. Plants and animals will shift from one location to another, tracking satisfactory habitat as much as they can. These changes may affect the human populations that depend on the ecosystems in their region for resources, raw materials and for something as basic as a sense of home.

Each biome, in short, will be affected by climate change in a different way. Some may expand in a warming climate, and those characterized by a cooler climate may disappear altogether. Similarly, the living organisms of each biome will feel the impact of climate change in different ways, with some species migrating, some becoming extinct, and still others thriving. From the frozen ground of the tundra to the equatorial rainforests, ecosystems will change and the livelihoods and settlements of the human beings who live there will be affected.

Temperature and Precipitation

It is impossible to predict the climate trajectory of any location. Although the average global trend is towards warming, some regions may become colder for periods each year. For example, thawing sea ice and heavier rainfall in the North Atlantic may spur the formation of cold fronts along the eastern seaboard of the United States, resulting in colder winters. Some regions may exhibit neither a strong warming nor a strong cooling trend. About 70 per cent of Antarctica is cooling; the rest is heating up. The East Antarctic ice sheet appears to be growing, while West Antarctica ice seems to be thinning.

Changes in precipitation are equally difficult to predict. In the United Kingdom, Mike Hulme of the Climatic Research Unit of the University of East Anglia found a relatively high likelihood that temperatures in southern Africa would rise between 1°C and 2°C (1.8°F and 3.6°F). Yet within this warming trend, he predicted three possible, very different outcomes: a 'dry' scenario, a 'wet' scenario and a 'core' scenario involving modest

RIGHT *Cumulonimbus clouds over the prairie in Saskatchewan. In this Canadian province and elsewhere, weather patterns are increasingly hard to predict as the climate changes continue.*
PREVIOUS PAGE *Clockwise from upper left: a creek runs through Oregon's Fort Clatsop National Memorial in a coastal coniferous forest biome; elephants lumber through the savannah grass in Kenya's Masai Mara National Reserve; Ama Dablam Mountain, Nepal, rises to 6,856 m (22,624 ft) and is typical of an alpine biome; a nomadic Tuareg drives his camels across the Sahara desert.*

drying over large parts of the region. The variability of these outcomes is part of what makes gauging climate change so challenging. In a warming world, weather patterns will probably be less predictable than they have been, not more so.

Models of Ecosystem Change

How ecosystems and the larger biomes in which they occur respond to the complexities of climate change is a knotty problem. Researchers use two main models to predict such responses: the ecosystem modification model and the ecosystem movement model.

In the ecosystem modification model, the assumption is that climate change will bring changes in species composition and dominance in a given location. In other words, some species will diminish in number or even become locally extinct, while others will increase in number. The evolution will be further affected by invasion from outside species and the ageing of existing populations. The ecosystems that result may be significantly different than those currently in existence.

Because of its complexity, the ecosystem modification model is not practical for forecasting. So most studies of possible climate change impacts on ecosystems have relied on the ecosystem movement model. In this model, the assumption is made that an ecosystem remains generally intact as it migrates to a new location that is similar to its current biome. This model allows

RIGHT *Meerkats thrive in South Africa's arid Kalahari Desert, but a change in regional climate could increase the threat to the animal from predators.*

scientists to project new ecosystem distributions under changed climate scenarios, using knowledge about the type of environment and climate that certain species favour today. However, this model runs into problems when applied to scenarios in real life because of differences in the ability of various species to migrate; their different life spans; their differing tolerance for climate shifts; and the nature of any invading species in the area. In the real world, ecosystems do not transfer neatly from one region to another.

Still, models of ecosystem change do help scientists assemble a picture of how the shifting climate will affect the world's different biomes and over time, a picture of each has begun to evolve.

Tundra

The tundra biome is large, covering about one-fifth of Earth's land, and is principally found above the Arctic Circle (66.66 degrees north). Its name comes from Lapland's Saami people, who call this treeless region home. Tundra is the coldest biome, lying above the highest latitudes at which trees thrive. Temperatures can drop below -30°C (-22°F) in winter and may exceed 25°C (77°F) in summer. Some areas at high altitude, such as the Himalayas, are tundra, as are parts

Extinctions on the Way

'Habitats at Risk', a 2002 report by the World Wildlife Fund (WWF), studied selected ecoregions – 'areas where the Earth's biological wealth is most distinctive and rich' – and found that more than 80 percent of them would suffer extinctions of plant and animal species as a result of global warming. The report also found that some of the most distinctive and diverse natural ecosystems might lose more than 70 per cent of the habitats on which their plant and animal species depend.

According to the report, changes in biomes from climate shifts will be more severe at high latitudes and altitudes than in lowland tropical regions. It estimated that many habitats will change about ten times faster than the rapid changes that took place after the most recent ice age.

Extinction rates would vary by region. If habitat loss in a region is restricted to 15 per cent, species loss may amount to two to three per cent of the local life forms. But proportionately more species will become extinct in regions where habitat loss is greater. British ecologist Chris Thomas and others reported in the scientific journal *Nature* that nearly one-quarter of species in the regions they studied would be doomed to extinction by 2050 under mid-range climate warming scenarios.

ABOVE *A three-week old green parrot, rescued during an endangered species control operation by the Costa Rican Ministry of Wildlife.*

of Antarctica. Tundra is the driest of biomes, annually receiving less than 25 cm (10 in) of rain. At the same time, swampy areas are common; moisture from rain and melted snow cannot penetrate the soil very deeply before being blocked by a layer of permafrost, ground that stays frozen all year.

Because of the harsh conditions and short growing season, only a few plants and animals can survive on tundra: lichens, mosses and dwarf birches and willows, along with caribou, polar bears, lemmings, Arctic hares, Arctic foxes and wolves.

Climate change is altering the tundra biome faster than most others. The polar regions, especially the Arctic, have seen their warmest temperatures in four centuries along with declining sea ice in spring and summer. Permafrost is melting across the Arctic as well as in higher reaches of mountainous regions such as Tibet. In many places in Asia's high latitudes, warming could cause permafrost to disappear.

As arctic permafrost thaws, undecayed organic matter that has been locked in the frozen soil starts to decompose and begins to release its carbon stores into the atmosphere in the form of methane. As a greenhouse gas, methane is 20 times more powerful than carbon dioxide and the melting of vast areas of permafrost will significantly accelerate the current warming trend. The melting of permafrost also has local consequences. Flooding will increase as coastal lands subside, and the ground will generally become more unstable, affecting any large object sitting or growing atop it. Houses and other structures may crumple as their foundations give way; arctic industries such as construction, mining and energy drilling will be affected.

When temperatures rise, the southern boundary of the arctic tundra will move toward the North Pole as boreal forests encroach from the south. Animal migrations are likely to change. Thin winter ice and rapid spring melts have already altered migration patterns and led to declines in some populations of the Peary caribou, a reindeer subspecies that lives on Greenland's tundra.

The northwards shift will not be uniform. Regional variations so far indicate that parts of Canada and Russia are likely to be relatively vulnerable to change, while Finland and the Scandinavian peninsula and western Alaska will be more stable.

High-altitude or alpine, tundra is likely to diminish increasingly and, in some cases, such as in the continental United States excluding Alaska, such areas may disappear altogether. Species that depend on tundra habitat have few options for relocation. Tundra species that will be affected include the polar bear, the dunlin (a waterbird), the white-tailed ptarmigan and the mountain marmot.

Evergreen Coniferous Forest

Also known as boreal forests or taiga, evergreen coniferous forests cover a large swathe of Alaska, Canada, northern Europe and northern Asia. This biome has long, cold winters and short summers – a growing season too short to support deciduous trees, which lose their leaves in autumn and must regrow them in the spring. Coniferous trees, or evergreens, which also abound in temperate regions of the Northern and Southern hemispheres and at high altitudes, have the advantage of preserving their leaves through the winter, with the result that they can begin photosynthesis as soon as seasonal temperatures are sufficiently warm.

There are several different types of evergreen coniferous forest. Furthest north, just south of the tundra, is the open taiga, with only sparse forests.

LEFT *Caribou migrate in Alaska's Arctic National Wildlife Refuge. In the northern regions of the world, shifting climate is changing the tundra faster than any other biome, as warming causes the melting of permafrost.*

TROPICAL RAINFORESTS

With high temperatures and heavy rainfall, tropical forests have more abundant life than any other biome. Tropical rainforest – a specific type of tropical forest – occupies only 6 to 7 per cent of the planet's surface but sustains more than half of the world's plant and animal species. Today the tropical forest biome is being transformed in response to several factors, including deforestation and climate change.

Warm and wet all year, the tropical rainforest biome hugs the equator. The areas in this biome receive more than 10 cm (4 in) of rain per month, sometimes more than 50 cm (20 in). Temperatures are relatively constant year-round and high – about 26.5°C (78°F). Tropical rainforests have a broad geographical range and are found in Mexico, Central America, the Amazon basin of South America, central Africa, western India, Sri Lanka, Myanmar, Thailand, Malaya, Indonesia, the Philippines and northern Australia. Tropical rainforests house a greater variety of species than any other biome on Earth and are lush with varieties of broadleaf evergreen trees, palms, tree ferns and climbing vines. The vast array of fauna includes bats, birds, lizards, monkeys, snakes and around 50 million species of invertebrates. A single tree may be home to 50 different species of ants alone. Other kinds of tropical forests include cloud forests, found at high altitudes in the tropics (see page 152–153) and tropical deciduous or seasonal, forests, found in areas where the dry season is long enough to cause some plants to lose their leaves. Tropical forests that are not rainforests can be found in eastern Mexico, Central America, northern South America and parts of Asia and Africa.

Climate change is expected to bring only modest temperature alterations in the tropics, but every degree of temperature change is likely to affect tropical forest species more drastically than would be the case in a temperate forest. Even small changes in temperature, rainfall, and evaporation and transpiration (release of moisture by plants) may lead to major changes in the tropical forest.

When coupled with increased precipitation, rising temperatures can make rainforests grow even faster. In Australia's Top End in the Northern Territory, researchers have found isolated rainforest patches that have doubled in size in the past 50 years as rainfall has increased. But in other cases, climate change may

ABOVE *A squirrel monkey foraging in the rainforest canopy, Costa Rica.*
UPPER LEFT *The Carara Biological Reserve rainforest in Costa Rica.* BELOW *A tree frog of the South American rainforest. Tree frogs are threatened by rainforest destruction.*

harm rainforests such as those found in South America's Amazon basin, which is expected to become much drier. In Africa, the effect of climate change may vary greatly by region. In the tropical rainforest of the Congo Basin, growth rates increased from 1982 to 1999, while the coastal tropical forests of western Africa grew at a slower pace. Human factors can cloud the picture. In central Africa, as in many parts of the world, the human population is stressing the tropical forests. Water is being diverted, huge

swathes of forest are being cut or burned and biodiversity is consequently on the wane. Climate change may inhibit the ability of such ecosystems to withstand or recover from these stresses. The United Nations Intergovernmental Panel on Climate Change (IPCC) predicts that if mean ambient temperatures increase beyond 1°C (1.8°F), significant changes will occur in land cover and the composition and distribution of species.

The impact of deforestation may be even worse in the rainforests of South America, where large areas of the Amazon Basin, site of the world's largest tropical rainforest, are expected to become drier not only because of higher temperatures that lead to increased evaporation, but also because the forests are being cut down and turned into pastureland. Forests are able to hold much more moisture than pastureland, so deforestation decreases humidity, which in turn reduces precipitation. Pastureland not only recycles much less water into the atmosphere, but also increases water loss through runoff. Production of cereal crops in these regions is likely to suffer as temperatures increase and rainfall diminishes.

In rainforest regions such as the Amazon, where precipitation declines and temperatures rise, drought and fire are likely to increase. Already, large areas of rainforest worldwide have been ravaged by fire. In Nicaragua in 1998, more

ABOVE *A Brazilian rainforest is burned to clear land for cattle ranching. The loss of such forests leads to decreased levels of precipitation and higher levels of atmospheric carbon, perpetuating the cycle of global warming.*

Extent of rainforest in 1980

Extent of rainforest today

Caribbean Sea

VENEZUELA

GUYANA

SURINAM

FRENCH GUIANA

COLOMBIA

Orinoco R.

Negro R.

Amazon R.

ECUADOR

PERU

Amazonia

Madeira R.

Tapajós R.

Xingu R.

Juruá R.

Ucayali R.

Andes

B R A Z I L

São Francisco R.

Mato Grosso

BOLIVIA

Paraná R.

PARAGUAY

LEFT *The magnitude of the deforestation in South America is shown on this map, which shows the difference between the extent of the rainforests in 1980 and their extent at present, after the heavy clear-cutting that took place during the 1980s and 1990s.*

URUGUAY

C H I L E

Andes

ARGENTINA

Patagonia

than 15,000 fires burned 890,308 hectares (2.2 million acres), including protected lands in the Bosawas Biosphere Reserve, which consists mainly of dense tropical rainforest. In Indonesia that same year, fires burned up to 809,371 hectares (2 million acres), including parts of the already greatly reduced forest habitat of the Kalimantan orangutan. In Nepal, forest fires may bring local extinction to rare species such as red pandas and leopards. In addition, as rainforests burn, they release large amounts of carbon dioxide, adding to the greenhouse gases already in the atmosphere and further accelerating climate change. Large parts of the present-day tropical rainforest may become replaced by more drought-tolerant forms of scrubby, open vegetation.

The taiga includes trees such as the black spruce and white spruce. To the south of that is boreal or mixed coniferous forest, which includes spruce, fir, pine and larch. Montane coniferous forests and coastal coniferous forests are marked by a milder, wetter climate than boreal forests. The animals in evergreen coniferous forests include moose, caribou, elk, wolves, bears, wolverines, badgers and otters.

Climate change is expected to have a major impact on evergreen coniferous forests. A study of Canadian national parks found that warming climate would lower the proportion of tundra and taiga/tundra areas in the system, while increasing the number of more southerly biomes – in some scenarios, doubling to quadrupling them. Throughout the Northern Hemisphere, there would be a poleward shift of belts of boreal as well as temperate mixed forests.

Trees cannot simply get up and move; their dispersal rates depend on many factors. Every year, spruce have historically shifted between 73.2 m and 457.2 m (240 and 1,500 ft) in response to climate variations and fir between 18.3 m and 270.43 m (60 and 900 ft). But climate models indicate that annually the evergreen coniferous forest range may have to shift at least 4,572 m (15,000 ft)

to keep up with climate conditions – more than ten times the historic migration rates of most evergreen coniferous species. Since climate is expected to change faster than some trees can spread to new ranges, the species composition of existing forest types is likely to change. In some regions, whole forest types may disappear and new communities of species arise.

Temperate Deciduous Forest

There are four distinct seasons and a moist climate in the temperate deciduous forest, with an average yearly rainfall that exceeds 80 cm (32 in). Winters are cold and summers warm, and the annual temperature averages about 10°C (50°F). The temperate deciduous forest biome is found in parts of North and South America, Europe, Asia and Australia. Broadleaf trees such as oak, maple and elm are typical. Animals include deer, raccoons, squirrels and many small birds.

The effects of a warming climate on temperate deciduous forests may vary widely, with some species declining and others flourishing. Deciduous forest covers more than half of the mid-Atlantic section of the United States – the eastern seaboard

from Virginia to New York. In this area, according to the Mid-Atlantic Regional Climate Change Assessment, part of the US National Assessment of the Potential Consequences of Climate Variability and Change, the mix of tree species can be expected to shift in a major way. The dominant maple-beech-birch forests will probably decline, giving way to oak-hickory forests and, to a lesser degree, southern pine and mixed oak-pine forests. The shift in forest types may reduce biodiversity and permit the spread of invasive species, affecting the local agriculture and timber industries.

Among animals as well as trees, diversity is expected to decrease. Some bird species might become more abundant, but important predators of insects such as the wood warbler and other perching species will probably decline, leading to unwanted changes in insect populations.

In temperate regions, including grasslands as well as forests, climate change is likely to bring milder, wetter, stormier winters and longer, drier, warmer summers to many areas. The climate will tend toward the Mediterranean (that of the chaparral biome). London's average August high temperature – presently 22°C (72°F) – will probably rise. By 2030, it may be similar to that of today's southern France at 27°C (80°F); by 2100, it may resemble that of present-day Athens at 31°C (88°F).

Growing warmth in the temperate deciduous forest will increase the likelihood of forest fires. Trees

LEFT *A boreal forest on Lake Superior, Ontario. In this area of Canada and around the world, animal species that depend on evergreen coniferous forests will become increasingly endangered as climate changes faster than trees can adjust their range.*

ABOVE *A red panda, or lesser panda, which is an endangered species living in a temperate deciduous forest in the Wolong Nature Reserve, China. Diversity among animal species will likely shrink in these biomes, as climate changes to a more Mediterranean pattern.*

weakened by droughts and insects are more likely to burn. Fires have already drastically reduced the amount of Asian ginseng in the northeast Asian deciduous forest biome. Yet in some coastal and central areas of China, temperate deciduous forests should expand north with a warmer, wetter climate. Still, conditions will vary, depending on how much the regional temperature changes and whether the local climate becomes wetter, drier or a mixture of both.

Temperate Rainforest

This biome can be found in coastal areas of western North and South America and parts of Asia, Europe, Tasmania and New Zealand. Every year, it rains between 152 and 500 cm (60 and 200 in) in these forests; a constant supply of moisture is ensured by their proximity to the ocean and to the mountains just inland, which halt the clouds blowing from sea to land

and thus keep rainfall plentiful on ocean-facing slopes. Fog also brings moisture, the annual equivalent of up to 2.5 cm (1 in) of rain.

Temperatures in temperate rain forests vary moderately with the season. Winters hover near freezing and are usually long and wet; summers are shorter but pleasant – drier, around 27°C (80°F). The world's tallest trees, which reach heights of 39 to 84 m (130 to 280 ft) and can live for 500 to 1,000 years, are found in temperate rainforests. Soil is rich and vegetation lush; with 250 tonnes of life per hectare, the biomass of this biome is more dense than any other. Redwood, giant sequoia, Sitka spruce, western hemlock and Douglas fir grow here; voles, squirrels, deer, elk and bears call these rainforests home.

moisture. The biome would narrow if climate dries. In addition, harvesting old-growth timber renders forests more vulnerable to drought, which comes with climate change; already, between 75 and 95 per cent of the old-growth forests in the US Pacific Northwest have been logged.

Different tree species in these forests will respond to climate change in different ways. For those that depend on winter chilling to produce seeds and confer frost-hardiness, warmer winters may reduce reproduction and survival. In some species, because of their long life, individual trees need to reproduce only once. But warming may make successful reproduction more elusive. The fate of the world's temperate rainforests is important in part because of the number of distinctive species inhabiting this biome. For instance, the temperate rainforest of Tasmania is home to some of Australia's most ancient flora, some dating back more than 60 million years, as well as some of the

HARVESTING OLD-GROWTH TIMBER RENDERS FORESTS MORE VULNERABLE TO DROUGHT. ALREADY, BETWEEN 75 AND 95 PER CENT OF THE OLD-GROWTH FORESTS IN THE US PACIFIC NORTHWEST HAVE BEEN LOGGED

Climate models disagree on whether temperate rain forest biomes will shrink or grow. A warming trend would let them stretch toward the poles, but expansion would be limited away from coasts, and thus away from the oceans where they receive their

world's most primitive invertebrates, such as the large land snail and the freshwater crayfish. Other unusual animals in this part of Tasmania include the pademelon, a hare-sized wallaby, and the spotted-tailed quoll, a small marsupial.

Temperate Grassland

Generally, arid and semiarid regions where annual precipitation runs between 25 and 75 cm (10 and 30 in) are the sites of temperate grassland. As the amount of rainfall increases, forests grow; with less precipitation, deserts take over.

Some grasslands are steppes, semiarid areas with short grasses; others are prairies, which have tall grasses and are more humid than the steppes. Besides grasses, other herbaceous plants are common; these include small, soft-stemmed plants such as alfalfa.

Temperate grasslands are essential habitats for migratory birds and also nurture antelope, pronghorn, bison, wolves and coyotes. The North American prairie pothole region, which extends northwest from northern Iowa in the United States into east-central Alberta in Canada, is the nursery for 50 to 80 per cent of the continent's waterfowl. Widespread conversion of grassland to farms may make it more difficult for grassland species to migrate and adapt. Bird species that depend on intact native prairie, such as Sprague's pipit and McCowan's longspur, may become extinct.

Grasslands that have been turned into pastures and fields form the basis for much of the world's agriculture. Vast steppes have been converted to agricultural uses in North America's Great Plains, Argentina's western pampas and parts of southern Africa's veldt. Productive former prairie lands are found in the midwestern United States, Argentina's eastern pampas, and parts of Hungary and northeastern China.

ABOVE *Agricultural fields near China's Inner Mongolian plains. Worldwide, many grasslands have been converted into fields; these have proved highly productive for crops and stimulate a country's agricultural economy. Balancing the benefits of agriculture are the problems which the loss of grassland causes for migrating bird species.*

In the short term, rising carbon dioxide levels may benefit agriculture in the grassland biome by producing faster plant growth and higher yields per unit of water required. Some regions may become too hot, but up to a point greater warmth may spur agricultural productivity. In subtropical Africa, for example, warmer winters would make it possible to grow crops at higher elevations than is now possible. However, agriculture and development generally damage natural grasslands, so that an increase in agriculture is not likely to benefit native species and ecosystems.

The fate of individual grasslands will depend on how much moisture they receive, and the seasonality of that moisture. In areas where grassland gets dry enough, deserts may spread. South of China's Gobi desert, for instance, grasslands are being replaced by desert as the climate warms and dries. Where precipitation increases, or becomes less seasonal, forests may replace grasslands. Conversely, higher temperatures may shift some forests toward grasslands. Land use practices interact with climate change. Fire suppression favours forests over grassland, and overgrazing favours desertification.

FROM THE CHAPARRAL TO THE TROPICS

Near the equator lie other biomes: chaparral, desert, tropical savannah and woodlands, and tropical forest.

Chaparral is a biome of tough, crooked shrubs and small trees, with mild, moist winters and hot, dry summers. Mediterranean climates are typical of the chaparral biome, which is also known as the Mediterranean forest, wood and scrub biome. Chaparral conditions also occur in California, southern Australia, Mexico, Chile and South Africa, but never on the east coast of continents.

Fires are common in chaparral because of the long summer droughts and because the oils in many local shrubs – among them scrub oak and manzanita – are highly flammable. The typically small, leathery leaves help the shrubs retain water during dry seasons. Coyotes, mules, deer and lizards inhabit the chaparral. Winter rains enable farmers to plant crops here.

By making the weather hotter and drier, climate change will make chaparral fires more common and farming more difficult. Many species will not survive. Along with the grassland biome, the chaparral biome is expected to experience the greatest proportional change in biodiversity by the year 2100. Plants in the Succulent Karoo – a South African ecosystem with a Mediterranean climate and over 6,000 plants, 40 per cent of them found nowhere else on Earth – are already showing signs of stress. Leaves are shrivelling, flowering is failing and, in some areas, plants such as proteas and Cape reeds are suffering local extinction. Climate model forecasts for the next 50 years predict that the region's winter rainfall may diminish by a quarter and temperatures rise by 2°C (3.6°F).

Climate change may convert areas of chaparral into desert. If more heat builds in equatorial regions, the upper-level winds that flow north and south from the equator are likely to strengthen and move farther before they slow and descend. The area where they now descend in the Northern Hemisphere is at about 30 degrees latitude, the latitude of many of the world's great deserts, which are formed by these hot, dry winds. If the winds travel farther, they could descend at 35 degrees north, the latitude of San Francisco and Europe's Mediterranean region. Much of this part of the chaparral biome is densely populated and desertification would have a significant impact on local populations.

Desert

Most of the world's major deserts are found in a belt in the subtropical zone north and south of the equator. As in Africa's Sahara and Kalahari deserts, heat often lasts year-round, but not always: in winter in the Gobi Desert in China and the Great Basin in the United States, the monthly mean temperature can fall below 6°C (43°F).

Aridity is universal, however, and the mean annual precipitation hovers under 25.5 cm (10.2 in). Nonetheless, deserts sustain abundant life, and many plant and animal species live only in deserts. There are cacti, sparse grasses and small-leaved shrubs. Lizards, snakes and small rodents live here as well.

RIGHT *A king protea in Cape Province, South Africa, its native habitat. The area is typical of the chaparral biome. Many plants in this region already show signs of stress from increasing heat and aridity.*

Aquatic Biomes

The world's oceans and bodies of freshwater have biomes of their own, but their relation to climate change is more complex than that of terrestrial biomes. Aquatic biomes are defined by the characteristics and depth of the water, the underlying sea floor or lakebed and the major plants and animals. The ocean's open waters are the pelagic zone, occupied by swimmers and drifters, such as whales, fish and jellyfish. Beneath the pelagic zone is the benthic zone, on or near the seabed. In the shallower waters of the continental shelf, occupants include fungi, seaweeds, worms, arthropods and fish. Most of the ocean bottom is made up of the abyssal plains, dark regions with no plant life. Between the sea and land lies the intertidal zone, subdivided into numerous habitats such as the rocky intertidal, mud flats, salt marshes and estuaries. The freshwater biomes include lakes, ponds, rivers, streams and wetlands.

Because climate is a characteristic of the atmosphere, not of water, aquatic biomes do not experience climate change directly. But large bodies of water are heat sinks, so changing temperatures in the atmosphere will affect temperatures in the water – which in turn will affect the organisms that live there. Climate change can also affect ocean circulation and nutrient dispersal, thereby increasing or diminishing biological productivity. Even the rising carbon dioxide levels that are responsible for climate change are taking a toll: as the carbon dioxide levels of the water in the ocean increase, the rate at which corals, clams and other animals can secrete calcium carbonate drops and their growth slows down.

ABOVE *A coral reef in New Caledonia, in the South Pacific. Soft corals, such as the sea fans pictured top left, are suffering from the effects of bleaching, a well-known phenomenon in hard corals that is associated with elevated water temperatures. Rising water temperatures have also been linked to outbreaks of a fungal disease that is destroying soft coral populations around the world.*

All these organisms are often exquisitely adapted to their rugged environment. Cacti, for example, store water in fleshy stems and protect themselves with spines against animals that would consume them. There may be a limit to how much heat and aridity these plants and animals can tolerate, however. For instance, Mexico's cerios, bizarre trees that are known also as boojums, are very susceptible to drought as seedlings and under current climate conditions reproduce successfully only every 100 years. If conditions become much drier in their already limited range, the species may perish.

Similarly, drought could endanger the desert animal species that live around oases. These water sources are often far apart, separated by tracts of desert. If lack of precipitation kills off the local vegetation on which they subsist, the animals cannot necessarily move to other oases.

In some regions, climate change could bring increasing moisture and this desert could be transformed into arid grassland. This may happen to much of North America's Sonoran Desert.

Around the edges of the world's deserts, poor land management has often degraded the soil of settlements in fertile areas. This land misuse, combined with rising temperatures, can desertify arable land in a matter of years, harming not only agriculture and herding, but also native plant and animal communities in this delicate transitional zone.

Climate change will not only influence deserts but will also affect marginal dry lands. In most cases, the dry lands will become even drier, increasing the trend toward desertification. In Asia, large areas of semiarid dry lands are expected to become desert because of climate change.

Tropical Savannah and Woodland

Found in the tropics, the savannah biome consists of grassland with widely scattered trees and shrubs. It has both a rainy season (summer) and dry season (winter). The dry season lasts five or six months, and during this season brush fires are common. The savannah separates the belt of deserts in Earth's middle latitudes from the tropical forests near the equator. Savannah grasses grow in clumps, punctuated by clusters of trees, such as acacias and baobabs. Giraffes, zebras, lions, hyenas and jackals live in the African savannah, which covers more than 40 per cent of the continent (specifically its central, southern and eastern regions).

Africa has the most developed savannah, but this region of the world is in jeopardy if mean temperatures climb more than 1°C (1.8°F). Many areas of savannah are threatened by land degradation and desertification and these are likely to worsen as the climate becomes hotter and drier. Rising temperatures are likely to increase the stress on hoofed species, including the Cape mountain zebra and grysbok, which roam the savannahs of South Africa. Both the distribution and diversity of these species are expected to fall in South Africa and in the arid highlands of Botswana, Namibia and Angola.

Some climate models predict that precipitation will drop even more in the Sahel and in southern Africa, two regions already stressed by lack of water. Precipitation is also expected to become more variable, making it even harder for farmers, who rely on the timing and wetness of the rainy season, to grow crops successfully. According to some climate models, rainfall will increase in the

ABOVE *This Masai, or Kenyan giraffe, feeds on the leaves of an acacia tree in a Kenya savannah area. Today, more than 40 per cent of Africa is savannah, but warming threatens to transform this biome into desert.*

highlands of east Africa and equatorial central Africa, which may make some marginal lands more productive than they are now. But overall, the trend is likely to be less favourable for most regions.

In India, where there are both dry and moist savannahs, it is estimated that climate change will shift about 60 per cent of dry savannah into xeric, or extremely dry, woodland; another 24 per cent likely to become xeric shrubland. Moist savannahs are more likely to change into tropical seasonal forests. In all these cases, increased carbon dioxide is expected to encourage the growth of trees, converting grassland into woodlands and woodlands into forests.

LEFT *The high veldt in Zululand, South Africa. If global warming continues, this region is likely to become hotter and drier. Worldwide, many animal species could be threatened, particularly in the savannah areas of South Africa and its neighbours.*

ABOVE *If climate change disrupts mountain habitats, serious erosion – such as that seen on this Nepal hillside – could ensue if rains should fall before new plant species have taken hold.*

Alpine

Found around the world, the alpine biome evidences a complex pattern of climate and vegetation types arranged according to altitude. At their bases, mountains share the climate of the surrounding region. But that climate changes with increasing altitude, just as it would with increasing latitude.

In the tropics, for example, a mountain's base will be tropical, with a hot climate and usually evergreen vegetation. Higher levels of the mountain can include both warm temperate zones and cool temperate zones. The next level up is the alpine zone, above the treeline, where tundra vegetation grows. Finally, just below the permanent snow line, a nival zone, similar to the Arctic biome, can be found. The nival zone includes lichens and a few scattered flowering plants.

The animal population varies by mountain zone, although mammals are generally sure-footed ones such as the Nubian ibex, a wild goat that can cross almost sheer rock faces in the mountains of Arabia and North Africa.

Species that live in mountainous areas are in special danger of losing their habitat as a result of climate change. Human settlements and agriculture have already stressed many mountain and highland ecosystems. In Asia, for example, vegetation cover and water supply are both being degraded in many mountain regions. Climate change is expected to further disrupt mountain ecology. With warming climate, mountain-dwelling plants and animals will seek higher elevations, where temperatures will be more similar to those once found at lower elevations. But species that live near mountaintops may have nowhere higher to go. Mountain habitats are also highly variable, with many differences in steepness and exposure. And many Asian mountain ranges are oriented east–west, so that species have little opportunity to move northwards in their search for a cooler habitat.

In the Himalayas, increased erosion is expected to limit the colonization of higher elevations by plants and animals. Weedy plants, which can tolerate a wide range of conditions, will be likely to have an edge over other species. In Japan, the area of habitat suitable for higher-elevation trees has diminished during the past three decades and the variety of alpine plants has decreased along with it.

The Altered Mosaic

From the poles to the equator, climate change is now altering the grand mosaic of Earth's biomes. Although it remains to be seen how the mosaic will look by 2050 or 2100, some trends are clear.

The tundra will shrink as permafrost melts and evergreen coniferous forests encroach on the region from the south. Even as the spruce, fir and pine trees of those forests expand northwards, their southern range will contract under pressure from the oak, maple and elm trees of the temperate deciduous forests. Those deciduous forests, like the temperate rainforests and temperate grasslands, will undergo a variety of changes, depending on their precise conditions of moisture, temperature, geography and ecology. But in many cases, these temperate regions will become more like the chaparral biome, with its moist winters and dry summers.

Some areas that are now chaparral will become desert. At the same time, some areas that are now desert will expand, while others will shrink, becoming arid grassland. The fate of the tropical savannah and tropical forest will vary depending on whether precipitation rises or falls in a particular region. In the alpine biome, species will shift to higher ground to escape rising temperatures. That shift will mirror the many transformations taking place around the globe in the face of climate change.

RIGHT *Each biome is found in several countries, wherever the climate supports certain life forms. Biome distribution is expected to change with the warming climate. Tropical biomes will move up into temperate regions, while the temperate biomes move closer to the poles. As rising temperatures accelerate desertification, deserts will expand, absorbing the biomes around them. Across the planet, the pattern of life will be altered.*

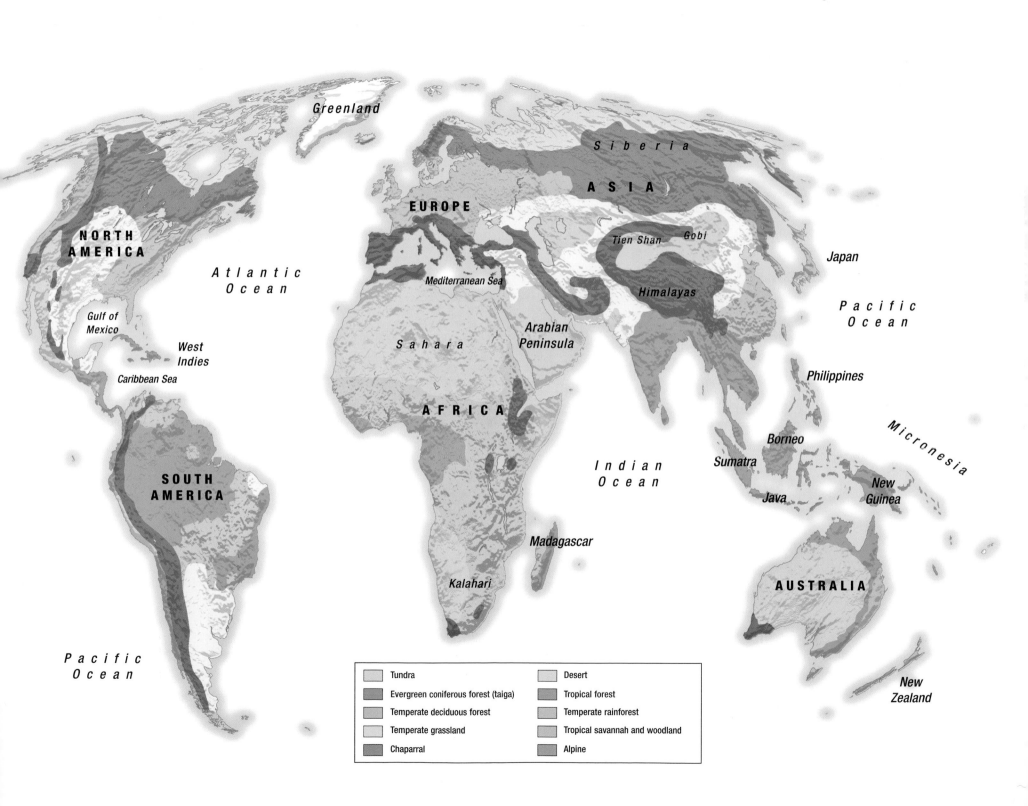

Greenland

Siberia

ASIA

EUROPE

Tien Shan   *Gobi*

Japan

NORTH
AMERICA

*Atlantic
Ocean*

Mediterranean Sea

Himalayas

*Pacific
Ocean*

Gulf of
Mexico

*Arabian
Peninsula*

West
Indies

Sahara

Philippines

Caribbean Sea

AFRICA

Micronesia

Borneo

*Indian
Ocean*

Sumatra

SOUTH
AMERICA

Java

New
Guinea

Madagascar

Kalahari

AUSTRALIA

*Pacific
Ocean*

New
Zealand

	Tundra		Desert
	Evergreen coniferous forest (taiga)		Tropical forest
	Temperate deciduous forest		Temperate rainforest
	Temperate grassland		Tropical savannah and woodland
	Chaparral		Alpine

LIMITING CLIMATE CHANGE

An Eye on Emissions

Most greenhouse gases generated by human society – that is, the so-called anthropogenic greenhouse gases – come from the production and use of energy. These gases enter the Earth's atmosphere in the form of emissions from motor vehicles and other means of transportation, from buildings and homes and from industrial activities such as the generation of electricity. By reducing such emissions, all of us – individuals, businesses and our governments – have the power to impose limits on climate change.

REDUCING EMISSIONS

Climate change is real and it is happening. Human society has a role in causing it and, regardless of debates over responsibility or blame, will also be affected by it. Fortunately, by changing the way they use energy, societies and individuals can play a part in limiting climate change. The key is to reduce the emission of greenhouse gases that contribute to climate change. That means using less energy and relying increasingly on sources of energy other than fossil fuels.

Some national governments and private enterprises argue that actions needed to limit climate change and reduce greenhouse gas emissions are too expensive. Other countries have already managed to reduce greenhouse gas emissions while growing economically. However, still more actions can be taken to limit climate change at the individual, community, regional, national and international levels. Some of these actions require no investment and no new technology. Others make use of energy-efficient and renewable energy technologies that are available commercially. It is not necessary to wait for technological advances or international agreements before taking steps to limit climate change.

People can learn to travel farther using less fuel and can find out more about how to lower energy costs at home and at work. They can also begin to use energy sources other than fossil fuels. As the growing market for renewable energy reveals, that is already beginning to happen.

The Kyoto Protocol

The Kyoto Protocol to the United Nations Framework Convention on Climate Change (UNFCC) commits developed countries to limiting their collective emissions of six key greenhouse gases to at least five per cent below 1990 levels by the period 2008 to 2012. Adopted in 1997, the Kyoto Protocol entered into force on February 16, 2005, after it had been ratified by at least 55 of the participating nations, including enough signatories from a group of specified industrialized countries to account for 55 per cent of the total emissions given off by that group in the year 1990.

The Protocol sets a group target designed to be achieved via a series of national emission reduction targets – including 8 per cent for the European Union (EU) and 6 per cent for Canada. Emissions can be traded internationally: nations that have cut emissions more than required by their targets can earn 'pollution credits' that can be sold to countries that have not.

Under the principle of 'common but differentiated responsibilities', developed countries are to take the lead in attacking the problem of climate change because, historically, they have contributed most to the problem and have more resources for dealing with it.

Within the Kyoto emissions-reduction target for the European Union as a group, the governments of Germany and the United Kingdom have made commitments to national targets of 21 per cent and 12.5 per cent, respectively. Between 1990 and 2002, they achieved emissions reductions close to or exceeding their targets. In Germany, this can be credited to efforts to increase the efficiency of power and heating plants and the restructuring of East Germany's ageing industries after reunification. Substantial reductions of emissions have been achieved in the United Kingdom by using natural gas instead of oil and coal to produce electricity. However, UK emissions of greenhouse gases rose between 2003 and 2004, raising fears that it might miss its Kyoto target.

RIGHT *Bicyclists ride through the streets of Shanghai. In China, the bicycle has long been a favoured form of transportation and emits no greenhouse gases.* PREVIOUS PAGE *Clockwise from upper left: the Sun, a key source of clean energy, shown setting behind power lines; north central Iceland's Godafoss Falls, a potential source of hydropower, which is efficient as an energy source but which can alter local ecosystems; a field of corn, convertible into ethanol, a greener alternative to gasoline; windmills, used for centuries to harness the power of the wind.*

ABOVE *Passengers board a bus in the city of Bangalore, India. Although not as fuel-efficient per vehicle as automobiles, buses can be an energy-efficient form of transportation; the more passengers there are for each trip, the lower the total carbon dioxide emissions per person.*

Differing Climate Policies

The Kyoto Protocol has not had runaway success. Among the countries that signed but did not ratify it are the United States, which is by far the world's biggest emitter of the greenhouse gases that lead to global warming, and Australia, which has the highest emissions per capita in the world. The governments of these two nations rejected the Kyoto Protocol on the grounds that it would be unacceptably expensive to carry out.

Although not legally bound to meet its emission targets, the Australian government has announced its intention to meet its Kyoto target of limiting total emissions to no more than 8 per cent above 1990 levels and has introduced voluntary initiatives, including grants and other incentives, to induce industry and households to become more energy-efficient. Legislation has also been adopted to force power generators to produce a small percentage of the electricity they use from sources that are renewable.

In the United States, the federal government has voted down a national climate policy, although some state governments have undertaken their own climate-related initiatives. In the western United States, 18 states are working together on strategies to increase renewable energy sources and efficiency in their electricity systems. In 2001, the New England governors, joined by premiers in eastern Canadian provinces, prepared a climate action plan that set emissions reduction goals. Canada's provinces are also working on their own. In 2000, the Canadian province of Manitoba established the Climate Change Action Fund to support community-based solutions to greenhouse gas emissions.

reductions to meet its national targets. Outside the framework of international agreements, some developing countries are adopting policies that would help limit climate change, usually in reaction to other environmental and political needs. For example, in the year 2004, as consumer car sales increased, China introduced its first fuel efficiency standards for cars.

Guiding Principles

Fuel efficiency, or getting more work from each unit of fuel consumed, is an important key to reducing emissions. In transportation, the most obvious means of improving fuel efficiency and lowering fuel costs per mile travelled is to maximize the number of people travelling in each vehicle at the same time.

Public transport does exactly that. Trains achieve this on a grand scale by moving hundreds of people over hundreds of kilometres simultaneously. Public buses shuttling passengers inside cities emit even less carbon dioxide per passenger kilometre travelled. Car-pooling, or ride sharing, lowers fuel costs and the amount of emissions for each passenger in private vehicles.

THE UNITED STATES, THE WORLD'S BIGGEST EMITTER OF GREENHOUSE GASES, SIGNED THE KYOTO PROTOCOL BUT DID NOT RATIFY IT.

Few developing countries have national climate policies. Under the Kyoto Protocol, some developing countries, such as Brazil and Honduras, are undertaking projects to lower their emissions. These projects are funded by developed countries and allow the investing country to use the emissions

Travelling without burning any fossil fuels is an even more basic way to lower fuel costs and reduce emissions. Bicycling and walking are as inexpensive and fuel-wise as they are healthy. Sidewalks and bicycle paths help build communities of people who use no fuel at all to get around town.

LEFT *France's high-speed TGV train near the city of Aix-en-Provence. While high-speed trains are responsible for higher emissions than low-speed ones, they are an efficient and popular form of public transportation.*

Traffic congestion and the consequent waste of fuel are becoming increasingly worse as cities continue to sprawl. Statistics show that the average American urban commuter was stuck in traffic for 46 hours in 2002. In central London, the speed of traffic seldom averages more than 16 km per hour (10 miles/h) throughout the day and about half the time commuters spend in traffic is passed without moving. Slow and stationary traffic emits a large amount of carbon dioxide for the short distance traveled. Traffic congestion also increases noise and reduces air quality.

Trains, Planes, Automobiles

Public transportation is an effective means of limiting congestion and pollution from private motor vehicles. In cities, underground trains are usually fast and not affected by congestion in streets. Trains, streetcars, trams and buses also consume less fuel and release less carbon dioxide for each passenger kilometre travelled.

The energy consumption and carbon dioxide emissions of passenger trains can vary greatly. Among the most energy-efficient vehicles are medium- and low-speed trains that are not fired by coal; their emissions are produced at relatively low intensities – that is, with a somewhat moderate amount of carbon dioxide per passenger kilometre. Producing the lowest emission intensity are city buses and trams that carry many passengers at once. Finally, high-speed trains that travel faster than 200 km per hour (125 miles/h), such as the Shinkansen in Japan and the TGV in France, consume more energy and

emit more carbon dioxide. But the amount of carbon dioxide emitted from electrically powered trains depends on how the electricity has been produced.

Many cities' public transportation networks are expanding and some of them even run on renewable energy. One example is the Light Rail system in Calgary, Canada, which started generating electricity from wind power, switching from coal, in 2001. In Europe, metro systems in Paris and Barcelona have been extended, as have similar underground systems in several cities in the United States. Tokyo's rail capacity has almost doubled since the mid-1960s and there are plans to extend the ten existing lines.

In some rural or suburban communities, where public transport frequently does not extend deeply into neighbourhoods, 'park and ride' programs make it easier for commuters to drive to a suburban station and continue their journey by public transport. Some US ski resorts have adopted this idea to reduce the amount of traffic and pollution in their valleys.

Outside urban areas, driving can appear to be a less expensive alternative to public transportation, particularly in North America, where fuel prices are relatively low. However, according to the Canadian Automobile Association, the true cost of owning and running a car can be ten times more than fuel costs when maintenance, insurance and depreciation are factored in.

The emission intensity of cars varies greatly. A small car carrying two or more passengers can emit as little carbon dioxide per passenger kilometre as a large bus travelling with few passengers. A light truck with only one passenger releases about five times more carbon dioxide per passenger kilometre than the bus – similar to the levels on a short-haul flight.

Walking and Cycling School Buses

Walking is good for one's health, a point that is appreciated by parents of schoolchildren around the world. In recent years, an idea called walking and cycling school buses has spread throughout Europe, New Zealand, Australia and in some states in the United States. A group of schoolchildren walk to and from school along a set route, accompanied by a minimum of two parents per 'bus'. One parent 'drives' at the front of the bus, while the other parent takes up the rear. The walking bus picks up its young passengers along the way at designated bus stops. The service is free, and all children are welcome to join the bus even if their parents cannot be drivers. A study in New Zealand showed that more than 60 per cent of the children who used walking school buses had previously been driven to and from school. Traffic outside schools is also reduced, making the area safer for the children. The cycling school bus runs on a similar principle, except children travel on bicycles rather than on foot.

ABOVE *Children in Britain form a walking school bus. The young passengers are picked up at designated stops along the way and walk to school in a chaperoned group.*

Commercial air travel is a more problematic form of public transport. In general, passenger aircraft emit twice as much carbon dioxide as trains or streetcars for each passenger kilometre travelled. And because aircraft consume large amounts of fuel during take-off and landing, short flights consume significantly more energy and emit more carbon dioxide per kilometre travelled than medium- or long-haul flights. Because air travel has become indispensable in the modern world, the development of airplanes with greater fuel efficiency is of paramount importance.

No-Fuel Solutions

By far, the cleanest and least expensive mode of transport is by human power. Besides walking, this can include cycling, roller skating, skateboarding, and even travelling by non-motorized scooter. Cities such as Amsterdam, Barcelona and Florence have instituted pedestrian-only zones to promote no-fuel transportation and thereby to reduce congestion and pollution; such initiatives also promote a sense of conviviality and community. Bikeways, including roadway traffic lanes reserved for bicyclists, have become a common feature of cities around the

BELOW *A corn field in Alsace, France. Corn is one of several crops that are used as biomass fuel, or fuel made from plant and animal materials.*

world. Leading the way are the Dutch, whose flat landscape is ideally suited to bicycles; the nation has created some 17,000 km (10,500 miles) of special lanes and paths for cyclists. This has encouraged the Dutch people to use their bicycles as regular transportation rather than simply for recreation.

Some cities have taken the promotion of bicycling a step further. Since the summer of 1995, the City Bike program in Copenhagen, Denmark, has been deploying up to 2,000 bicycles for public use in the city centre. The bicycles are paid for by sponsoring companies, which then display advertisements on the bikes. To use a city bicycle, a person drops a coin (about 2.50 euros, equivalent to about £1.75) into a slot on the handlebar and unlocks the bicycle from the rack. The bike can then be used within a designated area for as long as it is needed and can then be returned to any of 120 City Bike racks around town. When the bike is locked, the deposit is returned.

Postal services in several European countries have long relied at least partly on bicycles. The British Mail Department owns 35,000 bicycles and, like the French, Swiss and Finnish postal services, delivers letters by bicycle in some cities and their outskirts.

Green Driving

All cars run on some kind of energy. The vast majority run on fossil fuels – gasoline, diesel, or sometimes natural gas or propane – and emit carbon dioxide from their exhaust. Newer cars, called hybrids, run on a combination of gasoline and electricity, emitting less carbon dioxide. Still others run on biofuels, electricity, or hydrogen, emitting little or no carbon dioxide during operation. However, electricity is not necessarily emissions-free: most of the world's electricity is generated from the burning of fossil fuels. And current technology is not able to produce hydrogen or biofuels without generating carbon dioxide during the process. Therefore, whenever a car and its fuel is manufactured and used, carbon dioxide is emitted somewhere in the process. For that reason, the first step in lowering carbon dioxide emissions from a vehicle is to maximize the mileage from the amount of fuel used – that is, to use fuel most efficiently. Typically, smaller cars are more fuel-efficient than larger cars, trucks and so-called sport-utility vehicles.

Regardless of the vehicle, all drivers can take simple steps to improve fuel efficiency, save money and lower carbon dioxide emissions all at the same

time. The first step is to avoid speeding and aggressive driving. Fuel efficiency drops rapidly at more than 100 km per hour (just under 60 miles/h), and rapid acceleration and braking can increase fuel consumption by 40 per cent. In addition, idling an engine wastes gasoline. Driving with tires that are underinflated can diminish fuel economy by 0.4 per cent for every drop of 0.07 kg per sq cm (one pound/sq inch) in the pressure of all four tires. Replacing an air filter that is clogged or fixing a car that is noticeably out of tune or that has failed an emissions test can make its fuel combustion more efficient and improve its fuel economy.

When there is a choice of more than one car, it's smart to drive the one that is most fuel-efficient. Combining several errands into a single trip is another fuel-wise practice.

Sharing the Experience

Whether called *covoiturage*, *mitfahren*, ride share, lift share, or car-pool, driving with someone else is done almost everywhere in the world. As it reduces fuel costs and emissions per passenger kilometre, it also lets commuters share fuel costs and saves wear and tear on vehicles. Some employers run van-pool programs, even providing the vehicle; employees cover its operating expenses. In some cities and countries, vehicles transporting more than one or two people are allowed to use special lanes that speed travellers' drive time and slash per-passenger fuel consumption.

ABOVE *This minuscule Mercedes gets 25 km per litre (60 miles/gallon). Such tiny cars are popular in Europe and Asia; in those made by Smart, a company, strong structural bars reduce passenger injuries during collision, which is important given the car's length – just 2.5 m (8.2 ft).*

Smarter Cars

All these ways of saving fuel and decreasing emissions can be multiplied, of course, by the use of a more fuel-efficient vehicle.

Perhaps the most advanced of these vehicles now on the market is the hybrid-electric car, which gets between 17 and 25 km per litre (40–60 miles/gallon). This is two to three times more efficient than many of the conventional passenger cars used in the United States and four to six times more efficient than conventional sport-utility vehicles and pick-up trucks. In the near future, cars manufactured in Europe will have to average no fewer than 17 km per litre (40 miles/gallon) of gasoline and 19 km per litre (45 miles/gallon) of diesel fuel. These standards are twice as high as in the United States.

SAVING ENERGY AT HOME AND AT WORK

Lighting, heating, cooling and other electrical appliances account for most home and office energy consumption. Insulation, solar energy and wide eaves can lower energy costs and consumption and help reduce greenhouse gas emissions. In the United States, entire developments of zero-energy homes have been popping up in California, Arizona and other states. These homes incorporate energy-efficient designs and appliances and generate their own electricity using solar panels installed as roof tiles. At times, they generate more energy than they can use. According to a US Department of Energy study, property values of energy-efficient homes increased twice as much as those of conventional homes in the same region during the same period.

Small Steps That Add Up

Many energy-efficiency improvements do not require drastic remodelling – some tactics that provide immediate benefits are simple. For instance, heaters, radiators and air conditioners can be turned down or off when no one is home. In addition, because much heat is lost and gained through windows, blinds, shades and curtains can be opened by day in cold climates to admit heat or closed in warm climates to keep heat out, reducing the need for air conditioning. At night, curtains can be closed to help keep heat from escaping through the window. Sealing air leaks around doors and windows is effective. In addition, because glass is a poor insulator, installing double-paned windows can greatly reduce heat loss in cold climates; windows with special coatings can help reduce heat gain in warmer climates. Energy-saving appliances and light bulbs, although sometimes more expensive to buy than their conventional counterparts, save money on energy in the long run. Energy-saving light bulbs are a form of fluorescent lighting in which most of the energy is turned into light instead of heat, unlike ordinary light bulbs. For the same level of brightness, energy-saving bulbs consume one fifth the energy of ordinary bulbs and last eight to ten times longer.

Many smaller savings can add up. Taking showers rather than baths can reduce how much hot water is used. Working near windows in daytime eliminates the need for electric lighting. Task lighting that illuminates work areas directly is more efficient than area lighting. Heating and air conditioning can be limited when a space is unoccupied. In warm climates, light-coloured materials on a house's roof and facade can help reduce air conditioning needs by up to 15 per cent. In these regions, trees can add to the effect by providing shade. In cold climates, trees can act as windbreaks and reduce indoor heating needs.

Electricity Generation and Industry

Conventional electricity generation is inherently not very efficient. Only about one third of the fuel's potential energy is converted into usable energy. In co-generation, electricity and heat are generated simultaneously. For example, a steam turbine powered by natural gas can generate not only electricity but also heat from the steam. A similar process can be used with any method of electricity generation and can convert up to 90 per cent of the fuel into usable energy. Currently, co-generation plants make up over 10 per cent of the electricity generation capacity in Singapore. The European Union has a target to double its share of electricity produced from co-generation by 2010.

Creating industrial products is extremely energy-intensive. Improving the efficiencies of equipment and processes can save considerable energy. For example, by operating equipment at close to optimum levels, improving process controls and upgrading production equipment, the Australian aluminum industry reduced greenhouse gas emissions by 75 per cent. Some industries can also capture their waste heat for use in industrial co-generation systems. The resultant electricity can then be used on site or sold to a utility.

Creating Carbon Sinks

Trees and other vegetation are natural carbon sinks – they trap and use carbon dioxide that would otherwise be released into the atmosphere. When forested land is cleared for development, its value per hectare as a carbon sink declines. As the world's population increases and more land is converted from natural environments to buildings and agriculture, atmospheric carbon increases.

RIGHT *This zero-energy house uses two basic strategies to remain 'green'. First, it uses technologies that produce energy without emissions. Second, it is built to use less energy overall. By recycling, buying recycled paper goods, riding a bicycle to get around locally, driving a hybrid car to cover longer distances, using a push mower on a small lawn and drying clothes on a clothesline, the residents of this home can reduce both fossil-fuel energy use and greenhouse-gas emissions to almost nothing.*

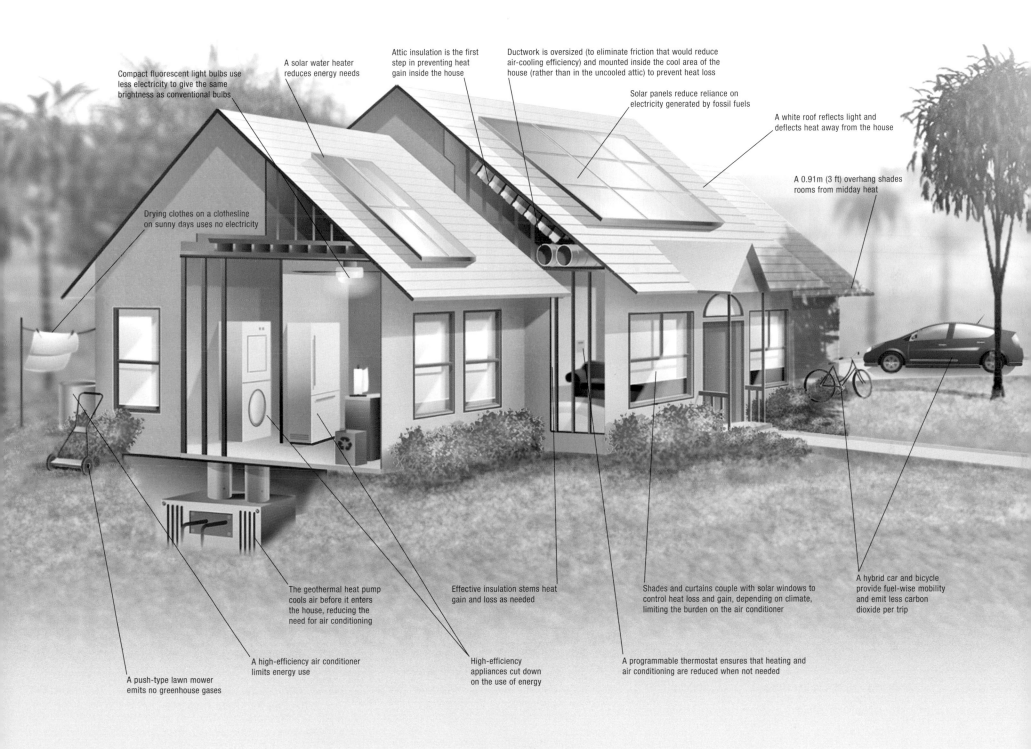

Compact fluorescent light bulbs use less electricity to give the same brightness as conventional bulbs

A solar water heater reduces energy needs

Attic insulation is the first step in preventing heat gain inside the house

Ductwork is oversized (to eliminate friction that would reduce air-cooling efficiency) and mounted inside the cool area of the house (rather than in the uncooled attic) to prevent heat loss

Solar panels reduce reliance on electricity generated by fossil fuels

A white roof reflects light and deflects heat away from the house

A 0.91m (3 ft) overhang shades rooms from midday heat

Drying clothes on a clothesline on sunny days uses no electricity

The geothermal heat pump cools air before it enters the house, reducing the need for air conditioning

Effective insulation stems heat gain and loss as needed

Shades and curtains couple with solar windows to control heat loss and gain, depending on climate, limiting the burden on the air conditioner

A hybrid car and bicycle provide fuel-wise mobility and emit less carbon dioxide per trip

A push-type lawn mower emits no greenhouse gases

A high-efficiency air conditioner limits energy use

High-efficiency appliances cut down on the use of energy

A programmable thermostat ensures that heating and air conditioning are reduced when not needed

ALTERNATIVE TRANSPORTATION

A quiet revolution is hitting the streets. Around the world, people are beginning to drive cars that run on electricity or on a combination of electricity and gasoline. Other alternatives to gasoline, from hydrogen-fuelled cars to electric bicycles, are also gaining attention. What these vehicles have in common is that they are more fuel-efficient than gasoline-powered cars and so reduce the emissions that contribute to climate change.

Electric Cars

The average gasoline-powered car is not at all efficient, burning copious amounts of fuel and generating much pollution for the speed and convenience it offers. Electric and hybrid cars are cleaner and more efficient, but they accomplish that in different ways. An electric car runs completely on a battery that is periodically recharged. In most cases, the energy for recharging comes from the regional power grid, which uses fossil fuels for at least part of its supply. Some people have electric cars they recharge at home from solar photovoltaic rooftop power and they may even achieve zero pollution. Though an electric car that draws on the resources of an old-fashioned power company is not pollution-free, it uses fuel more efficiently than a gasoline-powered vehicle and greatly reduces carbon dioxide emissions per kilometre.

Electric cars have disadvantages. They must be refuelled often – every 80.5 to 161 km (50–100 miles), compared to every 482 km (300 miles) for conventional cars. Refuelling takes more time than the usual gas-station stop; 30 minutes is considered quick. In addition, the cars' batteries must be replaced often and the vehicles have more limited acceleration and top speeds than conventional cars. Still, the technology is improving constantly and these weaknesses may soon disappear.

RIGHT *A solar-powered car from the Tasai-Techno team of Osaka Sangyo University, Japan, photographed in 2001 in front of the Shah Alam mosque near Kuala Lumpur, Malaysia. The car was an entry in the first World Solar Car Championship. Solar-powered cars can achieve zero pollution.* UPPER LEFT *Bicyclists on the streets of Wuxi, Jiangsu Province, China. The bicycle is an inexpensive and pollution-free form of transportation.*

ABOVE *Liquid hydrogen fuelling station. At left are photovoltaic solar cells, which power the process by which water is converted into liquid hydrogen. The liquid hydrogen is stored in the white tanks shown at right.*

Hybrid Cars

To eliminate the problems of electric-only automobiles, hybrids team an electric motor with a gasoline-powered internal combustion engine that is smaller than the one in a conventional car. To accelerate, both engine and motor provide power. On the highway, the gas engine works alone. When energy needs are minimal, as at a red light, the electric motor works alone. Electricity is supplied not only by the gas engine but also by regenerative braking – a process that charges the batteries with kinetic energy collected whenever the brakes are applied.

Hybrid cars – among them Toyota's Prius and Honda's Insight and Civic Hybrid – are gaining a following. The Prius, the first mass-produced hybrid, went on sale in Japan in 1997 and by late 2004, some 250,000 had been sold worldwide, with many prospective purchasers on waiting lists. Celebrity drivers such as Pierce Brosnan, Brad Pitt, and Cameron Diaz have given hybrids a high profile. Market demand for hybrids has been such that hybrid versions of sport-utility vehicles and pick-up trucks have started to appear.

In some countries, the government encourages the use of hybrid cars with tax breaks and reduced toll charges. For instance, Londoners driving alternatively fuelled cars do not have to pay the city's congestion fee, £8 per weekday.

Other Alternatives

In the near future, other alternatives to conventional cars may emerge. Automakers worldwide are developing cars that run on hydrogen-powered fuel cells. Instead of depending on batteries, these vehicles generate electricity by reverse electrolysis, a process that combines hydrogen with oxygen, with water as the only by-product. However, this technology is clean only if the fuel has a low-emissions source of energy. Hydrogen is made by electrolysis, in which water is split into hydrogen and oxygen, but that process requires a power source. Expense is an issue: at present, hydrogen costs about four times more than gasoline; and fuel cell technology has yet to be proved economical.

More cars in the future may run on fuels produced from biomass. The most widely used biofuel today is ethanol, which is produced from starch crops such as corn and involves fermentation. Biodiesel, which can be produced from most animal fats or vegetable oils, including leftover cooking oil, can be substituted for conventional diesel. A group of Canada-based filmmakers has made the news because of their plans to drive from Alaska to Argentina using only biofuel.

Further in the future may be solar-powered cars, although they are now used only in races meant to prove and promote the technology.

Another transportation option is the light electric vehicle. This category includes electric bicycles, electric scooters, electric motorcycles, or even neighbourhood electric vehicles (NEVs), which are similar to golf carts but faster and designed for public streets. Because many people use cars mainly within 16 km (10 miles) of home, a light electric vehicle often makes sense.

RIGHT *This two-seater electric car runs primarily on a rechargeable battery, but its energy is supplemented by solar power, collected by solar panels on its roof.*

To retain the natural carbon-trapping services of plants, land-use experts and urban planners advise increasing vegetation, even in city centres. Carbon sequestration is one benefit of green roofs, which are covered with low-maintenance plants such as sedum. Islands of trees and plants can be planned into parking lots and highway medians. Mature trees can be left on construction sites, not cut down. World forest conservation, in addition to protecting animal species and fragile ecosystems, also helps to limit carbon in Earth's atmosphere. Farms, although sequestering less carbon than forests, can demonstrate increased carbon retention capacity if harvested plant remains are left on top of soil, rather than tilled under.

Trapping Carbon Dioxide

Scientists advocate several different methods to isolate the carbon dioxide that would otherwise enter the atmosphere. In geologic sequestration, carbon dioxide is pumped underground, into oil or natural gas reservoirs, unminable coal beds, or deep salt formations. The petroleum industry has long used this technology to push more oil from depleted reservoirs. Another method is to sequester carbon dioxide in the ocean, by injecting liquefied carbon dioxide directly into the sea at a depth of at least 1,000 m (3,281 ft).

LEFT *An X-ray of an energy-saving lightbulb. This bulb converts more energy into light than into heat, so is four to six times more efficient than a filament lightbulb.*

At great depths, carbon dioxide is denser than seawater and it may be possible to store it on the bottom as liquid or deposits of icy hydrates. However, this approach may be problematic for a number of reasons. Although we know relatively little about life in the deep sea, recent exploration has made it clear that the deep ocean harbors more life than previously imagined; deep sea coral reefs are just one example. Pumping liquid carbon dioxide into this fragile, mysterious realm will create large dead zones, destroying a world we are only just learning about. The growth of tropical coral reefs is already limited by rising oceanic levels of carbon dioxide; adding more would make the problem worse. It is also unclear that underwater carbon dioxide would remain underwater: during the 1980s, concentrated carbon dioxide at the bottoms of two lakes in Cameroon was unexpectedly and violently released, perhaps because of activity in a nearby volcano. Nearly 2,000 people suffocated, as well as thousands of wild and domestic animals.

Another possible approach is to use the minute organisms in the ocean to store carbon. Iron stimulates growth of phytoplankton, single-celled algae that live in the ocean and absorb more carbon dioxide from the atmosphere as they grow more rapidly. As phytoplankton and the organisms that eat them die and sink to the bottom, this carbon dioxide sinks with them. However, overfertilizing the ocean may create dead zones like those in shallow areas like the Gulf of Mexico; here, excess nutrients from the Mississippi River have created a dead zone the size of New Jersey. Many oceanographers doubt that iron fertilization could increase oceanic carbon uptake enough to make a difference.

Can carbon dioxide be trapped permanently underground or in the ocean? How will marine animal and plant life in the shallow and deep ocean be affected by iron fertilization? How long does carbon remain sequestered using these various approaches? The current level of scientific understanding leaves many important questions unanswered about sequestration technology.

RIGHT *The Rocky Mountain Oilfield Testing Centre's Teapot Dome oilfield in Wyoming. Carbon sequestration procedures are being tested here by the US Department of Energy; the greenhouse gas will be injected into underground oil reserves in order to remove carbon from the atmosphere and slow global warming.*

Renewable Energy in the Real World

For the summer 2000 Olympic Games in Sydney, when the Australians built Athletes' Village, they made it the world's largest solar-powered community. Each of the 665 houses has solar panels on the roof to supply it with electricity. When the panels produce more electricity than is used, the excess power is fed back into the main distribution grid and credited to customers' accounts. When more electricity is needed than is available from the solar panels, the grid serves as a backup. After the Olympics, the houses were sold as part of the new suburb of Newington, which has become the largest solar-powered suburb on Earth.

In Denmark, some 5,000 wind turbines generate enough power to supply 20 per cent of the country's energy consumption. The Danes were pioneers in developing commercial wind power during the 1970s, and today, Danish manufacturers produce almost half of the world's wind turbines. The wind energy industry brings about 3,000 million euros (over £2,000 million) into the nation's economy and employs some 20,000 people.

In Iceland, geothermal energy provides the heating for nearly 90 per cent of households. The oldest known household geothermal system was created in the early fourteenth century in Chaudes-Aigues Cantal, a village in France. This system distributed warm water

ABOVE *A bus fuelled by coconut oil in the Pacific nation of Vanuatu. The oil is mixed with diesel fuel; many vehicles here have been modified to run on this inexpensive, efficient and renewable source of energy.*

through wooden pipes and remains in use even today. In other cities all over the world, many homes get their heat from geothermal energy.

In the Pacific nations of Vanuatu, Samoa and the Marshall and Cook islands, residents use coconut oil to fuel diesel engines. Coconut has long been a basic foodstuff here and residents discovered during World War II that engines could run on a mix of diesel and coconut oil. Local coconut oil not only costs four times less than imported diesel but also, as locals report, has a 'beautiful smell'. Others have had similar ideas. In 2004, four US college students created a portable biodiesel manufacturing setup that they could operate in campgrounds, then drove across North America using biodiesel made from waste oil collected in restaurants along the way.

LEFT *Danish wind turbines on a breakwater. The wind energy industry is an important source of income for Denmark, which manufactures almost half the wind turbines in use worldwide.*

GREEN POWER

Renewable energy comes from natural sources that replenish themselves over short periods of time. These resources include the Sun, the wind, moving water, organic plant and waste material (biomass), and Earth's heat (geothermal energy). Renewable energy emits little or no carbon dioxide and its environmental impacts are generally less significant than those of conventional power generation.

Nuclear power is not a renewable energy source. It is fuelled by uranium, which comes from an ore that is mined. In production of nuclear power, very little carbon dioxide is emitted, but highly toxic radioactive wastes are produced. To date, no country has found a safe way to dispose of them. And while some countries are promoting nuclear energy as a replacement for fossil fuels, others have decided to phase out use of nuclear power.

For consumers, electricity that comes from renewable energy – so-called green energy – is often more expensive than energy produced from fossil fuels or by nuclear power plants. This is because the price of conventional energy does not include its full costs to society – for example, the costs of dealing with its negative effects. Burning fossil fuels produces air pollution that can lead to asthma and other medical problems. It also pollutes lakes and streams through acid rain and mercury and contributes to the problems that are connected with atmospheric warming. These associated medical and environmental costs are not included when calculating the price of electricity; it is assumed that the consumer, individually or through the state, will pay for them. To be competitive in a market that, both directly and indirectly, subsidizes dirty energy, renewable ener-

ABOVE *A man checks solar panels on the roof of a school in rural San Ramon, Honduras. The village became Latin America's first solar-powered village in 1998, with assistance from UNESCO and the Honduran energy council, COHCIT.*

gy will continue to require government financial support in the short term.

The costs of renewable energy, however, are likely to drop as its technology advances and market demand increases. In the United States, the cost of wind energy declined 88 per cent between the 1980s and the 2000s. At the same time, as governments continue to force conventional power suppliers to cut the amount of pollution they produce, as existing power stations age out of service, and as the effort and expense involved in extracting fossil fuels from deeper and more remote reservoirs increases, the costs of conventional energy are expected to rise. In short, the price difference between fossil and renewable energy sources is likely to continue to narrow over time.

LEFT *A wind farm in Helix, Oregon, USA The aerodynamic turbine blades drive a generator that produces electricity. This source of power is especially beneficial in the Pacific North-west; electricity here is also generated by hydropower, but the region's dams threaten salmon migration, while wind farms have little negative environmental impact.*

Sun and Wind Power

Humans have long harnessed the elemental forces of nature to provide energy. Technology now provides more sophisticated, and more efficient, means.

Besides making the most of direct sunlight for heat and light, a simple way to take advantage of the Sun's energy is to use it to heat water by means of solar panels. These easy-to-manufacture panels expose a concentrated network of plumbing directly to the Sun's rays. Photovoltaic cells, a more advanced technology, convert solar energy into electricity. Panels made from photovoltaic cells are now sometimes used as roof tiles to produce a building's electricity.

Like the windmills used since medieval days to power mills to grind flour and pump water, modern wind turbines convert wind to rotary motion; the blades, now sleek and aerodynamic, drive a generator that produces electricity. Usually found in non-urban areas or offshore, modern wind turbines are large: a tower for a 250-kilowatt turbine stands some 40 m (130 ft) high, with a blade sweep of 30 m (98 ft). Each turbine requires about 0.4 hectares (1 acre) plus wind speeds averaging approximately 24 km per hour (15 miles/h) at a height of 50 m (160 ft).

Power from Water and the Earth

The most common form of hydropower uses dams on rivers to create large reservoirs of water, which is funnelled out through turbines, causing them to spin. The turbines are connected to generators that produce electricity. However, dams alter the ecology of a region. The water below the dam is often colder than would normally flow down the river, so fish sometimes die. The water level of the river below the dam

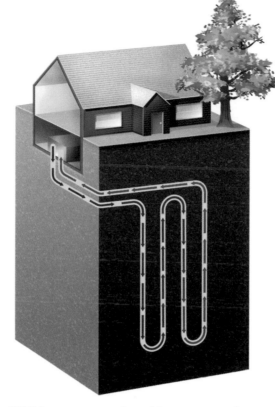

ABOVE *In summer, a geothermal heat pump transfers heat from building to ground. Cool water pre-cools air in the house and warm water carries out the heat.*

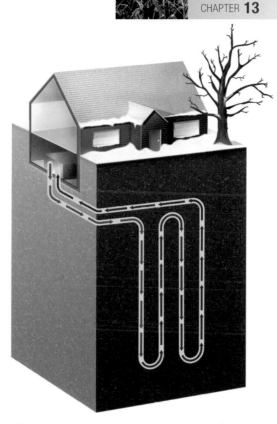

ABOVE *In winter, heat is transferred from ground to building. Warm water preheats air in the house and leaves cooled after heat exchange with the air inside.*

can be higher or lower than its natural state and the flow regimes less natural, which affects plants along the riverbanks.

The centre of the Earth has a temperature of more than 6,000°C (10,000°F). As a result, in temperate regions, even the upper 3 m (10 ft) of the Earth's surface stay at a nearly constant 10°C to 16°C (50°F to 60°F) year-round. In regions of high seismic or volcanic activity, with natural hot springs, geothermal heat can be tapped directly for heating buildings or for producing electricity. In addition, in most areas worldwide, the stable ground surface temperature can be used to warm and cool buildings. Air for ven-

tilation can pass through a series of underground pipes before entering the building. These pipes, part of a geothermal heat pump, transfer heat from the ground to ventilation air in winter and transfer unwanted heat from the ventilation air to the cool ground in summer. A geothermal heat pump is one of the most efficient ways to heat and cool buildings.

Ocean Power and Biomass

The ocean harbours tremendous amounts of energy, from the motion of its waves to the twice-daily movements of the tides. Wave energy can be captured by a type of dams fixed to the sea floor or by devices floating at the surface. By compressing an air chamber inside a dam, or by making part of the floating device move, wave motion can be harnessed to turn a turbine to generate electricity. Tidal motion can be channelled to generate electricity with the use of large dams, called barrages, which span estuaries. As the tides move in and out, water flowing through tunnels in the dam is made to turn a turbine.

The world's first commercial wave power station opened on the island of Islay off the west coast of Scotland in the year 2000. Denmark also has an offshore facility supplying electricity to the national grid system. The world's largest tidal power station is in the Rance estuary in northern France. It generates 640 million kilowatt hours per year and has been commercially successful for more than three decades.

Wave and tidal power are not without disadvantages. Building and maintaining such facilities is expensive. Dams and pipes can disrupt other activities such as fishing and shipping. Tidal barrages flood some areas and deprive others of the sediments they would normally receive from upstream. These changes can affect local ecosystems that are often rich with birds and fish.

Biomass is plant and animal waste, including wood chips, cornstalks and animal manure. It can be burned to produce heat or electricity, or refined into a fuel that can be directly used in vehicles. Biomass fuels provide energy while minimizing the amount of waste that needs to be disposed of. But since all organic matter is made up of carbon and hydrogen, carbon dioxide is released when biomass or biofuels are burned. For biomass energy to achieve zero

ABOVE *Tidal barrage on the Rance River, France. The barrage harnesses the energy of the tides to generate electricity. When the tide rises, the sluice gates open, and water flows into the estuary, which acts as a reservoir. When the tide falls, water is released only through the 24 turbine generators. These generators produce a total of 640 million kilowatt hours per year. Tidal power is being studied as an efficient and environmentally viable form of energy, but tidal barrages can have significant impacts on local ecosystems.*

net carbon dioxide emissions, it is necessary to grow more plants to absorb the amount of carbon dioxide emitted.

Buying Green Power

For many power consumers, both commercial and residential, it is now possible to purchase renewable energy from utility companies. In Geneva, Switzerland, individual customers can choose to buy 100 per cent of their electricity generated from hydropower, or request that at least 2.5 per cent of it come from solar power. In Grenoble, France and ten states in the United States, corporate consumers can choose to buy 10 to 100 per cent of their electricity generated from wind, biomass, solar, or hydropower. In the United States, the Johnson & Johnson corporation has been purchasing 11 megawatts of wind power per year from Texas and East Coast states.

Globally, more and more countries are increasing their capacity and use of renewable energy. The European Union aims to supply 22 per cent of its electricity from renewable energy by 2010. Australia, the United Kingdom and some US states are requiring electric utilities to increase the amount of renewable energy resources in their electricity supplies. Under some arrangements, utilities that have exceeded their renewable-energy targets can sell the excess to the utilities that have not yet done so. Germany and Spain have gone further by setting a price for renewable energy, in an effort to encourage further investment.

Educating Governments

During the past decade, with the publication of the expert IPCC reports and the entry into force of the Kyoto Protocol, the public is gradually coming to accept the fact that climate change is very much a reality. Still, significant distance remains between accepting its reality and taking action to limit its

Businesses, the media and individuals all have their roles to play in limiting climate change. Businesses can produce products or offer services to help individuals as well as other businesses reduce greenhouse-gas emissions. Individual citizens, in turn, can create demand to force the businesses they patronize to focus even more on these products, as in the case of hybrid vehicles. As part of the process, the media has the power to influence public opinion and to define fashion and trends. Today, the Canadian government is making use of advertisements on television and other media to promote a 'one tonne challenge' – that each Canadian should reduce his or her annual carbon dioxide emissions by one tonne.

BETWEEN THE IPCC REPORTS AND THE KYOTO PROTOCOL, THE PUBLIC IS ACCEPTING THAT CLIMATE CHANGE IS REAL. STILL, SIGNIFICANT DISTANCE REMAINS BETWEEN ACCEPTING ITS REALITY AND TAKING ACTION TO LIMIT IT

effects. In many nations, climate change policies are directed at the national level and local governments tend to need more education, as well as considerable political incentive, to translate the national targets into action. In other countries, such as the United States, state and local governments and even industries have taken the lead in initiating climate change programs. In this case, education, as well as increased political pressure, needs to be directed toward the national government.

The younger generation holds the key to the future of climate change. The choices that young people make individually and as a group will have an important influence on atmospheric greenhouse gas levels for the next century, and they will bear the impact of climate change for many years. In classes and extracurricular programs, schools can give children an introduction to real-world problems and solutions, such as waste management, energy choices and city planning for energy efficiency.

Turn Off the Lights

There is a common belief that electrical appliances consume next to no energy in standby mode and that computer hard drives and monitors are damaged by frequent switching on and off. In fact, photocopiers and printers consume about half as much electricity during standby as when they are in full ready mode. Modern computer hard drives are not significantly affected by frequent shutdowns. Turning off computers at night and on weekends may actually make them last longer, while monitors should be shut down even during short periods of inactivity. Lights that are left on in unoccupied rooms simply waste energy. In offices, lights are often left on in lounges, meeting rooms, restrooms and other common areas on weekends and after hours.

A study conducted by the US Environmental Protection Agency showed that turning lights off when rooms are not used can lower office energy consumption by 20 to 60 per cent.

ABOVE *Office buildings in Chicago illuminated at night. Energy consumption can be greatly reduced by turning off lights and electrical appliances whenever they are not in use.*

ADAPT WE MUST

Living with Climate Change

Human-induced climate change has already passed the point where it can be completely stopped. As temperatures and sea levels rise, societies must adapt to the changes that cannot be averted – or suffer the consequences. Some adaptations must be designed to protect cities and towns: building must be limited on vulnerable coastlines and new kinds of crops developed. Still other adaptations must focus on helping ecosystems to resist or habituate to the changing climate: pollution and habitat destruction must be sharply limited and room provided for species to shift their ranges.

PROTECTING HUMAN COMMUNITIES

'Mitigate we might, adapt we must' are the words of William Nordhaus, a US economist and Yale University professor who has written extensively about the economics of climate change. When climate change first came to public attention in the late 1980s, the focus was on mitigation – limiting climate change. The hope was to reduce emissions from factories and cars and convert to forms of energy that do not produce greenhouse gases, such as solar power.

Even though such measures are indispensable in the effort to limit climate change, enough greenhouse gases have already been emitted to make some degree of climate change unavoidable.

Rising temperatures and sea levels, an increase in extreme weather, melting glaciers, droughts and floods, altered ecosystems – all are already happening, to a greater or lesser degree, and are likely to continue happening, no matter how successfully society reduces greenhouse gas emissions. Whatever human beings do to mitigate climate change, they will have to adapt to it – to protect both their own communities and the natural ecosystems on which those communities depend. The more wisely societies can adapt, the greater the possibility of reducing the harm that climate change can cause and the greater our ability to take advantage of possible opportunities.

The inevitability of climate change is no reason to give up on mitigation, however. Adapting to an altered climate will be easier and will cost less if changes are held at bay. In the United States, University of Texas–Austin researcher Camille Parmesan has estimated that if global warming is limited to 2°C (3.6°F), it will be possible for societies to adapt effectively. But warming beyond that threshold would make adaptation measures increasingly expensive and less effective.

In adapting to changes in climate, flexibility is absolutely essential. Because no two regions will be affected in precisely the same way, there is no single best way for an area to adapt. The first important step is for every society to assess how its own area is most likely to be affected and subsequently to evaluate what the members of that society most want to protect.

This is no simple task. Priorities will not be the same in all communities, or even for groups within a single community. In one case, foreign aid workers built extensive shoreline protection in front of a village, only to be told later by the inhabitants that they would rather have protected their graveyard. After all, they observed, the living could relocate their houses; the dead could not.

Some adaptations will be of a scale that can be undertaken only by governments. Others are in the hands of individuals. Individuals can prepare for climate change first and foremost by being informed. They can consult reliable books, articles and Web sites to find out as much as possible about what to expect. They can prepare for probable changes in their region by buying appliances that are more energy-efficient or water-efficient or by not building in coastal areas prone to flooding. They can support organizations that deal with fundamental problems such as rapid population growth, poor land-use practices and poverty – all factors that make societies and their ecosystems more vulnerable to the stresses of climate change. Individuals can exert pressure on all levels of government to undertake the collective actions that only governments are in a position to implement.

RIGHT *Waves crash over a seawall in Winthrop, Massachusetts, USA. Such barriers will be more important as sea levels rise and weather patterns become less predictable.* PREVIOUS PAGE *Clockwise from upper left: aerial view of a rain catchment and water storage system on the uninhabited island of Kahoolawe, Hawaii, USA; a woman works in rice paddies in Goa, India, where farmers are being encouraged to plant saline-tolerant crops; a seawall built of hollow concrete hexagons prevents coastal erosion; artificial islands in Valemount, British Columbia, Canada, attract migrating birds such as avocets to the wetlands. Wetlands restoration is important to limiting water runoff.*

SURFACE AIR TEMPERATURE CHANGE

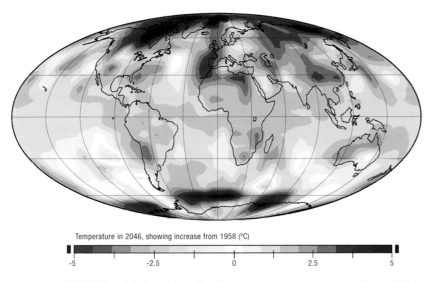

Temperature in 2046, showing increase from 1958 (°C)

-5   -2.5   0   2.5   5

ABOVE *This global model predicts how surface air-temperature will rise or fall by 2046. Most areas will become warmer, but to different degrees. Areas near the poles will experience the greatest warming, while some areas will cool.*

The Coming Millennium

What Earth's climate will be like one thousand years from now, or even one hundred, is impossible to predict. Signs point to a warmer world and rising sea levels, yet other drastic scenarios have been painted, including a 'century of hunger' and a global deep freeze resulting from the shutdown of the thermohaline circulation (see page 26). How likely are these?

LEFT *Pivacharam, in the southern Indian state of Tamil Nadu, where this fisherman stands among mangrove trees, has managed to retain its mangrove forests, a natural coastal buffer. During the tsunami of 2004, the presence of the trees saved many lives and preserved about a dozen villages, some as close as 100 m (110 yards) from the sea. Mangrove restoration projects are now underway in much of South-east Asia.*

Scores of computer modellers have tried to create realistic predictions for Earth's climate. In 2001, working with results from several of these models, the United Nations Intergovernmental Panel on Climate Change (IPCC) predicted that during the next century, the temperature could increase between 1.4°C and 5.8°C (2.5°F and 10.4°F). However, the IPCC did not estimate the probability of these numbers.

The uncertainty derives at least in part from science's limited understanding of the myriad forces shaping global climate. While it is known, for instance, that increasing carbon dioxide levels trap more heat closer to Earth, very little is known about whether increased cloudiness would warm or cool the planet. Just as important, scientists have no way of knowing what will happen politically, socially, or technologically to alter future climates. If the rest of the world begins to consume fossil fuels at the same rate as the United States, for example, the future will very probably be much warmer. On the other hand, if everyone drastically limits fossil fuel use because of a breakthrough in transportation technology, the prospects change.

Scientists at the Massachusetts Institute of Technology, in the United States, predict that in the absence of any reduction in greenhouse gas emissions, there is a 50 per cent chance that over the next 100 years the global mean surface temperature will increase by 2.4°C (4.3°F). If an aggressive approach to emissions reduction is taken, the probability changes to a 50 per cent chance of a 1.6°C (2.9°F) increase. As for longer-term predictions, the IPCC estimates that even if every country in the world stops emitting greenhouse gases today, climate change will continue for quite some time: atmospheric carbon dioxide levels and global temperatures could stabilize in a century, or perhaps two or three centuries, while sea levels would not stop rising for at least a millennium.

Limiting Coastal Erosion

As rising sea levels and bigger storm surges threaten coastal communities, a key concern is limiting loss of land and livelihood from erosion and degradation of coastal environments. The usual response of coastal communities or governments to this threat is to build offshore structures like breakwaters, or to protect beaches and channels with projecting jetties that reduce along-shore currents. Another strategy is armouring the shore with seawalls of concrete, rocks and other hard material and building dykes and levees to limit flooding. Although effective in the short term, these structures can severely harm natural ecosystems, disrupting the periodic flooding that nourishes marshlands and, ironically, increasing the loss of shoreline by keeping sediment from coming in.

In recognition of these problems, some governments and organizations around the world are attempting new strategies. The United Kingdom's governmental Environment Agency is seeking to help 'nature protect the coastlines'. The agency's concept is to do this by removing some existing defences.

INUNDATED BANGLADESH

The nation of Bangladesh is accustomed to flooding. At the head of the Bay of Bengal, straddling the mouths of the mighty Ganges River, the country consists almost entirely of a flat alluvial plain, formed from soil deposited by rivers. The rivers often overflow and most of the land is less than 15 m (50 ft) above sea level. A third of the country floods every year, affecting the lives of millions of Bangladeshis who live and farm in the flood-prone areas.

More than a third of Bangladeshis subsist below the poverty line, and 40 per cent are unemployed or underemployed. The increased severity and frequency of flooding expected to result from climate change threatens even greater disaster. During the next 100 years, 15 to 17 per cent of Bangladesh's land is expected to be lost due to flooding or salinization as a result of inundation by seawater. Since two-thirds of Bangladeshis depend on agricultural work to survive, this loss of land will have dire consequences.

The CARE Project

Into this seemingly desperate situation stepped CARE Bangladesh and its 'Reducing Vulnerability to Climate Change' project. Launched in 2001, the project focused on six districts of south-west Bangladesh: Bagerhat, Gopalganj, Jessore, Narail, Khulna and Satkhira. Its stated goal was 'to build the capacity of

Bangladeshi communities in the southwest to adapt to the adverse effects of climate change. The project will reduce vulnerability and increase adaptive capacity at the household and community levels'.

The project began with local community meetings to raise awareness about climate change, in part by using traditional songs and theatre. Through these meetings, the project also explored what aspects of climate change were of greatest concern to local people. Flooding, salinity, waterlogging of soil and

BELOW *Villagers navigate a boat through flooded paddy fields south of Dhaka, Bangladesh, after a flood in August 2004. The region saw widespread loss of crops due to flooding and salinization; some farmers have now switched to growing a profitable, salt-tolerant grass called mele, which is used for weaving.*
UPPER LEFT *A fisherman in Bangladesh.*

ABOVE *A Bangladeshi woman and girl cross a makeshift bridge at Sirajhonj, to the north of Dhaka. Annual floods in this area are becoming more severe and more frequent – and this condition can be attributed to climate change.*

earned more from growing and harvesting the mele than they once earned from the harvesting of traditional crops.

Gardens that Float

Still another simple but powerful idea was to build gardens that would float during floods and when not inundated would rest firmly on the ground. CARE worked with local communities and development workers to train people to build *dhaps* – floating beds of compost and water hyacinth on bamboo platforms. Water hyacinth is a highly invasive, free-floating plant native to northern South America; as it decomposes, it provides structure and nutrients for crops as diverse as potato, cucumber, tomato and eggplant. Each *dhap* is about 7.5 m (25 ft) in length and 1.5 m (5 ft) in width and lasts up to a year.

ABOVE *Water hyacinths, which are native to South America, are being introduced in Bangladesh as part of an adaptive measure against rising flood levels. The plant is used to create* dhaps, *bamboo platforms for floating crop gardens.*

The beds are ready for planting after nearly a week and farmers can start harvesting vegetables as early as three weeks after planting.

Along the Kapataksha River, the floating gardens are doubly adaptive because they take advantage of what had been a serious local problem: water stagnation along large stretches of the river. The near-dead river is now full of water hyacinth that is being cleared and used to grow vegetables.

The decomposed water hyacinth was so useful as fertilizer that farmers did not need to use any additional chemical fertilizer. After harvesting crops from their *dhaps,* farmers could then sell the vegetables profitably in the local market and in neighbouring towns. The brevity of the interval between planting and harvesting meant improved economic conditions for farmers. In this way, the *dhaps* have allowed many marginal farmers to make a living without abandoning farming or permanently leaving their homes. Far from being a sacrifice forced by climate change, these floating gardens address for Bangladeshis both their current problems and an anticipated changing landscape.

inconsistent availability of water for irrigation all were major concerns. Many approaches to adaptation were developed as part of the project. Some were techniques to secure food and income in the midst of a changing climate: duck rearing, homestead gardening and cage aquaculture, which employs submerged cages in which fish and plants can be securely farmed. Other techniques were aimed at conserving resources, such as rainwater harvesting and community pond conservation for drinking and irrigation.

Residents of the village of Mandra began cultivating mele, a salt-tolerant relative of grass, on land that had been so infiltrated by salty water that it could not be used to grow other crops. The mele is used to make mats, baskets and other popular handcrafted products. Farmers were pleased with the results because they

ABOVE *Erosion barriers at a resort hotel in Cancun, Mexico. Such protective means of 'armouring' the shore have short-term benefits but may not be effective in the long term, as they can alter and damage local ecosystems. More adaptive measures are called for in cases such as this.*

Rhode Island has banned armouring along much of its coast; on the West Coast, residents of Santa Barbara, California, raised private and public funds to purchase 27 shorefront hectares (69 acres) to serve as an erosion buffer. Rather than fighting erosion, this approach allows the natural processes of shoreline erosion and rebuilding to continue while protecting local real estate. In the aftermath of the late 2004 tsunami in South-east Asia, Sri Lanka established a designated no-build zone that extended up to 200 m (660 ft) from the mean high tide line.

Any one who has sat out a tropical storm nestled in the shelter of a mangrove forest can attest to the ability of these trees to lessen the destructive powers of wind and wave. For this reason, a mangrove forest is an excellent buffer against coastal erosion. During the tsunami, areas with intact mangrove forests suffered less damage than those where mangroves had been cleared for aquaculture or development of tourist facilities. Today, after a half century during which nearly half of the world's

mangrove area has disappeared, communities have finally begun to appreciate the importance of the trees. Coastal villages that have lost their mangroves are seeing not only increased rates of erosion but also a decline in marine species owing to loss of habitat.

In South-east Asia, the Vietnam Red Cross Society initiated a large-scale mangrove restoration project in 1994 to protect northern provinces from the typhoons and storms that regularly break through dykes, damage property and flood fields with salty water. Although such storms are a normal occurrence, climate change is expected to worsen their effects on coastal communities. The program's success is already clear: by 2002, mangroves had been planted in nearly 12,000 hectares (29,600 acres), and decreases in erosion and increases in marine biodiversity had been documented there.

Limiting Runoff

Although climate change will continue to increase the frequency and severity of flooding and droughts, human influences on the natural water cycle are making the situation worse. Wetlands and marshes in their natural state act as sponges, absorbing water during heavy rains and feeding underground aquifers. Similarly, floodplains slow floodwaters and allow them to soak into aquifers. Floodplain soils are often rich because of the regular influx of nutrients.

Yet wetlands around the world have been filled or drained and floodplains have been covered with networks of dams, levees, dykes and canals to control flooding and irrigate fields. Restoring and protecting wetlands and floodplains can provide immediate benefits by reducing flood damage, increasing water quality and decreasing an area's vulnerability to

Sand and silt could be moved from eroding cliffs into the sea, where cross-shore currents can transport the sediment to beaches. Seawalls might be set back farther from the shore. In the United States, the state of

LEFT *Wetland reclamation of old farmland in Canada's Riding Mountain National Park, Manitoba. Wetlands are an important link in water preservation, as they enable the land to absorb rainwater and limit the volume of runoff into streams or rivers.*

the effects of climate change. Already this approach has been taken up around the world. Wetlands International, a non-profit organization that is based in the Netherlands, runs programs to sustain and restore wetlands in more than 120 countries.

As developers build in new areas and more land is paved or hardened, the land is less able to absorb rainwater, the volume of runoff into streams and rivers increases and the likelihood of flooding rises. One approach to reducing runoff, implemented in countries worldwide, makes use of so-called 'rain gardens' – swales, or depressions and ponds that have been created specifically to capture rain or runoff and allow water to seep back into the soil. Many communities in New South Wales, Australia, have built flood-retarding basins that also help to reduce erosion. In the Pacific North-west region of the United States, Seattle's Street Edge Alternatives program all but eliminated runoff in some areas by narrowing the streets and filling swales created on either side with boulders, plants and porous soil similar to that of forest floors.

If modifying the landscape is not a viable option, even something as simple as rain barrels can make a significant difference. Rain barrels not only reduce the amount of water that pours into storm drains; they also provide water that people can use in their gardens. Rainwater harvesting in this fashion has been used in many low-rain and rural areas around the world for centuries. Today, modified and updated rain barrel programs have been adopted in cities ranging from Vancouver, British Columbia, in Canada, to Austin, Texas, in the United States.

Agricultural Adaptation

The twentieth-century rise of international agri-business has steadily shifted agriculture away from the domain of the small farmer, who knew which crops were well adapted to the local environments and made a point of growing them.

In the modern world, crops are increasingly grown from a limited number of cultivars, or culti-vated varieties, sold by companies that are often based far away from where the crops will be planted. As a result, much of the genetic diversity and adap-tation to local conditions that once existed is disappearing. As climate change worsens droughts and floods, salinizes coastal farmland and makes weather more unpredictable, many have looked to traditional cultivars for solutions.

AS CLIMATE CHANGE WORSENS DROUGHTS AND FLOODS, SALINIZES COASTAL FARMLAND, AND MAKES WEATHER UNPREDICTABLE, MANY LOOK TO TRADITIONAL CULTIVARS FOR SOLUTIONS.

In India, a number of traditional rice varieties can be grown without irrigation and these have been a mainstay for farmers who lacked funds to build wells and irrigation systems. With increasing droughts, the preservation of such species may be key to the suc-cess of rice farming in some localities. In 1996, a dry year in India's Tamil Nadu province, farmers grow-ing irrigation-dependent, high-yield plants suffered losses, while those who had planted traditional vari-eties survived the drought without excessive losses.

The use of plant varieties adapted to local condi-tions is not only found on small farms. In Canada, the British Columbia government runs a Provincial Tree Improvement Program, which maintains genetically diverse trees adapted to varied conditions. By provid-ing these varieties to tree farmers province-wide, the program helps increase yields and strengthen the abil-ity of the province's forestry industry to adapt.

In response to climate change, some communities are modifying the crops they grow. In Tanzania, some farmers now mix fast-growing crops with more water-intensive crops as a means of preparing for either drought or heavy rains. As rising sea levels increase the salinity of farmlands, agricultural agencies around the world, including India's Halophyte Biology Laboratory, work to identify salt-tolerant cash crops. One tomato native to the Galapagos Islands can be watered with seawater. Several salt marsh plants, such as salicornia (pickleweed) or atriplex (saltbrush), are already in use as food crops and salt-tolerant wheat relatives have been identified.

LEFT *Women cultivating the land near Lushoto, Tanzania. Some farmers in Tanzania are experimenting with more varied crops, in anticipation of changing precipitation patterns brought on by climate change.*

Rainwater Harvesting

The collection or diversion of rainwater for human use, called rainwater harvesting, has been practiced for millennia in areas of the world where there is inconsistent access to clean water. As early as 2,600 BCE, the Harappan people developed rainwater harvesting systems in the Indus Valley (in modern-day Pakistan). Rainwater wells and collection jars were used in China 2,000 years ago. In India's Thar Desert, people have been using kunds, or catchment areas of up to 2 hectares (5 acres) with a collection tank in the middle, for almost 1,000 years. As European colonialism spread, disrupting indigenous cultures, many traditional rainwater harvesting techniques fell into decline. Now, with water resources less predictable, these techniques have been reintroduced and redesigned.

In 1995–1996, the government of China's arid Gansu Province assisted farmers by developing the so-called '1-2-1 project' – one water collection field with two water-storage tanks and one area for cultivation of fruits and vegetables irrigated by water that had been collected in the field and stored in the tanks. By large-scale replication of this system, a consistent water supply was ensured for more than a million people and their livestock; many families were able to develop orchards and vegetable gardens as a result. In the United Kingdom, the Rainwater Harvesting Association encourages people to install water-storage tanks in their yards to catch rooftop runoff. In Tanzania and Kenya, some Masai communities that traditionally sealed their roofs with mud and cow dung are switching to corrugated metal or plastic, which permits rainwater harvest. Their systems are not costly, and they are easy to install. As a result, Maasai women are spared many hours of walking to and from the nearest river every day to obtain water.

Like many other techniques for adapting to climate change, rainwater harvesting not only lessens the vulnerability of communities, but it also immediately increases quality of life. The technology is simple and inexpensive and it can be implemented on many scales, both in individual households and as community-wide projects.

One of the challenges is that farmers will need to use strategies that work not just for drought, but for flooding and increased storms as well, while respecting a region's natural water cycle.

In Honduras, traditional Quezungal farming techniques address both problems: crops are planted under trees, which anchor the soil and prevent the erosion resulting from slash-and-burn agriculture. In this practice, farmers mow grass and chop down trees to prepare land for farming. When erosion then leaves it unfit for crops, they move to a new plot and repeat the process. The Quezungal use of both terraces and trees has the additional benefit of conserving water; soil nutrients are replenished by digging trimmings from the trees back into the soil. The effectiveness of this approach was demonstrated in 1998 when Hurricane Mitch devastated the region: farmers using Quezungal techniques lost only 10 per cent of their crops. The technique is well suited to the small hillside plots that form 70 per cent of the country's tillable land. The Quezungal method, like rainwater harvesting, traditional cultivars, rain gardens, floating gardens and mangrove restoration, helps protect communities from the effects of climate change.

ABOVE *Terraced rice paddies in Indonesia are good examples of anticipatory adaptation. Terracing allows farmers to use their land and water resources with greater efficiency.*

PROTECTING THE ECOSYSTEMS OF PLANET EARTH

In addition to affecting human societies, climate change is also having marked effects on biological communities. The animals and plants of all the world's biomes, from the tundra to deserts and tropical forests, are valuable for many reasons: their genetic diversity, their scientific interest, their possible usefulness as resources for new drugs and chemicals and simply for their beauty. Preserving Earth's natural ecosystems in the face of climate change will almost certainly require innovative approaches to conservation.

Ecosystems around the world are already labouring under a variety of stresses, foremost among them pollution, habitat destruction and an influx of invasive species. Climate change aggravates all of these scenarios. Curbing air and water pollution, curtailing harmful land-use practices and reducing the damage to and fragmentation of natural habitat are all ways of helping ecosystems remain healthy and therefore better suited to withstand climate change.

Protecting Species

As plant and animal species worldwide shift toward the poles or to higher altitudes in response to warming, they will require adjoining habitat that enables them to move – habitat that often no longer exists or that is now in danger. Wild animals and plants protected in nature reserves are symbolic of this problem. Many of these reserves are isolated islands of wilderness, surrounded by development. The Bukit Timah Nature Reserve, situated within Singapore's city limits, only 12 km (7 miles) from the city centre, is one extreme example. As the changing climate alters the environment of many of these isolated reserves, the resident species will be unable to migrate to safety. In a few cases, they may be aided by the establishment of reserve networks, or multiple protected areas linked by migration corridors that cut across political boundaries. For instance, the Selous-Niassa wildlife corridor links two game reserves in Tanzania and Mozambique, allowing free movement of elephants and other species. Preservation of such habitats will be essential to the survival of countless species.

Identifying and protecting populations that appear more resilient in the face of climate change may be another, different approach to conservation – a sort of ecological triage. For example, on occasions when there has been mass bleaching of coral reefs, some patches of coral have been less affected or have recovered faster. By targeting these hardier patches for protection, the likelihood that resistant and resilient individuals will spread is increased.

Conserving natural ecosystems in the face of the uncertainties of climate change will require flexibility and responsiveness. Rather than rely on one or two strategies, scientists and conservation managers will need to develop a broad array of options and be ready to abandon or modify methods that fail.

Climate and Humanity's Future

The involvement of humanity is what makes the present epoch of climate change so crucially different from all the previous instances when Earth's climate was transformed. In the vast stretch of time from the formation of the planet's first climate some 4,000 million years ago to the relatively recent past, human beings either were not on the scene or did not affect the climate on any meaningful scale. But since the mid-nineteenth century, when greenhouse gases produced by industrial activity began to accumulate in the atmosphere, climate has been changed by human intervention.

Anthropogenic climate change began without human awareness that it would happen. It was a by-product, rather than an intended consequence, of the use of fossil fuels. But now the consequences of humanity's actions have become clear and we have the power, for the first time, to consciously and deliberately influence Earth's climate. By our own actions, we can lessen the climate change that would otherwise occur. And we can deliberately and consciously adapt to the climate change that cannot be avoided, making the best of what is inevitable.

Our future can now be glimpsed. Worsening forest fires, more severe droughts, increased coastal flooding, threatened wildlife – these are but a few of the dramatic and ominous consequences of the climate change that we can now see all around us. These are consequences that we helped bring about, and that we can continue to influence, for better or worse, for ourselves and for future generations.

RIGHT *Dawson City, a former gold mining town on the Yukon River, in far north-western Canada. Industry and expansion here and elsewhere have brought human settlement to nearly every corner of the globe and have introduced conditions that lead to climate change. Thus, our own future and that of the planet are intimately interconnected.*

A CHRONOLOGY OF CLIMATE

4.6 billion
Earth and the rest of the solar system originate in a nebula of dust and gas.

4 billion
With oceans, atmosphere, and the beginnings of continents, Earth starts to have a climate.

3.8–3.4 billion
Life appears on Earth.

2.7–2.3 billion
One or more episodes of glaciation occur worldwide.

543 million
The Cambrian explosion, a great diversification of life, takes place at a time when global climate is warmer and more uniform than at present.

443 million
At the end of the Ordovician period, an ice age brings on a mass extinction.

354–290 million
A warm, moist climate during the Carboniferous period encourages the growth of dense, swampy forests, source of many coal deposits.

260 million
During the Permian period, the formation of the supercontinent Pangaea leads to a drier climate that favours the proliferation of reptiles.

248 million
The end of the Permian period brings the largest mass extinction in Earth's history. Glaciation and climate change due to volcanic eruptions may have contributed.

206–144 million
During the Jurassic period, the breakup of the supercontinent Pangaea leads to wetter, greener conditions; new kinds of dinosaurs evolve.

100 million
In the mid-Cretaceous period, global climate reaches its warmest point in Earth's history, with an average global temperature of 6°C to 12°C (10.8°F to 21.6°F) warmer than at present.

65 million
At the end of the Cretaceous period, an asteroid 10 km (6 miles) in diameter strikes Earth. The impact causes climatic chaos and the extinction of many of Earth's species, including all the dinosaurs.

50 million
The Cenozoic climate decline, a cooling trend characteristic of the rest of the Cenozoic era (65 million years ago to present), begins.

13 million
The southern ice cap spreads rapidly, covering all or most of Antarctica with the Antarctic ice sheet.

8–5 million
In Africa, the first hominids evolve. The emergence of these bipedal primates and human ancestors may be linked with the spread of savannah as the climate becomes cooler and drier.

2.5 million–10,000
A series of glacial and interglacial periods holds sway worldwide. Most of them occur during the Pleistocene epoch (1.8 million to 10,000 BCE).

2.5 million
The hominid genus *Homo* emerges, perhaps in response to variable climate in Africa.

1.8 million
A new hominid species, *Homo erectus,* evolves in east Africa. It goes on to colonize many parts of Europe and Asia.

200,000–100,000
Homo sapiens man evolves in Africa and begins to migrate to the rest of the world during the warm spell before the most recent glacial period.

71,000
The eruption of Mount Toba in Sumatra darkens the world; 1,000 years of the coldest temperatures of the late Pleistocene begin.

40,000
Lower sea levels as a result of glaciation narrow the gap between continents and allow humans to cross from Asia to Australia.

23,000–12,000
Because of lower sea levels, a land bridge opens from Siberia to Alaska; humans populate the Americas.

14,000–9500
A wave of extinctions of animals, including woolly mammoths, woolly rhinoceros and giant deer, may be caused by a warming trend.

10,900–9600
In the Younger Dryas event, the North Atlantic thermohaline circulation stops, resulting in a period of glaciation.

8000
Warming climate speeds the development of agriculture and civilization in northern Mesopotamia.

6200–5800
The Mini Ice Age brings cold, dry conditions from Europe to the Middle East. As far south as Indonesia, the sea surface cools by about 3°C (5.4°F).

5600
As melting glaciers contribute to rising sea levels, the Mediterranean overflows into Euxine Lake, transforming it into the Black Sea.

5000–4000
The present interglacial period reaches its peak with the Climatic Optimum, a time when global average temperatures are at their highest point for the Holocene epoch.

4000–3000
In the American Southwest, increasing dryness coincides with the disappearance of nascent agricultural societies.

3500
The Neoglaciation begins. This trend toward a cooler and drier climate continues, with fluctuations, to the present.

3500
The Sahara, formerly wet and green, begins to become a desert.

3500–2000
The Sumerians in Mesopotamia adapt to the dry climate by building canals to irrigate their fields.

2200
In Mesopotamia, prolonged drought in the Akkadian empire contributes to its collapse. The drought is connected to more frequent El Niño–Southern Oscillation (ENSO) events.

2184
Egyptian civilization nearly collapses, perhaps because of droughts related to El Niño that drastically reduce the flow of the River Nile.

2000
Severe floods in China on the lower Yangtze and Yellow rivers contribute to the decline of the Longshan and Liangzhu cultures.

1700
Drought and changing river patterns play a part in the collapse of the Harappan culture of western India and Pakistan.

1470–50
A massive volcanic eruption on Thira, or Santorini, an island north of Crete, produces a dense cloud of debris that has a widespread cooling effect. The eruption destroys the community on Thira and contributes to the collapse of the Minoan civilization of Crete.

900–300
A decline in solar radiation may be responsible for a cooler, wetter period in higher and middle latitudes. The last two centuries are especially cold.

200 BCE–100 CE
The climate in North Africa is more moist than at present, making it possible for the Romans to grow crops extensively in that region.

CE

300–400
Drought conditions prevail in the Mediterranean, North Africa and western Asia.

400
Global warmth and sea levels, both rising since the last few centuries BCE, are at an apex. Sea levels are at or slightly above present-day levels.

400–500
The climate in the Roman world becomes cooler, wetter, and more variable, reducing cereal production in Gaul (France and some adjoining areas) and contributing to the fall of the western Roman empire. Climate in Europe remains generally colder and more variable until 900.

535 or 536
Probably due to a volcanic eruption, solar radiation is blotted out for some period of time; this disastrous climate event brings political turmoil worldwide. A dense, dry fog covers Europe, China, and southwest Asia.

500s
When drought hits their homeland, in Asia, the nomadic Avars invade Europe.

750–900
Severe drought stresses the Classic Maya civilization of Mesoamerica and contributes to its collapse.

900–1300
In the Little Optimum or Medieval Warm Period, the North Atlantic climate becomes warmer, with summer temperatures in Europe becoming 1°C (1.8°F) higher than today.

1000

Taking advantage of the warmer climate, the Vikings found the colony of Vinland in what is now Newfoundland, Canada.

1300

Drought coincides with the collapse of the Anasazi civilization of the American Southwest; the Anasazi abandon their cliff dwellings.

1314–17

A series of cold wet summers, connected to a general cooling trend, leads to disastrous harvest failures in Europe.

Late 1300s

As the North Atlantic climate becomes frigid, the Norse settlements in Greenland are lost.

1332

Floods in China cause mass migrations of wild animals, including rats. Some of those rats may carry bubonic plague, or the Black Death, to Europe, which suffers an epidemic of the disease from 1347 to 1352.

1450

Drought contributes to the collapse of the Tiwanaku state in South America. Its distinctive raised fields dry up and its cities are abandoned.

1450–1850

A period of cold known as the Little Ice Age prevails around the world.

1600–1800

Timbuktu, a powerful trading city in what is now Mali, in Africa, suffers from both drought in the summer and periodic winter floods.

1600

As the Little Ice Age redefines the northern limit of where corn can be cultivated, corn growing is abandoned at a site near Winnipeg, Canada.

1645–1715

The Maunder minimum, a period of very low sunspot activity, corresponds to the coldest part of the Little Ice Age in Europe.

1693–1700

Poor harvests in Scotland lead to famine, one of the factors that prompt Scotland to unite with England in the 1707 Act of Union.

1783

The eruption of Laki in Iceland may have contributed to a severe winter in northern Europe in 1783–1784. The dust cloud emitted by the volcano dims the Sun in Paris for months.

1789

The French Revolution is sparked in part by the rising grain prices and widespread hunger arising from bitter weather and crop failures in the eighteenth century.

1815

The eruption of Mount Tambora in Indonesia hurls large quantities of debris into the stratosphere and is linked to exceptionally cold conditions over the next two years.

1845

Cool, wet summer conditions lead to the spread of potato blight fungus in Ireland, ruining the potato crop and starting a famine.

1850

After the Little Ice Age, a general trend toward greater warmth begins that continues, with fluctuations, to the present.

Mid-1800s

Emission of carbon dioxide and other greenhouse gases rises dramatically as a result of escalating industrialization. The result is a discernible human impact on Earth's climate.

1863

Irish physicist John Tyndall discovers what will become known as the greenhouse effect.

1883

The eruption of Krakatau in Indonesia spews out dust that is linked to a worldwide drop in temperature for five years.

1900–2000

Sea levels, which have risen and fallen throughout Earth's history, rise at a rate of 1 to 2 mm (0.04 to 0.09 inches) per year.

1900–2000

Global average surface temperature, which has fluctuated throughout Earth's history, increases by 0.6°C (1.1°F).

1930s

The invention of the radiosonde allows measurements of air temperature, humidity, and pressure throughout Earth's atmosphere.

1934–38

The Dust Bowl: the southern Great Plains of the United States are devastated by dust storms brought on by drought, high winds and poor agricultural practices.

1940s–1970

The worldwide warming trend of the 1920s–1940s is replaced by a cooling trend; the cool period is succeeded by a warming trend that continues until the present day.

1941–42

An unusually cold winter slows the advance of Hitler's army into the Soviet Union, assisting Soviet forces in their defeat of the Germans in the Battle of Stalingrad (1942–1943), a major turning point in World War II.

1950s–1990s

Annual worldwide economic losses from extreme climate events increase tenfold, from $4 billion per year to $40 billion per year.

1970s–present

The average thickness of Arctic ice at the end of the summer season decreases 1.3 m (4.3 feet), or as much as 40 per cent. Between 1979 and 2004, Arctic sea ice declines in extent by about 8 per cent.

1970

A tropical cyclone in the Bay of Bengal on November 12–13 kills 300,000 people in East Pakistan. The Pakistani government's failure to help the victims leads to civil war and independence for East Pakistan as Bangladesh.

1976–97

Cotopaxi, a snow-capped volcano in Ecuador, loses 31 per cent of its ice cover.

1980s–present

Tropical and subtropical ocean evaporation rises 10 per cent.

1982–1983
During these El Niño years large waves occur, stripping substantial portions of California's beaches from Malibu, and Santa Monica, to Venice Beach.

1988
The United Nations Environment Programme (UNEP) and the World Meteorological Organization (WMO) establish the Intergovernmental Panel on Climate Change (IPCC) to provide independent scientific advice on the issue of climate change.

1988
American scientist James Hansen testifies before the U.S. Senate about mounting evidence that accumulation of greenhouse gases is responsible for global warming.

1991
Mount Pinatubo erupts in the Philippines, sending about 20 million tonnes of sulfur dioxide into the atmosphere and reducing the average global temperature by 0.2°C–0.5°C (0.36°F to 0.9°F) for a few years.

1992
The Earth Summit, or United Nations Conference on Environment and Development, takes place in Rio de Janeiro, Brazil. The United Nations Framework Convention on Climate Change is drafted. This is the first intergovernmental attempt to address the problem of climate change.

1994–1999
The Gobi desert grows by 52,000 sq km (20,000 square miles), advancing closer to China's capital, Beijing.

1995
Long periods of high ocean temperature off Nova Scotia, Canada, lead to almost complete green urchin mortality due to parasitic amoebas.

1996
A panel consisting of members of the World Health Organization, WMO, and UNEP emphasizes the seriousness of the threat of climate change to human health.

1996
China's government reports that unusually heavy flood damage costs more than 3,000 lives and more than $20 billion.

1996
Spring in the Northern Hemisphere is coming seven days earlier than in the mid-1970s, as measured by climatic and biological phenomena. Among the events happening earlier are plant flowering, tree budding, frog breeding, and arrival of migrant birds and butterflies.

1997
The Kyoto Protocol to the U.N. Framework Convention on Climate Change (UNFCC) is adopted. This international agreement, once ratified, will commit developed countries to limiting their collective emissions of six greenhouse gases to at least 5 per cent below 1990 levels by 2008–2012.

1997
The Toyota Prius appears on the Japanese market; it is the first mass-produced hybrid petrol-electric vehicle. By the end of 2004, a quarter million have been sold worldwide.

1998
1998 (with 2002, 2003 and 2004) is the hottest year since instrumental records began nearly 150 years earlier.

1999
An Antarctic ice core shows that levels of greenhouse gases are higher than at any time in the past 420,000 years.

Late 1990s
The American West begins to experience a period of prolonged drought.

2000s
Saint Mark's Square in Venice floods ten times more often on average than it did in 1900. The average water level in Venice is 23 cm (9 inches) higher than a century before.

2001
The IPCC predicts a 1.4°C to 5.8°C (2.5°F to 10.4°F) temperature rise over the next century.

2002
Heavy rainfall brings floods in Germany and other European countries.

2002
Tuvalu, in danger of being flooded by rising seas, plans to file a lawsuit against the United States and Australia for their greenhouse gas emissions.

2002–2003
Australia suffers its worst drought in 100 years.

2002
Coral bleaching affects 60 to 95 per cent of the Great Barrier Reef, with some areas experiencing 90 per cent mortality.

2003
Analyzing data from studies covering several decades, American researchers Camille Parmesan and Gary Yohe show that rising temperatures are affecting 84 per cent of 334 species studied.

2003
A heatwave in Europe takes 30,000 lives and costs 16 billion euros in damage. About 600,000 hectares (1.5 million acres) of European forests are burned.

2004
Weather-related events take over 10,000 lives and cost the insurance industry $40 billion.

2004
Florida suffers more hurricanes in one season than any US state in more than a hundred years. Beginning on 13 August, Florida is hit by Hurricanes Charley, Frances, Ivan and Jeanne.

2004
In the Yukon, Canada, fires burn more than 1 million hectares (2,470,000 acres) of forest.

2004
The Arctic Climate Impact Assessment (ACIA) reports that the Arctic is experiencing 'some of the most rapid and severe climate change on Earth' and that the change is likely to accelerate over the next 100 years.

2006
For every square metre (1 square yard) of surface area, Earth is now absorbing 0.71 watts more solar energy than it is radiating back to space as heat.

CHARTING CLIMATE CHANGE

Climate studies and models make it clear that humans are changing the climate of Earth. Scientists, even global warming sceptics, have not been able to explain how recent climate changes could have occurred without significant human (anthropogenic) influences. The following images record or predict changes in many of the elements that affect climate: the shrinking polar cap, rising carbon dioxide levels, and temperature and precipitation levels. Although there is always a level of uncertainty in predictions of the future, taken together, these images help to demonstrate that climate change is a reality, and that it will be dangerous for human society to continue allowing carbon dioxide levels to rise.

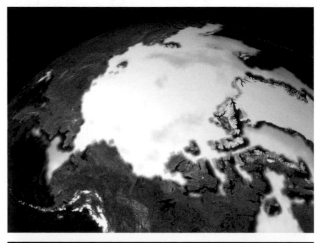

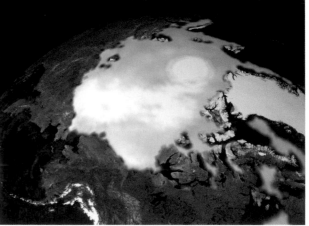

ABOVE *Observed Arctic sea ice extent in 1990, top, and 1999, below. These images were created using data from the Defense Meteorological Satellite Program's Special Scanning Microwave Imager, an instrument employed by the National Aeronautics and Space Administration (NASA) in the United States.*

1,000 YEARS OF CHANGES IN CARBON EMISSIONS, CO_2 CONCENTRATIONS, AND TEMPERATURE

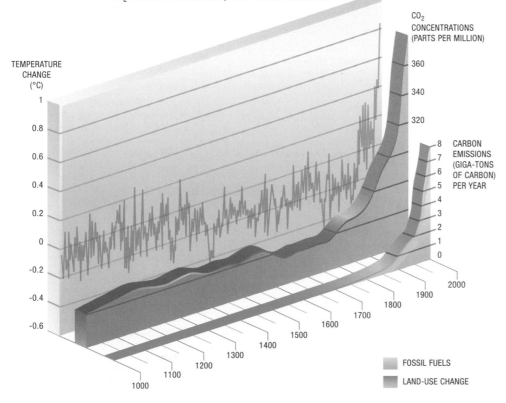

FOSSIL FUELS

LAND-USE CHANGE

LEFT *This 1,000-year record tracks the rise in carbon emissions due to human activities and the subsequent increase in atmospheric carbon dioxide concentrations and air temperatures. The earlier parts of this Northern Hemisphere temperature reconstruction are derived from historical data, tree rings and corals, while the later parts were measured directly. (Recreation of a graph released by the Arctic Climate Impact Assessment.)*

PROJECTED CHANGE IN PRECIPITATION

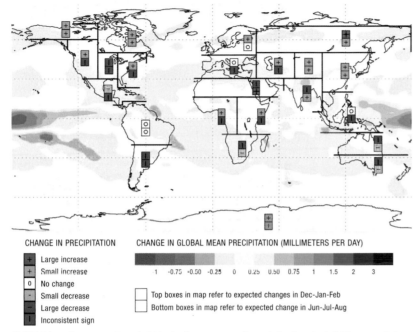

CHANGE IN PRECIPITATION

- Large increase
- Small increase
- No change
- Small decrease
- Large decrease
- Inconsistent sign

CHANGE IN GLOBAL MEAN PRECIPITATION (MILLIMETERS PER DAY)

1 -0.75 -0.50 -0.25 0 0.25 0.50 0.75 1 1.5 2 3

Top boxes in map refer to expected changes in Dec-Jan-Feb

Bottom boxes in map refer to expected change in Jun-Jul-Aug

ABOVE *Map shows predicted shifts in the amount of precipitation (rainfall) around the world, in both winter and summer, by the end of the 21st century as a result of climate change. This map was produced using the Intergovernmental Panel on Climate Change's B2 scenario. The B2 is a 'medium low' scenario that predicts a moderate rise in global population and fossil fuel emissions by the end of the 21st century.*

PROJECTED CHANGE IN TEMPERATURE

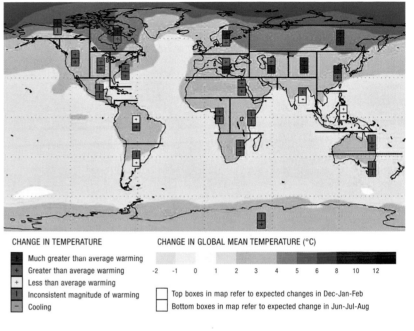

CHANGE IN TEMPERATURE

- Much greater than average warming
- Greater than average warming
- Less than average warming
- Inconsistent magnitude of warming
- Cooling

CHANGE IN GLOBAL MEAN TEMPERATURE (°C)

-2 -1 0 1 2 3 4 5 6 8 10 12

Top boxes in map refer to expected changes in Dec-Jan-Feb

Bottom boxes in map refer to expected change in Jun-Jul-Aug

ABOVE *Map published by the Intergovernmental Panel on Climate Change (IPCC) predicts where and how much temperature change is likely to occur over the coming century, as predicted by a climate model using the IPCC's B2 scenario. The B2 scenario projects that average global temperatures will rise by about 2.2°C (3.9°F) between the late 20th century and the end of the 21st century.*

What You Can Do To Limit Climate Change

TRAVEL SMART:

- Take public transport, carpool, walk or bike.
- If you have a vehicle get your engine tuned up for best fuel consumption.
- Start a no-idle rule in your school or carpool lane.
- When purchasing a vehicle, make fuel efficiency a priority.

BE ENERGY EFFICIENT:

- Turn off lights, computers and electrical appliances. Unplug adaptors and chargers from the wall when not in use.
- Replace incandescent light bulbs with compact fluorescent bulbs.
- Buy high-efficiency "Energy Star" appliances.
- Use clean energy: solar, geothermal heat pump or wind power.
- Call your local utility and sign up for renewable energy sources.
- Weatherize your home: caulk and weather-strip doorways and windows, insulate roofs, use a light coloured roof and overhangs to shade rooms.

- Manage heat through your windows: use shades to shield or let in sunlight for cooling or heating. Install double-pane solar windows.
- Install a programmable thermostat. Programme it to turn down a few degrees in winter, or up a few degrees in summer, while you are asleep or out.
- Encourage energy-efficient practices in public places. When you are in a building that is over-air-conditioned, or wasteful of electricity, complain!

CREATE INCENTIVES:

- Agree to local carpool lane incentives. Offer free parking for hybrid vehicles.
- If you are a business owner, offer clean-energy incentives to employees.
- Buy from responsible companies. Let the companies know how you feel by writing letters.

USE RECYCLABLE PRODUCTS:

- Use post-consumer products, including paper towels, toilet paper, tissues.
- Recycle your paper, glass, metal and plastic.

TREES AND PLANTS:

- Replace some of your lawn with natural vegetation. Cut lawns less frequently. Leave grass clippings on top of your lawn after cutting. Do not overfertilize your lawn.
- Plant more greenery and trees. Encourage your community to install islands of greenery in parking lots, and put green roofs on public buildings.
- Participate in forest conservation and tree-planting community campaigns.

MAKE YOUR VOICE HEARD:

- Join a national or local environmental group that is fighting the climate crisis every day, so its numbers increase and its voice cannott be ignored.
- Write to your elected officials and tell them that addressing climate change must be a high priority. Ask your friends to do the same.
- Do not invest in companies that have poor records on climate change.
- Write to your local newspapers and call radio shows and tell them you are concerned about climate change.
- Get these issues into the curriculum in schools.

GLOSSARY

ablation The process by which ice and snow waste away because of melting and evaporation.

ACIA Arctic Climate Impact Assessment. An international project of the Arctic Council and the International Arctic Science Committee (IASC) to evaluate knowledge on the effects of climate change in the Arctic.

adaptation An evolved characteristic of an organism that results in a better fit with its environment. Also, an adjustment made in response to, or in anticipation of, actual or expected change to reduce harm or maximize benefits.

aerosol Solid or liquid particles suspended in a gas, such as mist in the atmosphere. Also refers to a liquid substance packaged under pressure, to be dispersed in a fine spray. Use of aerosol cans, which release greenhouse gases, can contribute to global warming.

air mass A large body of air, usually at least 1,600 km (1,000 miles) across and several kilometres thick, characterized by similar temperature and moisture throughout.

albedo The reflectivity of a surface; the proportion of incoming **radiation** that a surface reflects.

allergen A substance that can produce a hypersensitive reaction in some people.

alpine Of or relating to high mountains.

anthropogenic Induced by humans, as in anthropogenic **climate change**.

AOGCM Atmosphere–Ocean General Circulation Model; a sophisticated mathematical model that shows global climate behaviour.

aquifer An underground layer of porous rock, sand or gravel that bears water; a source of groundwater for wells and springs.

asthenosphere A zone of Earth's mantle beneath the **lithosphere** that is made up of hot, deformable rock.

atmosphere The gases or air surrounding a planet and bound to it by gravity.

AVHRR Advanced Very High Resolution Radiometer. An instrument carried on US polar orbiting satellites that measures radiation to determine cloud cover and surface temperature.

backscattering The phenomenon whereby some solar **radiation** is reflected or scattered back by Earth's surfaces or its **atmosphere**.

biological community A group of interdependent organisms living together in one area.

biome A large-scale region distinguished by its climate and dominant vegetation.

biosphere The regions of Earth and its atmosphere capable of sustaining life.

biota Living organisms of a particular region, including flora (plants) and fauna (animals).

boreal forest see **taiga**

breakwater An offshore wall or barrier that protects a coast or harbour by breaking the force of waves and diminishing their power to cause erosion.

carbon cycle The continuous processes by which carbon circulates between living organisms and the environment. Plants convert atmospheric carbon dioxide to organic compounds, which are consumed by plants and animals; the carbon is then returned to the biosphere through respiration and decay of living organisms.

carbon dioxide (CO$_2$) A colourless, odourless greenhouse gas produced by burning fossil fuels as well as by natural processes. Each molecule consists of one carbon atom and two oxygen atoms.

carbon sequestration The isolation and storage of carbon dioxide in the oceans, forests, soil, or in former oil and gas reservoirs underground. These processes are intended to prevent carbon from entering the atmosphere and contributing to climate change.

chaparral A **biome** of tough, crooked shrubs and small trees, with mild, moist winters and hot, dry summers.

climate change A statistically significant alteration in the long-term behaviour of weather that persists for an extended period of time.

climate forcing A change in a mechanism outside the Earth's climate system, such as a change in **thermohaline circulation** or in Earth's orbit, that can effect, or force, a change in climate.

climate model A mathematical way of representing the interactions of the **climate system**, used to understand the processes that govern climate and predict climate change.

climate system All the elements that contribute to the climate, including **atmosphere**, **hydrosphere**, **cryosphere** and **biosphere**.

cloud forest An evergreen tropical forest nearly always immersed in mist and clouds, usually found in **montane** environments.

cloud seeding The introduction, usually by aircraft, of substances such as silver iodide or dry ice into a cloud to promote **precipitation**.

coccolithophore A single-celled marine plant that lives in large numbers throughout the upper layers of the ocean.

co-generation The simultaneous generation of electricity and heat in a combination heat-and-power plant.

convection The phenomenon in which the warmer, lighter part of a fluid (such as air or water) rises and the cooler, denser part of it sinks.

coral bleaching The process by which corals turn white from the loss of beneficial algae that inhabit them.

Coriolis effect The phenomenon, resulting from Earth's rotation, in which currents of air or water in the Northern Hemisphere are deflected to the right and currents in the Southern Hemisphere to the left.

cryosphere The frozen portion of Earth's surface, including ice and snow.

cyanobacteria A type of bacteria that uses photosynthesis. Also known as blue-green algae.

deforestation Large-scale burning and clear-cutting of forested land to provide fields for agriculture or to graze animals. Also, the clearing of forested areas to build homes, businesses or factories.

desertification The progressive transformation of fertile land into desert, either by overgrazing, soil depletion or climate change.

drought A period of abnormally dry weather long enough to cause serious shortages of water.

dry land Arid land that is at risk of becoming uninhabitable through **desertification**.

eccentricity A measure of how elliptical Earth's orbit is.

ecosystem A community of interacting organisms and their physical environment.

El Niño A periodic warming of the surface water of the eastern and central Pacific Ocean, occurring approximately every 4–7 years, which disrupts typical weather patterns worldwide, bringing heavy rainfall to coastal South America and drought to Eastern Australia and Indonesia. This warming is associated with the atmospheric phenomenon called the Southern Oscillation, hence the El Niño–Southern Oscillation (**ENSO**).

electromagnetic radiation Forms of radiation, including ultraviolet, visible light and infrared, requiring no medium for transmission and moving in a vacuum at the speed of light (300,000 km or 186,000 miles per second).

emissions In the context of climate change, the release of **greenhouse gases** and **aerosol** into the atmosphere, especially from industry, transport and agriculture.

emissions trading A market-based system designed to reduce total global emissions of greenhouse gases or other pollution. Generally, nations, states or other regulatory bodies allow regulated parties a given amount of emissions or pollution, called a cap, and the parties are allowed to trade or sell credits for any unused amounts of emissions or pollution under that cap.

energy budget The balance by which the amount of energy the planet receives from the Sun equals the amount it radiates back into space.

ENSO see **El Niño**

ESA European Space Agency.

eustatic forces Processes, such as the melting of continental glaciers, that contribute to changes in global sea levels.

evaporation The conversion of a liquid to a gas.

evapotranspiration The combined process of **evaporation** and **transpiration**.

feedback A process resulting from **climate change** that either magnifies the original change (positive feedback) or weakens it (negative feedback).

fossil fuels Carbon-based fuels, including coal, oil and natural gas, that are extracted from the remains of organisms such as plants and animals that lived millions of years ago.

front A boundary where two **air masses** of different origins and characteristics meet. Changes in weather usually occur along fronts.

general circulation The large-scale motions of the atmosphere and ocean over Earth's surface as a result of differences in heating by the Sun.

geothermal Of or relating to Earth's internal heat.

glaciation The process by which **glaciers** expand across land, or a period when glaciers are expanding.

glacier A thick, moving mass of ice that forms on land when snow, accumulated from year to year, compacts and recrystallizes.

global warming A term often used to refer to the present ongoing change in climate that began in the twentieth century as a result of industrial activities. It is notably represented by the increase in average global surface temperature since the late 1970s. However, this **anthropogenic climate change** is manifested in different forms, depending on region, with not all places becoming warmer.

Gondwana or **Gondwanaland** A prehistoric supercontinent of the Southern Hemisphere that broke up into India, Australia, Antarctica, Africa and South America.

great ocean conveyer belt see **thermohaline circulation**

greenhouse effect The warming of a planet as a result of **carbon dioxide** or other atmospheric gases that allow solar **radiation** to pass through the **atmosphere**, but at the same time slow the release of infrared radiation from Earth back into space.

greenhouse gas An atmospheric gas, such as **carbon dioxide** or **methane**, that traps infrared radiation when it leaves a planet's surface.

heat island An urban area with temperatures higher than the surrounding region because of factors such as the high absorption of solar energy by asphalt.

Heinrich event A phenomenon in which icebergs are deposited rapidly and massively into the North Atlantic ocean, causing **climate change**.

high A centre of high pressure around which winds blow outward, clockwise in the Northern Hemisphere and anticlockwise in the Southern Hemisphere. Also called an anticyclone. A high is associated with fair, dry weather.

hominid Any member of the family of human beings and prehistoric human-like species.

hurricane An intense tropical storm system with sustained winds of 119 km/h (74 mph) or more.

hydropower Energy generated by the flow of water from a higher to a lower place.

hydrosphere All of the water present on Earth, including oceans, rivers and lakes.

hydrothermal vent A superheated opening in the ocean floor usually at mid-ocean ridges and capable of supporting organisms that draw their energy from chemicals in the hot water.

ice age Any of several periods in Earth's history when **glaciers** covered large regions of the planet.

ice sheet A large mass of land ice that covers most of the underlying bedrock.

ice shelf A flat sheet of ice cap that extends over the sea.

infrared radiation Heat energy emitted by solids, liquids, and gases. Earth absorbs solar radiation and emits it as infrared, which is the form of energy trapped by **greenhouse gases**.

intertidal zone The shore area between the high tide and low tide marks.

invasive species A species that is able to expand its range quickly when introduced into a new habitat.

IPCC Intergovernmental Panel on Climate Change. An international body established in 1988 to assess scientific information relevant for the understanding of climate change

isostatic The state in which pressures from every side or direction are equal. The term is often used in relation to segments of the earth's crust, which maintain an equilibrium based on their thickness and density.

Kyoto Protocol An amendment to the UN Framework Convention on Climate Change (UNFCC), an international treaty on **global warming**, in which signing countries agree to reduce their emissions of carbon dioxide and other **greenhouse gases**, or engage in **emissions trading**. It was negotiated in Kyoto, Japan, in December 1997, and came into force on 16 February 2005.

lithosphere The outer shell of Earth, composed of **tectonic plates**.

Little Optimum A warming period from roughly 900 to 1300 CE, in which summer temperatures in Europe became 1°C (1.8°F) higher than today. Also called the Medieval Warm Period.

low A centre of low pressure around which winds blow inward, anticlockwise in the Northern Hemisphere and clockwise in the Southern Hemisphere. Also called a cyclone or depression. A low is associated with clouds, storms and **precipitation**.

mangrove A salt-tolerant tree that grows in wetlands along tropical coasts. Mangrove forests provide a buffer against coastal erosion.

Maunder minimum A period of very low sunspot activity from 1645 to 1715, associated with low temperatures and severe winters in western Europe.

Mediterranean climate A climate with warm to hot, dry summers and mild, rainy winters, typical of lands bordering the Mediterranean Sea.

MetOp Meteorological Operational Weather Satellite, a series of three polar-orbiting weather satellites launched by the **ESA**, beginning in 2006.

methane A colourless, odourless, flammable gas (CH_4); this **greenhouse gas** is a major constituent of natural gas.

migration The movement, usually seasonal, of animal species from one location to another.

Milankovitch cycles Three cyclical variations in Earth's orbit, described by Serbian mathematician Milutin Milankovitch (1879–1958), all of which may affect climate.

Mini Ice Age A cooling period lasting from roughly 6200 to 5800 BCE.

monsoon A seasonal wind that blows persistently in one direction during one season, but changes direction in the next, occurring primarily in the Indian Ocean and southern Asia.

montane Relating to or inhabiting mountain areas.

morphology The science of the form and structure of animals and plants.

MSG Meteostat Second Generation, a series of **ESA** weather satellites in use since 2002.

NAO North Atlantic Oscillation, a large-scale, cyclical fluctuation of air pressure between the subtropical high and the polar low, driving surface winds and wintertime storms from west to east across the North Atlantic. The NAO affects climate from North America to Europe, then eastwards to Siberia and southwards to western Africa.

NASA National Aeronautics and Space Administration, a governmental agency of the U.S.

Neoglaciation A series of relatively small advances made by mountain glaciers since the Pleistocene ice age.

NOAA National Oceanic and Atmospheric Administration, an agency of the US Department of Commerce that conducts environmental research.

obliquity The degree of tilt of Earth's axis.

ocean current A movement of seawater propelled by wind (surface currents) or by differences in water density related to temperature and salt content (**thermohaline circulation**).

pack ice Ice floating on the ocean's surface that has been driven together into a mass.

paleoclimatologist A scientist who studies past climates.

particulate A minute solid suspended in air or water.

pathogen A disease-producing microorganism.

permafrost Ground that is perennially frozen.

phenology Seasonal or recurrent changes in plants and animals such as migration or blossoming.

photosynthesis The process by which green plants use sunlight to convert carbon dioxide and water into carbohydrates (food).

phytoplankton Tiny, free-floating aquatic plants.

plankton Small organisms that drift or swim weakly in oceans and other bodies of water. They include phytoplankton, which use photosynthesis, and zooplankton, which do not.

precession The wobble of Earth on its axis.

precipitation Any water, liquid or solid, that falls from the atmosphere, including rain, snow, sleet, hail and dew.

radiation Emission of energy in waves or rays or particles.

radiosonde A balloon-borne instrument used to make and transmit measurements of air temperature, humidity and pressure throughout Earth's atmosphere.

rain garden A pond or low-lying ground area created to capture rain or runoff and allow the water to seep back slowly into the soil.

rainwater harvesting The collection or diversion of rainwater for human use.

renewable energy Power that comes from natural resources that replenish themselves over short periods, including the Sun, wind, moving water, organic plant and waste material (biomass), and Earth's heat.

runoff Rainfall that flows across the surface of the ground to streams, rivers and lakes.

savanna A tropical or subtropical grassland.

stilling well A gauge for measuring water levels that dampens wave activity.

storm surge A rapid, temporary increase in local sea levels produced when strong winds drive ocean waters ashore.

stromatolite A mound or dome left by cyanobacteria as they formed in mats in shallow water.

subduction The sinking of one part of the earth's crust, or **tectonic plate**, beneath another as two plates collide.

sunspot A dark area on the surface of the Sun that is thousands of degrees cooler than its surroundings. The number of sunspots varies over an 11-year cycle, related to a 22-year cycle in the reversal of the Sun's magnetic field.

taiga An evergreen coniferous forest, also known as a **boreal forest,** with long, cold winters and short summers.

tectonic plate One of several vast sections of rock that form Earth's **lithosphere** and move around individually on a hot, deformable layer called the **asthenosphere**. The continents are embedded in the tops of tectonic plates.

thermal expansion A component of sea level rise resulting from the increase of ocean volume as the water warms.

thermohaline circulation A large-scale system of water movement in the world's oceans stemming from differences in water density related to temperature and salt content; also known as the **great ocean conveyor belt**.

thermokarst Pits or depressions that are caused by the thawing of **permafrost** and the subsequent collapse of the ground surface.

tide The regular rise and fall of sea levels that occurs as a result of the gravitational pull of the Moon and, to a lesser extent, the Sun.

tornado A storm consisting of a funnel-shaped column of air that rapidly rotates under a large thundercloud or developing thundercloud.

transpiration The **evaporation** of water from plants, primarily through the leaves.

troposphere The lowest layer of the **atmosphere**, containing about 80 per cent of the atmosphere's total mass and reaching an average height of about 11.8 km (7.4 miles).

tundra A cold, dry **biome** where trees do not grow and the ground has a layer of **permafrost**.

UNEP United Nations Environment Programme. UN organization that makes information about climate change understandable to the public and to decision-makers.

UNFCCC United Nations Framework Convention on Climate Change. UN organization focusing on climate change.

upwelling The marine phenomenon in which cooler water rises from below to replace warm surface water driven away by winds.

vernal pool A seasonal wetland area that is dry for much of the year.

water vapor Water (a molecule composed of two hydrogen atoms and one oxygen atom) in its gaseous state; a greenhouse gas.

wetland An area, such as a marsh or swamp, where the water level remains near or above ground level for most of the year.

WMO World Meteorological Organization, a United Nations agency concerned with the international collection of weather and climate data, and with related geophysical sciences.

WWF World Wildlife Fund (outside North America, the World Wide Fund for Nature).

Younger Dryas event A period of rapid **climate change** from roughly 10,900 to 9600 BCE, in which the North Atlantic **thermohaline circulation** stopped, resulting in a reglaciation.

zooplankton Minute drifting or weakly swimming marine animals.

FURTHER READING & ORGANIZATIONS

Books

Alley, Richard B. *The Two-Mile Time Machine: Ice Cores, Abrupt Climate Change, and Our Future.* Princeton: Princeton University Press, 2000.

Arctic Climate Impact Assessment. *Impacts of a Warming Arctic: Arctic Climate Impact Assessment.* Cambridge: Cambridge University Press, 2004. Available at www.acia.uaf.edu.

Buddemeier, Robert W., et al. *Coral Reefs and Global Climate Change.* Arlington, Va.: Pew Center on Global Climate Change, 2004.

Burroughs, William James. *Climate Change: A Multidisciplinary Approach.* Cambridge: Cambridge University Press, 2001.

Calvin, William H. *A Brain for All Seasons: Human Evolution and Abrupt Climate Change.* Chicago: University of Chicago Press, 2002.

Cloudsley-Thompson, John. *Ecology.* Lincolnwood, Illinois: NTC Publishing Group, 1998.

Cowen, Richard. *History of Life. 2nd ed.* Boston: Blackwell Scientific Publications, 1995.

Crowley, Thomas J., and Gerald R. North. *Paleoclimatology.* New York: Oxford University Press, 1991.

Douglas, Bruce C., et al., eds. *Sea Level Rise: History and Consequences.* San Diego, California: Academic Press, 2001.

Easterling, William E., et al. *Coping With Global Climate Change: The Role of Adaptation in the United States.* Arlington, Virginia: Pew Center on Global Climate Change, 2004.

European Climate Forum. *What Is Dangerous Climate Change? Results of an International Symposium on Key Vulnerable Regions and Climate Change.* 2004.

Fagan, Brian. *The Little Ice Age: How Climate Made History, 1300–1850.* New York: Basic Books, 2002.

ibid. The Long Summer: How Climate Changed Civilization. New York: Basic Books, 2004.

Gelbspan, Ross. *The Heat Is On: The Climate Crisis, the Cover-Up, the Prescription.* Reading, Massachusetts: Perseus Books, 1998.

Hansen, Lara, et al. *Buying Time: A User's Manual for Building Resistance and Resilience to Climate Change in Natural Systems.* Berlin: WWF Climate Change Program, 2003.

Intergovernmental Panel on Climate Change. *Climate Change 2001: Impacts, Adaptation, and Vulnerability.* Cambridge: Cambridge University Press, 2001.

ibid. Climate Change 2001: Mitigation. Cambridge: Cambridge University Press, 2001.

ibid. Climate Change 2001: The Scientific Basis. Cambridge: Cambridge University Press, 2001.

ibid. Climate Change 2001: Synthesis Report. Cambridge: Cambridge University Press, 2001.

Jones, Steve, et al. *The Cambridge Encyclopedia of Human Evolution.* Cambridge: Cambridge University Press, 1992.

Lamb, H.H. *Climate, History and the Modern World, 2nd ed.* London: Routledge, 1997.

Lynas, Mark. *High Tide: The Truth About our Climate Crisis.* New York: Picador, 2004.

Lynch, John. *The Weather.* Toronto: Firefly Books, 2002.

Malcolm, Jay R., et al. *Habitats at Risk: Global Warming and Species Loss in Globally Significant Terrestrial Ecosystems.* Gland, Switzerland: World Wide Fund for Nature (WWF), 2002.

Mayewski, Paul Andrew, and Frank White. *The Ice Chronicles: The Quest to Understand Global Climate Change.* Hanover, New Hampshire: University Press of New England, 2002.

Parmesan, Camille, and Hector Galbraith. *Observed Impacts of Global Climate Change in the U.S.* Arlington, Virginia: Pew Center on Global Climate Change, 2004.

Philander, S. George. *Is the Temperature Rising? The Uncertain Science of Global Warming.* Princeton: Princeton University Press, 1998.

Pielou, E.C. *After the Ice Age: The Return of Life to Glaciated North America.* Chicago: University of Chicago Press, 1991.

Potts, Rick. *Humanity's Descent: The Consequences of Ecological Instability.* New York: William Morrow and Co., 1996.

Robinson, Peter J., and Ann Henderson-Sellers. *Contemporary Climatology, 2nd ed.* Harlow, UK: Pearson Prentice Hall, 1999.

Ross, Tom, and Neal Lott. *A Climatology of 1980–2003: Extreme Weather and Climate Events. National Climatic Data Center Technical Report 2003–01.* Asheville, North Carolina: National Climatic Data Center, 2003.

Schneider, Stephen H., and Terry L. Root. *Wildlife Responses to Climate Change: North American Case Studies.* Washington, D.C.: Island Press, 2002.

Southwood, Richard. *The Story of Life.* Oxford: Oxford University Press, 2003.

Stevens, William K. *The Change in the Weather: People, Weather, and the Science of Climate.* New York: Delta, 2001.

Tarbuck, Edward J., and Frederick K. Lutgens. *Earth Science, 9th ed.* Upper Saddle River, New Jersey: Prentice Hall, 2000.

World Health Organization. *Climate Change and Human Health – Risks and Responses.* Geneva: World Health Organization, 2003.

Articles

Davis, Curt H., Yonghong Li, Joseph R. McConnell, Markus M. Frey, and Edward Hanna. 'Snowfall-Driven Growth in East Antarctic Ice Sheet Mitigates Recent Sea-Level Rise.' *Science Express Reports* (May 2005) DOI: 10.1126/science.1110662.

deMenocal, Peter B. 'African Climate Change and Faunal Evolution during the Pliocene-Pleistocene.' *Earth and Planetary Science Letters* 220 (2004): 3–24.

'The Heat Is On.' *National Geographic,* September 2004, 2–75.

Kolbert, Elizabeth. 'The Climate of Man', 3-part series. *The New Yorker,* 25 April 2005, 56–71; 2 May 2005, 64–73; 9 May 2005, 52–63.

Malcolm, Jay R., et al. 'Estimated Migration Rates under Scenarios of Global Climate Change.' *Journal of Biogeography* 29 (2002): 835–849.

Parmesan, Camille, and Gary Yohe. 'A Globally Coherent Fingerprint of Climate Change Impacts across Natural Systems.' *Nature* 421 (2003): 37–42.

Ruddiman, William F. 'How Did Humans First Alter Global Climate?' *Scientific American,* March 2005, 46–53.

Serreze, Mark C., et al. 'Observational Evidence of Recent Change in the Northern High-Latitude Environment.' *Climatic Change* 46 (2000): 159–207.

Thomas, Chris D., et al. 'Extinction Risk from Climate Change.' *Nature* 427 (2004): 145–148.

Walther, Gian-Reto, et al. 'Ecological Responses to Recent Climate Change.' *Nature* 416 (2002): 389–395.

Organizations

AUSTRALIA
Australian Greenhouse Office
Department of the Environment and Heritage,
GPO Box 787, Canberra ACT 2601
www.greenhouse.gov.au
Government agency charged with delivering most of the programmes related to Australia's climate change strategy.

CSIRO (Commonwealth Scientific and Industrial Research Organisation) Marine and Atmospheric Research
GPO Box 1538, Hobart TAS 7001
www.cmar.csiro.au
Research unit that studies atmospheric and Earth systems and predicts climate, weather and ocean processes.

BANGLADESH
Climate Action Network – South Asia (CANSA)
Secretariat House 4 (1st Floor), Road 11 Dhanmondi
Dhaka – 1205
www.can-sa.net
South Asian arm of Climate Action Network, a worldwide network of nongovernmental organizations (NGOs) working to limit human-induced climate change.

BELGIUM
The European Climate Change Programme
European Commission, Environment DG Information Centre, Office: BU–9 01/11, B – 1049 Brussels
europa.eu.int
Intergovernmental programme to identify and develop a European Union strategy to implement the Kyoto Protocol.

BRAZIL
Inter American Institute for Global Change Research (IAI)
Av. dos Astronautas, 1758, São José dos Campos
São Paulo CEP 12227-010
www.iai.int
Brazil-based intergovernmental organization of 19 Western hemisphere countries, collaborating to increase scientific understanding of phenomena related to global change.

CANADA
Canadian Institute for Climate Studies
C199 Sedgwick Building, University of Victoria,
P.O. Box 1700, Sta CSC Victoria, BC V8W 2Y2
www.cics.uvic.ca
Organization that provides science-based information on climate change and other issues to decision-makers in business, industry and government.

Climate Change Knowledge Network (CCKN)
c/o International Institute for Sustainable Development
161 Portage Avenue East, 6th Floor, Winnipeg, Manitoba R3B 0Y4
www.cckn.net
Network of organizations aimed at increasing the exchange of knowledge and research expertise between developed and developing countries on climate change.

ICLEI – Local Governments for Sustainability
City Hall, West Tower, 16th Floor, 100 Queen Street West, Toronto, Ontario M5H 2N2
www.iclei.org
International association of local governments and national and regional local government organizations committed to sustainable development. Through its Cities for Climate Protection (CCP) Campaign, ICLEI promotes reduction of greenhouse gas emissions.

Taking Action on Climate Change
Public Works and Government Services Canada, Ottawa, Ontario K1A 0S5
www.climatechange.gc.ca
Government website that provides information, tips and resources in support of Canada's commitment to the Kyoto Protocol on combating climate change.

FRANCE
National Observatory on the Effects of Climate Warming
39, rue Saint Dominique, 75007 Paris
www.ecologie.gouv.fr
French government agency dedicated to gathering and disseminating information on climate change.

GERMANY
Bavarian Cooperation for Research on Regional Climate Changes (BayFORKLIM)
Universität München, Theresienstraße 37,
D-80333 München
www.bayforklim.uni-muenchen.de
Interdisciplinary association of state universities, state agencies and federal research institutes studying climate changes in Bavaria.

Climate Alliance
European Secretariat, Galvanistr. 28, D-60486 Frankfurt am Main
www.klimabuendnis.org
This organization has as its focus the reduction of greenhouse gas emissions in industrialized countries.

European Climate Forum
c/o Potsdam Institute for Climate Impact Research (PIK)
P.O. Box 601203, 14412 Potsdam

www.european-climate-forum.net
Organization that brings together different parties concerned with the climate problem, including industry, technology and major energy users.

Federal Ministry for the Environment, Nature Conservation and Nuclear Safety
P.O. Box 12 06 29, 53048 Bonn
www.bmu.de
Government ministry charged with carrying out Germany's commitment to the Kyoto Protocol.

United Nations Framework Convention on Climate Change (UNFCCC)
P.O. Box 260124, D-53153 Bonn
unfccc.int
UN organization founded during the 1992 Earth Summit at Rio de Janeiro, Brazil. Its focus is on climate change, with complete text of the Kyoto Protocol on its website.

INDIA
TERI (The Energy and Resources Institute)
Darbari Seth Block, IHC Complex, Lodhi Road, New Delhi   110 003
www.teriin.org
Indian organization committed to sustainable development and concerned with the impacts of climate change.

JAPAN
Asia-Pacific Network on Climate Change (AP-Net)
Shibakoen Annex 7th Floor, 1–8, Shibakoen 3chome, Minato-ku, Tokyo 105-0011
www.ap-net.org
On-line clearing house for the Asia–Pacific region on climate change issues.

KENYA
African Centre for Technology Studies (ACTS)
P.O. Box 45917, Nairobi
www.acts.or.ke
International research and outreach organization dedicated to strengthening the capacity of African countries to use science and technology for sustainable development. One of its projects concerns adaptation to climate change.

NETHERLANDS
Greenpeace International
Ottho Heldringstraat 5, 1066 AZ Amsterdam
www.greenpeace.org
International organization, with a presence in 40 countries, that focuses on confronting threats to Earth's biodiversity and environment, including the threat of climate change.

NEW ZEALAND
New Zealand Climate Change Office
PO Box 10362, Wellington
www.climatechange.govt.nz
An organization that is responsible for coordinating and implementing New Zealand's programmes and policy developments in response to climate change.

NORWAY
United Nations Environment Programme (UNEP)/GRID-Arendal
Longum Park, Service Box 706, N-4808 Arendal
www.grida.no
Organization that makes science-based knowledge,

including information about climate change, understandable to the public and to decision-makers. Website includes full text of the IPCC Third Assessment Report.

SAMOA
Pacific Regional Environmental Programme
Secretariat of the Pacific Regional Environment Programme (SPREP), P.O. Box 240, Apia
www.sprep.org.ws
Regional organization established by the governments and administrations of the Pacific region; among its areas of action are climate change and sea-level rise.

SENEGAL
Enda Energy
c/o ENDA Tiers Monde, 4 & 5 rue Jacques Bugnicourt ex Kléber, BP 3370, Dakar
www.enda.sn
Organization focusing on energy use in Africa, with an emphasis on the links between energy and development. Climate change is among its areas of concern.

SOUTH AFRICA
South African Weather Service (SAWS)
Private Bag X097, Pretoria 0001
www.weathersa.co.za
Government agency that forecasts weather and climate and provides information on climate change.

Sustainable Energy Africa
The Green Building, 9B Bell Crescent Close, Westlake Business Park, Tokai, 7945
www.sustainable.org.za
This organization promotes the use of sustainable energy practices through information, research and lobbying.

SWITZERLAND
Intergovernmental Panel on Climate Change (IPCC)
IPCC Secretariat, c/o World Meteorological Organization, 7bis Avenue de la Paix, C.P. 2300, CH 1211 Geneva 2
www.ipcc.ch
Organization established by the World Meteorological Organization (WMO) and the United Nations Environment Programme (UNEP) to assess scientific and other information relevant to climate change.

IUCN: The World Conservation Union
Rue Mauverney 28, CH Gland 1196
www.iucn.org
Worldwide, multicultural conservation network that aims to assist societies in conserving natural resources. Among the issues it addresses is climate change.

WWF International
Avenue du Mont Blanc, CH 1196 Gland
www.panda.org
Global conservation organization active in promoting solutions to climate change.

UNITED KINGDOM
Carbon Trust
www.carbontrust.co.uk
The trust runs a range of schemes to help business and the public sector cut carbon emissions.

Climatic Research Unit
University of East Anglia, Norwich NR4 7TJ
www.cru.uea.ac.uk

Department for Environment, Food and Rural Affairs (Defra)
Information Resource Centre, Lower Ground Floor, Ergon House, c/o Nobel House, 17 Smith Square, London SW1P 3JR
www.defra.gov.uk
Government environmental department that addresses climate change and informs the public on the subject.

Hadley Centre for Climate Prediction and Research
Met Office, FitzRoy Road, Exeter, Devon EX1 3PB
www.meto.gov.uk
Part of the government's Met Office, the centre studies climate change and develops climate models.

Responding to Climate Change (RTCC)
Victoria Chambers, 16–18 Strutton Ground, London SW1P 2HP
www.rtcc.org.uk
RTCC is a not-for-profit organization that encourages partnership between the private and public sectors in tackling climate change. They are an official observer to the UN Climate Change negotiations.

Soil Association
Bristol House, 40–56 Victoria Street, Bristol BS1 6BY
www.soilassociation.org.uk
The Soil Association is the UK's leading campaigning and certification organization for organic food and farming.

WWF–UK
Panda House, Weyside Park, Godalming, Surrey GU7 1XR
www.wwf.org.uk
The UK branch of this conservation organization is campaigning for action on climate change

UNITED STATES
Arctic Climate Impact Assessment (ACIA)
Center for Global Change, University of Alaska Fairbanks, P.O. Box 757740, Fairbanks, AK 99775-7740
www.acia.uaf.edu
International project of the Arctic Council and the International Arctic Science Committee (IASC) to evaluate and synthesize knowledge on the effects of climate change in the Arctic.

Climate Institute
1785 Massachusetts Avenue NW, Washington, DC 20036
www.climate.org
Organization that aims to increase international awareness of climate change and identify practical ways of achieving emissions reductions.

Conservation International
1919 M Street, NW, Suite 600, Washington, DC 20036
www.conservation.org
Environmental organization that undertakes initiatives to address climate change and its impact on biodiversity.

Global Environmental Facility (GEF)
GEF Secretariat, 1818 H Street, NW, Washington, DC 20433
www.gefweb.org

International organization that helps developing countries fund projects and programs that protect the global environment, including initiatives related to climate change.

The Global Warming International Center
International Headquarters, 22W381 75th Street, Naperville, IL 60565-9245
www.globalwarming.net
An international body disseminating information on global warming science and policy.

Lamont-Doherty Earth Observatory (LDEO)
P.O. Box 1000, 61 Route 9W, Palisades, NY 10964-1000
www.ldeo.columbia.edu
Research institution affiliated with Columbia University and dedicated to the study of Earth, including climate change.

National Climatic Data Center (NCDC)
Federal Building, 151 Patton Avenue, Asheville, NC 28801-5001
www.ncdc.noaa.gov
The world's largest active archive of weather data. It is part of the Department of Commerce, National Oceanic and Atmospheric Administration (NOAA), and the National Environmental Satellite, Data, and Information Service (NESDIS).

The Nature Conservancy
4245 North Fairfax Drive, Suite 100, Arlington, VA 22203-1606
www.nature.org
International environmental organization dedicated to preserving the diversity of life on Earth, in part through its Global Climate Change Initiative.

Pew Center on Global Climate Change
2101 Wilson Blvd., Suite 550
Arlington, VA 22201
www.pewclimate.org
Organization that brings together business leaders, policy-makers, scientists and other experts to address the issue of climate change.

US Global Change Research Program (USGCRP)
1717 Pennsylvania Ave., NW, Suite 250, Washington, DC 20006
www.usgcrp.gov
Government programme that supports research on changes in the global environment, including climate change.

World Resources Institute (WRI)
10 G Street, NE, Suite 800, Washington, DC 20002
www.wri.org
Environmental think tank aimed at protecting Earth's environment by providing objective information and practical proposals. Among its goals is addressing climate change.

URUGUAY
South American Climate Change (SACC) Consortium
c/o Programa de Ciencias del Mar y de la Atmósfera Facultad de Ciencias, Universidad de la República Iguá 4225, 11400 Montevideo
glaucus.fcien.edu.uy
International consortium for the study of global and climate changes in the western South Atlantic.

INDEX

Page numbers in italics indicate illustrations.

A

abiotic forces, 181
ablation, *40,* 117, 121, 125
abyssal plains, 234
Accra, Ghana, 203
accretion, 18
ACIA (Arctic Climate Impact Assessment), 114, *124,* 127
acidic rain, 44
acqua alta, 201
adaptation, 256–68
  agricultural, 266
  limiting coastal erosion, 261–64
  limiting runoff, 264–65
  rainwater harvesting, 265, 267
  species protection, 268
  varieties of, 267
Adelie penguins, *114,* 122, *122*
Admiralty Inlet, Canada, *93*
Advanced Very High Resolution Radiometer (AVHRR), 90
aeolus, 70
aerosols, 93
Afghanistan, 146
Africa
  desertification, 142–43, 213
  formation of, 45, 48, 50, 54
  human evolution in, 54, 56–57, 60, 62
  savannah, 54, 56, 235
agnathans, 38
agriculture
  adaptation, 266, *266,* 267
  albedo and, 86
  desertification and, 142–43
  development of, 65–66
  extreme weather and, 214
  Mediterranean spread and, 181
  rising carbon dioxide levels and, 89, 232
  sea level rise and, 162
  water supply and, 132, 134, 138, 140, 169
air- conditioning, 246
air masses, 212
air pollution, 100, 101, 206, 243
air travel, 244
Akkadians, 69, 194
Alaska, U.S.A., 63–64, 128, 160, 190, 227
albedo, 22–23, *30*
  heat islands and, 86, 88–89
  sea levels and, *117,* 117–18
Alberta, Canada, 232
Alexandria, Egypt, 196

algae, 145, 177, 181
allergies, heat's effect on, 100
Alliance of Small Island States (AOSIS), 106
Allosaurus, 48
alpine ecosystems, 129, 227, 236
alpine (mountain) glaciers, 58, *58,* 92, 117, 119, 125
alpine tundra biome, 227
Alsace, France, *244*
altered interactions, 178–81, 184–85
alternative energy sources, 240, 246, 251–55
alternative transport, 245, *245, 248,* 248–49
altimeters, 161
Ama Dablam Mountain, *224*
Amazon Basin, 185, 228–29
Amboseli National Park, Kenya, *13*
Ambulocetus natans, 51
American pikas, *182,* 182–83, *183*
amoebas, 188
amphibians, 42–43, 44, 46
amphipods, 186
Amsterdam, Netherlands, 93, 244
Anasazi Indians, 74–75
Andes Mountain, 125, 202
Andreas Stone, *70*
angiosperms, 49, 50
anoxia, 44
Antarctica, 54, 114, 123–24, 224
  effects of global warming on, 108, 122
  formation of, 48, 50, 51
  sea-level rise and, 160
Antarctic ice sheets, 58, 114, 123–24, 224
Antarctic paradox, 123
Anthropocene epoch, 86
anthropogenic climate change *see* human-generated climate change
anticipatory adaptation, 267
anticyclones, 210
Antizana glacier, 202
AOGCM (Atmosphere-Ocean General Circulation Model), 91
apes, 54, 56–57
Appalachians, 40
appliances, energy-saving, 246, *247*
aquatic biome, 234, *234*
aquifers, 134, 140, 144, *144*
Archaeopteryx, 48
Archean period, 32
Arctic Circle, tundra biome, 119, 122–23, 226–27
Arctic Climate Impact Assessment (ACIA), 114, *124,* 127
Arctic foxes, *107,* 108
Arctic ice, thawing of, 111, 114–29
  albedo and sea levels, *117,* 117–18
  effects on industry and shipping, 111, 128–29
  impact on lives, 125, 128–29

  impact on native cultures, 111, 128
  impact on species, 10–11, 122, *126,* 126–27, *127*
  ocean currents and, 118
  permafrost and tundra, 119, 122–23
  signs of melting, 92, 114, 117, *124,* 127, *127,* 158, 160
Arctic National Wildlife Refuge, *227*
Arctic Ocean, 111, 119, 128–29
Argentina, 207, 232
Argentinosaurus, 48
arsenic, 144
arthropods, 40
artificial weather (cloud seeding), 88, *88*
asphalt, and heat islands, 86, 88–89
Assateague Island, USA, *162*
asthenosphere, 28
Atlanta, Georgia, USA, 206
Atlantic Fault, *41*
Atlantic Ocean, 59, 118
atmosphere, 12, 20, 33–34
  ocean surface and, 44, *44*
atmosphere-ocean general circulation models (AOGCMs), 91
Attenborough, David, 120
Aukurun wetlands, Australia, *166*
Australia
  bush fires, 214
  climate policies of, 165, 242
  droughts, 8, 141, 154–55
  extreme heat events, 105
  formation of, 48, 50
  hotspots, 150, *150*
  human migrations, 64
  rainforests, 228
  renewable energy, 254–55
  sea-level rise and, 134, 221
Australian Bureau of Meteorology, 154
Australian Greenhouse Office, 155
Australopithecines, 57, 58, 60
Australopithecus, *56,* 57, *60*
Austria, 214
autonomous adaptation, 267
AVHRR (Advanced Very High Resolution Radiometer), 90
axis of Earth, 18, 30, *30*

B

backscattering, 22
bacteria, 32–34, 38
Baffin Island, Canada, *114*
Bahamas, 76
Baiyangdian Lake, China, 146
Baltica, 40

Baltic Sea, 77, 166
Banda Aceh, Indonesia, *197*
Bangalore, India, *242*
Bangkok, Thailand, 196
Bangladesh, 262–63
  CARE Bangladesh, 262–63
  drinking water, 144, *144*
  floods (flooding), *132, 158, 171,* 201, *201,* 262–63
  founding of, 81
Banjul, Gambia, 204
Barcelona, Spain, 243, 244
bark beetles, 110, 180, *188,* 188–91
barnacles, 181, *181*
barn swallows, 176
bats, 51
bears *see* grizzly bears; polar bears
beech trees, 178, *178*
Beever, Eric, 182–83
Beijing, China, *86,* 202–3
Bengal tigers, 169–71, *170*
benthic zone, 234
Bergen, Norway, 203
Beringia, 63–64
Bering Land Bridge, 63–64, 76
Bering Sea, 110
Bering Strait, 64
Bhopal, India, *98*
Bhutan, 137
bicycles (bicycling), 240, 242, 244
bighorn sheep, 151
biology, heat affects on, 98–101, 105
biomass fuels, 245, 254
biomes, 91, 222–36
  distribution of, *236*
  models of ecosystem change, 226
  temperature and precipitation, 224, 226
biosphere, 12
biotic forces, 181
bipedal stance, 56, 60
birds, 48, 50–51
  migration timing and, 174, 176–77
Black Death, 75
Black Sea flood, 66–67
blue-green algae, 33, *34*
blue tits, 177, *177*
body temperature, 100
Bolivia, 202
Bonamia ostreae, 190, *190*
Bonelli eagles, 101
bonobos, 57, *57*
bony fish, 41, 50
boreal forests (taiga), 98, *107,* 227, 230, *230*
Bosawas Biosphere Reserve, Nicaragua, 229
Both, Christian, 176

bottlenose dolphins, 110
Brazil, 110, 185, 191, *194, 196,* 197, 206, 207, 242
breeding, 174, 176–77
Brest, France, 93
Britain *see* United Kingdom
British Columbia, Canada, *86,* 190–91, 202, *258,* 266
British Post Office, 244
British Meteorological Office, 91
brittle stars, *184*
Bronze Age, 65–66
Brueghel, Pieter, *77, 79*
Brunei, *194*
Bryce Canyon National Park, U.S.A., *179*
bubonic plague, 75
buffers, 162, 166, 168, 262–63, 264–65
Bukit Timah Nature Reserve, Singapore, 268
bulk water exports, 145–46
buses, 242, 243, 244, 251, *251*
business interests, and global warming, 95
butterflies, 107, 108, *174,* 176, *178,* 179
Buzzard's Bay, Massachusetts, U.S.A., *161*
Bwindi Impenetrable Forest National Park, Uganda, *153*

C

Cabot, John, 76
cacti, 234
Calcutta, India, 202
Caledonian orogeny, 40
Calgary, Canada, 243
California, USA, 101, 111, 151, 180, 194, 264
Cambrian period, 38, *42,* 46, 47
camels, 51
Cameroon, 148, 250
Campobello Island, Canada, *29*
Camuffo, Dario, 200
Canada
    agricultural adaptation, 266
    bulk water exports, 145–46
    climate policy, 240, 242, 255
    landslides, 122
Canadian Automobile Association, 243
Canadian Forest Service, 190–91
Canaletto, 200, *200*
Canary Islands, 76, *147*
Cancun, Mexico, *264*
Candobolin, Australia, *155*
Cape Floristic Province, South Africa, 151, *233*
Cape Hatteras, North Carolina, USA, *210*
Cape Verde Islands, *147*
Caracas, Venezuela, 203
Caracol, Mexico, 73
Carara Biological Reserve, Costa Rica, *228*
carbon credits, 221, 240

carbon cycle, *30,* 30–31, *31,* 43
carbon dioxide (CO$_2$), 20, 29–31, 89, 92–95
    greenhouse effect and, 20, 29–31, 89
    ocean chemistry and, 102–3
    photosynthesis and, 26
    reducing emissions. *See* Emissions reduction
carbonic acid, *31*
carboniferous period, *42,* 42–43, 46
carbon monoxide, 89
carbon sequestration, 250, *250*
carbon sinks, 89, 110, 119, 185, 246, 250
CARE Bangladesh, 262–63
Caribbean tectonic plate, 58–59
carnivores, 48, 50, 60, 63
carpooling, 242, 243, 245
cars, 243–44, 245, *245, 248,* 248–49
Carroll, Alan, 190–91
cars, 243–44, 245, *245, 248,* 248–49
Cartier, Jacques, 76
cascading effects, 179–80
Castelo Branco, Portugal, *8*
catastrophic weather events *see* extreme weather
    events
caterpillars, 177
Cauvery River, 146
cave paintings, *62, 67*
Cenozoic era, 38, 46, 50–51
cephalopods, 38
Cerios, 234
Chac, *73*
Chacaltaya glacier, 202
Chad, Lake, 148, *148*
chaparral biome, *233, 233*
checkerspot butterflies, *174*
chemical weathering, *31*
Chicago, Illinois, U.S.A., *255*
Chichancanab Lake, Mexico, 73
Chichen Itza, Mexico, 73
Chicxulub crater, 49, *50*
Chile, *114,* 134, 180, 206, *206*
chimpanzees, 57, *57*
China
    agriculture, 65, 146, 267
    desertification, 132, 142, 143, 149
    extreme weather, 213
    floods (flooding), 71, 75, 78, 137
    glacier melt, 125
    land restoration projects, 143, *143,* 149
    sea-level rise and, 196
    water supply, 146, 149, 169, 202–3, 267
cholera, 138, 203, 206
Christmas Carol, A (Dickens), 79
chronic warming, effects of, 107–11
Chukchi culture, 128

Chumash Indians, 74
cities (urban areas), 192–207
    on the coast, 196–201
    disease in, 206–7
    heat islands and, 86, 88–89, 203, 206
    as part of the land, 194, 196
    rich and poor, 196–97
    sea-level rise and, 196–201
    transport, 243, 244
    in the 21st century, 204–5
    varieties of impact on, 202–7
    world's largest, 194, 196, *196*
Clark's nutcrackers, *179,* 179–80
climate (climate system)
    chronology of, 270–73
    first, 20
    humanity's future and, 268
    overview of, 24–31
    vs. weather, 11–12
climate change, 26–31, 92–95
    business interests and, 95
    carbon cycle and, 30–31
    as catalyst in history, 68–69, 71
    changing ecosystems and, 172–91
    charting, 274–75
    continent movements and, 28–29
    contrarians and, 93–95
    Earth's orbit and, 30
    effects of rising temperatures, 96–111
    effects on cities, 192–207
    effects on evolution, 54, 56–57
    extreme weather and, 208–21
    greenhouse effect and, 29–31
    human-generated *see* human-generated climate
        change
    ice and snow thawing and, 112–29
    investigation of past, 35
    limiting *see* emissions reduction
    ocean's role in, 26, 27, 44, *44*
    sea-level rise and, 156–71
    signs of, 92–93
    water supply and, 130–55
    your role in, 275
Climate Change Action Fund, 242
'climate envelope' model, 107, 178
climate models, 91, 94, 107, 138, 218–19
climate observations, 35, 64, 90–91
climate policies, 242, 255
climate refugees, 164–65
Climatic Optimum, 65, 67, 70
climatologists, 90–91
cloud forests, 152–54, 228–29
clouds, 20, *22, 26*
    albedo of, 22–23

cloud seeding, 88, *88*
coal, 42, 89
coastal coniferous forests, 230
coastal wetlands, 166, 168, 264–65
coastline, 196–201, 261–64
    erosion of, 162, 163, 166, 261–64
    flooding of, 171, 196–201
    populations and storm impact, 171, 261–64
    protection of, 162, 166, 168, 261–64
    sea-level rise and, 106, 160, 162, 163, 166, 168, 196
coconut oil, 251
Coe, Michael T., 148
coelophysis, 45
coelurus, 48
Columbia Glacier, *114*
Columbia River, *135,* 135–36
Columbus, Christopher, 76, *76*
Commonwealth Scientific and Industrial Research
    Organization, 154
compaction, 162
computers, in standby mode, 255
condensation, *26*
Congo Basin, 228
coniferous forests, 227, 230, *230*
conservation, 243–55
continental glaciers (ice sheets), 58, *58,* 59, 117,
    123–24, 158, 160
continents
    formation of, 20, 28–29, 38, 42, 43, 45, 48, 50,
        51, 54, 58–59
Cook, James, 76
Copán, Mexico, 73
Copenhagen, Denmark, 244
coral bleaching, 10, 101, *102,* 102–3, *103*
coral reefs, 92, 102–3, 162, 166, 168, 234, 250
Corangamite Lake, Australia, *8*
core of Earth, 18
Coriolis effect, 24, 25, 210
corn, 77
coronal mass ejections, 23, *23*
Corpus Christi, Texas, USA, *216*
Costa Rica, 153–54, *185,* 203, 228
Cretaceous period, *28, 42,* 48–49
Cretaceous/Tertiary (K/T) extinction, 49, 50, *50*
Crete, 71
Crick, Humphrey, 176
crocodiles, 45
Cro-Magnons, *56,* 62–63
crust of Earth, 20
Crutzen, Paul, 86
cryosphere, 12, 91
cryptosporidiosis, 203
Cuba, *158*
Cumberland Island, Georgia, USA, *158*

cyanobacteria, 33, *34*
cyclones, 210, 212, 213–14, 216–17
cynodonts, 43

D

Dakar, Senegal, 196–97
Dakhla Oasis, Egypt, *143*
Dal Lake, India, *11*
Davis, Curt H., 123
Dawson City, Yukon Territory, Canada, *268*
Dayton, Ohio, U.S.A., 89
Dead Sea, 146
deciduous forests, 230–31, 236
deep sea cores, *31*, 35, 64, 90–91
deforestation, 86, 89, 121, 229, *229*
dengue fever, 132, 206, 207
Denmark, 221, 251, *251*, 254
Deque, Michel, 205
Dermo, 190, *190*
desertification, 142–43, 152
deserts, 45, 233–34
Devonian period, *28*, 41–42, *42*
Dhaka, Bangladesh, 201, *201*, 203, *210*, *262*
dhaps, 263
Diamond, Jared, 57
diapsids, 45
Diatryma, 50
Dickens, Charles, 79
Dinosaur National Monument, Utah, USA, *47*
dinosaurs, 29, 45–49
diseases, 186–91
  cities and, 206–7
  heat's effect on, 100
DNA, 47
dogs, 65
doldrums, *25*, 76
dolphins, 110
domesticated animals, 65
Douglas, Bruce, 160
Douglas firs, 110
dragonflies, 42–43
Drake, Francis, 76
Drake Passage, 51
drinking water (freshwater), 125, 132, 134,
    144–46, 203
  bulk water exports, 145–46
  declining quality of, 144–45
  disputes over, 146
  rainwater harvesting, 265, 267
drought, 140–43
  agriculture and, 132, 134
  in Australia, 8, 141, 154–55
  cost of, 214

during Little Optimum, 74–76
long-term damage from, 154–55
in the Sahel, 152
warmth and, 140
widespread, 141
Dryas octopetala, *64*
Dryas pollen, *64*
dust bowl of the 1930s, USA, 80–81, *81*, 142, 143
dust storms, 147, 148, 149, 154–55
Dutch elm disease, 188–89

E

Earth
  axis of, 18, 30, *30*
  formation of, 16–20
  life begins on, 32–51
  magnetic field, 43, *43*, *90*
  predictions for climate, 261, *261*
  variations in orbit, 30, *30*
East Antarctic ice sheet, 58, 114, 123–24, 224
echolocation, 161
ecosystem modification model, 226
ecosystem movement model, 226
Ecuador, 68, 152, 202, 205, *205*
Ediacaran fauna, 34–35
Egypt, 67–68, *68*, 149, *149*, 160, 194
electric cars, 245, 248, *249*
electricity generation and industry, 246, 251, 252–55
elm bark beetles, *188*, 188–91
El Niño, 67–68, 91, 134, 166
El Niño Southern Oscillation (ENSO), 68, 69, 91
emissions reduction, 238–55, *248*, 261
  alternative transport, 245, *245*, 248–49
  differing climate policies, 242, 255
  energy-saving at home and work, 246–47, 250
  guiding principles of, 242–43
  Kyoto Protocol and, 240, 242
  no-fuel solutions, 244
  renewable energy, 246, 251–55
  your role in, 275
emissions trading, 221
Emperor penguins, 122, *122*
energy corporations, 95, 254
Energy Department, US, 246
energy-efficient technologies, 240, 246–47, *247*,
    250, 255
England *see* United Kingdom
ENSO *see* El Niño Southern Oscillation
Environmental Protection Agency, US (EPA), 197,
    255
Eocene epoch, 50–51
eoraptors, 45
Epic of Gilgamesh, 66, *66*

epidemics, 75
Epstein, Paul, 100
equatorial currents, 25, *26*
Erasmus, Barend F. N., 108
erosion barriers, 264, *264*
Eryops, 43
ESA (European Space Agency), *90, 91*, 123
Eta Carinae, *18*
Ethbaal, king of Tyre, 71
Ethiopia, 54, 56
Eukaryotes, 34
Euphrates, River, 69
Eurasia, 50
Europe
  formation of, 50
  heatwave (2003), 101, 105, 203, 214
  timing of ecological events in, 108
European Space Agency (ESA), *90, 91*, 123
European tectonic plate, *41*
European Union (EU)
  Kyoto Protocol and, 240, 242
  renewable energy, 254–55
Euxine Lake, 66–67
evaporation, 26, *26*, *31*, 125
evaporites, 35
evergreen coniferous forests, 227, 230, *230*
Everest, *120*, 120–21, *121*
evolution, 32–35 *see also* human evolution
  Cenozoic era and, 50–51
  climate's role in, 54, 56–57
  Mesozoic era and, 45–49
  Paleozoic era and, 38, 40–44
  plate tectonics and, 28–29
  timeline of, *42*
exoskeletons, 38, 47
extinctions, 226
  climate change and, 56–57, 65, 105–6, 150–51
  Cretaceous, 49, 50, *50*
  Devonian, 42
  Jurassic, 48
  Permian, 44
  Pleistocene, 65
  Proterozoic, 34, 35
  Triassic, 45
extreme heat events, 100–101, 105, 219, 220–21
extreme weather events, 11, 208–21
  costs of, 214, 220, 221
eye of a hurricane, 216–17

F

famine, 75, 77, 79
farming *see* agriculture
feedback *see* negative feedback; positive feedback

figs, 189, *189*
fig wasps, 189, *189*
Fiji, 8, 10, 108, 160, 163
Finland, 128
fish, 41–42, 110, 129
fisheries, 111, 162
flatworms, 186
Fleming, Charles, 106
floodplain reclamation, 264–65
floods (flooding), 134, 137–38
  of coastlines and cities, 171, 197–201
  cost of, 214, 221
  legends about, 66–67
  storm surges and, 163, 165, 171, 214, 221
Florence, Italy, 244
Florida, USA, 8, 109, 169
flycatchers, *176*, 176–77
Foley, Robert A., 56–57, 148
foraminifera, *64*
forest fires, 8, 101, 105, 230–31
forests, 178, *178*, 266
  as carbon sinks, 89, 110, 185, 246, 250
  death of, 110
  in a changing climate, 185
formation of earth, 16–20
Fort Clatsop National Memorial, Oregon, USA, *224*
fossil fuels, 89, 95 *see also* greenhouse gases
fossils, 46–47, *60*
  determining age of, 46–47, 91
foxes, 107, *107*, 108
France, 78, 93, *244*, 254
  heatwave (2003), 101, *101*, 105, 203, 205, *205*, 214
Frankenstein (Shelley), 79
Fraser River, 202
French Revolution, 78
freshwater *see* drinking water
freshwater biome, 234
frogs, 177
fronts, 212
frost, 213
fuel efficiency, 95, 242–43, 245

G

Galapagos Islands, 266
Gama, Vasco da, 76
Ganges River, 202, 262
Gangtori glacier, 202
gases *see* greenhouse gases
Gaul, 73
'Generation X' water measurement, 161
Geneva, Switzerland, 254
geologic sequestration, 250
geothermal energy, 246, *247*, 251, 253, *253*

Germany, 81, 88–89, 110–11, 214, 240
Ghana, 203
giardia, 203
Gila River, 151
Gilgamesh, 66, 66
Gill, Richardson Benedict, 73
glacial isostatic adjustment, 160
glaciation, 28, 58–64, 117
    Devonian, 42
    hominids and, 60–64
    Ordovician, 38, 40
    Proterozoic, 32, 34–35
    Younger Dryas event, 26, 63, 63, 64
glacier ablation, 40, 92, 117, 121, 125
Glacier National Park, U.S.A., 117, 129
glaciers
    albedo of, 22–23, 40, 117, 117–18
    formation of, 28, 40, 40
    melting of (glacier melt), 92, 114, 114, 117–18,
        158, 160, 202
    types of, 58
global climate change see climate change
global mean surface temperature change, 261, 261
global population density, 202
global warming, 12, 86, 92–95
    albedo and, 23
    business interests and, 95
    changing ecosystems and, 172–91
    contrarians of, 93–95
    effects of rising temperatures, 96–111
    effects on cities, 192–207
    effects on living things, 96–111
    extreme weather and, 218–21
    greenhouse effect and, 29–31
    human-generated see human-generated climate
        change
    ice and snow thawing and, 112–29
    limiting emissions see emissions reduction
    sea-level rise and, 156–71
    signs of, 92–93
    water supply and, 130–55
global winds, 24–25, 25, 76, 210
    North Atlantic Oscillation, 78–79
goats, 65
Gobi Desert, 142, 142, 143, 233
Godafoss Falls, Iceland, 240
Golden Age stories, 70
golden-striped salamanders, 108
golden toads, 153–54, 154
Gondwana, 38, 40, 42, 43, 45, 48
grasslands, 54, 224, 232
gravity, 18, 20
Gray, William M., 217
Great Barrier Reef, 10, 102, 103, 105, 165

Great Britain see United Kingdom
Great Lakes, Canada–USA, 58, 145–46, 213
great ocean conveyor belt see thermohaline
    circulation
Great Plains, USA, 80–81, 81, 142, 143, 144, 232
Great Red Spot (Jupiter), 21
Great Rift Valley, Africa, 54, 54, 56, 60, 62, 132
Greeks, ancient, 71
green driving, 245, 245, 248–49
green homes, 246, 247, 250
green power, 246, 251, 252–55
greenhouse effect, 20, 29–31, 30
greenhouse gases, 29–30, 86, 89, 92
    reducing emissions see emissions reduction
Greenland, 50, 58, 74, 74, 76, 78–79
Greenland ice sheet, 80, 106, 114, 117, 119,
    123–24, 124, 199
Greenland sharks, 125
green sea turtles, 109, 109
green urchins, 188
Gregory, Jonathan, 106
Grenoble, France, 254
grizzly bears, 174, 179, 179–80
Guatemala, 95
Gulf Stream, 26, 28, 44, 44, 59, 118

H
Habitats at Risk (2002 report), 226
Hadean eon, 18
Hadley cells, 25, 25
Hadley Centre, 91
hadrosaurs, 48
hailstorms, 213, 214
Hainan Province, China, 132
Haiti, 163
halfmens, 151, 151
Halophyte Biology Laboratory (India), 266
Hamelin Pool, Shark Bay, Australia, 32
Hansen, Lara, 105
Harappan culture, 71, 267
Harley, Chris, 181
Harvell, Drew, 186
Hawaii, USA, 203, 258
heat see also extreme heat events
    in the cities, 203, 206
    effects on human health, 98–101, 105
heat exhaustion, 100, 220
heating systems, 246
heat islands, 86, 88–89, 203, 206
heat sinks, 25, 34
heatstroke, 100, 206, 220
heatwaves, 100–101, 105, 105, 203, 214, 220–21
Heinrich events, 63

Helix, Oregon, USA, 252
Hemingway, Ernest, 120
herbivores, 48–49, 56, 63
high pressure, 24, 210
high tides, 29
Himalayas, 28, 50, 59, 92, 114, 117, 137, 226, 236
Hitler, Adolf, 81
Holdridge life zones, 91
Holocene epoch, 65–66, 86
home, energy-saving at, 246, 247, 250
hominids, 51, 54–58, 60, 60–64
Homo erectus, 56, 60, 60, 62–63
Homo ergaster, 60
Homo habilis, 54, 56, 60
Homo heidelbergensis, 60
Homo neanderthalensis, 60, 62–63
Homo sapiens, 60, 60, 62–63
Honda Civic Hybrid, 249
Honda Insight, 249
Honduras, 152, 242, 267
Hong Kong, 194, 203, 206
Hopi Indians, 144
hotspots, 150, 150–51
huge impacts era, 18, 20
Hulhumale, Maldives, 204
Hulme, Mike, 224
human evolution, 52–65
    climate's role in, 54, 56–57
    after ice ages, 65–76
    during ice ages, 58–64
    progression of skulls, 56
    savannah hypothesis, 54, 56–57
    sea levels and migrations, 63–64
    turnover pulse hypothesis, 56–57
    variability selection hypothesis, 56–57, 60
human-generated climate change, 12, 86–95
    business interests and, 95
    changing ecosystems and, 172–91
    charting, 274–75
    contrarians of, 93–95
    effects of rising temperatures, 96–111
    effects on cities, 192–207
    fossil fuels and, 89
    heat islands and, 86, 88–89
    ice and snow thawing and, 112–29
    sea-level rise and, 156–71
    signs of, 92–93
    water supply and, 130–55
human physiology, heat's effects on, 98–101, 105
human power transportation, 244
Hunters in the Snow (Brueghel), 77, 79
hurricanes, 8, 213–14, 216, 216–17, 217, 220
    formation of, 216, 216–17
Hurricane Allen, 163

Hurricane Andrew, 216
Hurricane Charley, 8
Hurricane Felix, 106
Hurricane Frances, 8
Hurricane Isabel, 210, 217
Hurricane Ivan, 8, 214
Hurricane Jeanne, 8
Hurricane Mitch, 138, 267
hybrid cars, 245, 248, 249
hydrocarbons see carbon dioxide
hydrogen, 89, 245, 249, 249
hydrologic cycle, 26, 26
hydropower, 135, 136, 136, 137, 203, 240, 253
hydrosphere, 12
hydrothermal vents, 32, 34

I
Iberian lynx, 101
ice see also Arctic ice, thawing of
    albedo of, 22–23, 40, 117, 117–18
    signs of melting, 92, 114, 117, 124, 127, 127,
        158, 160
    thawing of, 111, 112–29
ice ages, 58–64
    evidence of, 64
    Little, 23, 54, 77–79
    Mini, 66, 69
    plate tectonics and, 28
    Pleistocene, 52, 54, 56, 58–64
    positive feedback and, 22–23
    Proterozoic, 32, 34–35, 38
icebergs, 80, 80, 125
ice cores, 35, 35, 64, 90–91, 92
Iceland, 240, 251
ice sheets, 58, 58, 59, 117, 123–24, 158, 160
ice shelves, 123
ichthyosaurs, 45
Ilulissat Icefjord, 123–24
Incas, 70
India
    agriculture, 266
    extreme heat, 100
    floods, 138
    formation of, 48, 50
    savannah, 235
    sea-level rises and, 118
    volcanism and, 79
    water supply, 146, 202
Indian Ocean, 210, 216, 219
Indonesia, 68, 79, 134, 196, 197, 229, 267
Industrial Revolution, 69, 80, 86, 89
Indus Valley, 71
infiltration, 26

infrared radiation, *30*
Inner Mongolian plains, *232*
inorganic carbon cycle, 30–31
insects, 41–43, 44, 48, 122
instrumental records of climate, 35, 64, 90–91
Intergovernmental Panel on Climate Change
(IPCC), 12, 91, 92, 93, 94, 106, 142, 146,
158, 196, 261
interstadials, 59
intertidal zones, 33, 181, 234
intertropical convergence zone, 76
Inuits, 70, 128
invasive species, 186–91
invertebrates, 38, 186, 188
IPCC *see* Intergovernmental Panel on Climate Change
Iraq, 65, 66, 69
Ireland, 79, *79*
iridium, 49
iron, 33
Iroquois Indians, 70
isotope dating, 64
Israel, 145, 160
Isthmus of Panama, 58–59
Italy, 199–201, 244
heatwave (2003), 101, 105, 203, 214

J

jaguars, 151, *151*
Jakarta, Indonesia, 196
Japan, 65, 197, 206, 236, 243
Japanese oysters, 190
Jason-1 satellite, *161*
Jelonka, Poland, *221*
Johnson & Johnson, 254
Jones, Gregory, 110–11
Jordan, 145
Jordan, River, 145
Jupiter, 21
Great Red Spot, *21*
Jurassic period, 29, 45, 46, 48

K

Kahoolawe, Hawaii, USA, *258*
Kalahari Desert, *226, 233*
Kalimantan, *98*
Kansas, U.S.A., 80–81
Kapataksha River, 263
Karoly, David, 154
Keeling, Charles, 29
kelp, 188
Kenya, 54, 56, 70, 267
Kenyan giraffes, *235*

Kerala, India, *214*
Khumbu icefall, *120*
Kilimanjaro, 92, 117, *120,* 120–21, *121*
Knoxville, Tennessee, USA, 206
Kodiak Island, Alaska, USA, *136*
Krakatau, Indonesia, 79
K/T extinction, 49, 50, *50*
Kulkarni, Vivek, 168
Kunds, 267
Kyoto Protocol, 240, 242, 255

L

Lagos, 214
lake bed sediments, 64, 91
land bridges, 63–64
landmasses, rebounding, 160, 162
landslides, 122, 138, *139*
land use practices, and water supply, 147
La Niña, 68
Larsen B Ice Shelf, *93*
Lascaux caves, France, *62*
La Tigra cloud forest, Honduras, 152
Laurasia, 40, 42, 43
Laurentia, 38, 40
Laurentide ice sheet, 65
layers of Earth, 18, 20
leatherback turtles, 109, *109*
Leatherman, Stephen, 171
Lebanon, 169
Leonaspis maura, 38
Leuk, Switzerland, *98, 101*
lianas, 185
Liangzhu culture, 71
Libya, *91*
lightbulbs, energy-saving, 246, *250*
light electric vehicles, 249
lightning storms, *213*
lights, turning off the, 255
Lima, 134
limited dispersal ability, 179
lithosphere, 28
Little Ice Ages, 23, 54, 77–79
Little Optimum, 54, 74–76, 79
lizards, 45
locusts, *191*
lodgepole pines, 180
loggerhead turtles, 109, *109*
London, United Kingdom, 243
Longshan culture, 71
Longwang, 70
Los Angeles, California, USA, 194
Louisiana, USA, 162
low depressions, 210, 212

low pressure, 24, 210
Lushoto, Tanzania, *266*

M

Magellan, Ferdinand, 76
magnetographs, *90*
magnetosphere, 43, *43*
Maine, USA, 188
Majuro Atoll, 163
Makalu, Nepal, *59*
Malaria, 100, *100,* 132, 207
Malawi, 134
Malaysia, 109, 152
Malcolm, Jay R., 107
Male, Maldives, 204, *204*
Mali, 77
Mallory, George, 120
mammals, 45, 48, 50–51, 56
mammoths, 65
mangrove swamps, 162, 168, *169,* 169–71, 264
Manicouagan impact, 45
mantle of Earth, 18
Maori, 70
Mars, 20, 21
South Polar Cap, *21*
Marshall Islands, 163
marshes, 166, 168
marsh reclamation, 264–65
Masai culture, 267
Masai Mara National Reserve, Africa, *224*
Massospondylus, 45
Maunder, E. Walter, 23
Maunder minimum, 23, 78
Mayan civilization, *69,* 73, *73,* 74
measurement of climate change, 35, *35,* 64, 90–91
Medieval Warm Period, 54, 74–76, 140
Mediterranean climate, 72, 180–81
Mediterranean Sea, 54, 66, 72
Mediterranean spread, 180–81
meerkats, *226*
megaliths, 66
Melbourne, Australia, 204
Mentuhotep I, 68
Mercury, 21
Mesopotamia, 65–66, 69
Mesozoic era, 38, 45–49
metabolism, heat's effect on, 100
metazoans, 34–35
meteorites, 18, 20, 32, *38,* 42, 45, 49, *49*
Meteosat Second Generation (MSG), *91*
methane, 50, *50,* 89, *89,* 119
MetOp (Meteorological Operational Weather
Satellite), *90*

metro systems, 243
Mexico, 65, 73, 108, 234
Mexico City, Mexico, *194,* 206
microalgae, 102
Micronesia, 118
Mid-Atlantic Regional Climate Change Assessment, 230
Middle Ages, 74–76
Middle East, 67, 146
migration, 71, 107–8, 129
altered interactions, 178–79
sea levels and, 63–64
timing of, 174–77
Milankovitch, Milutin, 30
Milankovitch cycles, 30, 59, 65, 119
Miller, Ronald, 94
Minas Basin, New Brunswick, Canada, *29*
Ming Dynasty, 78
Mini Ice Age, 66, 69
Ministry of Water Resources (China), 203
Minoans, 71
Miocene epoch, 54
Mississippi River, 250
Mississippi River delta, 199, 214
mitigation, 256–68
agricultural adaptation, 266
limiting coastal erosion, 261–64
limiting runoff, 264–65
rainwater harvesting, 265, 267
species protection, 268
Moller, Anders, 176
monkeys, 51, 54, *228*
monsoons, 201, *201, 214,* 219
Montana, U.S.A., 140
montane forests, 152–54, 230
Monteverde Cloud Forest, Costa Rica, 153–54,
185
Moon, 18
tides and, 29
Mosel Valley, Germany, 110–11
mosquitoes, 100, 132, 138, 207, *207*
mountain glaciers, 58, *58,* 92, 117, 119, 125
Mount Everest, *120,* 120–21, *121,* 183
Mount Kilimanjaro, 92, 117, *120,* 120–21, *121*
Mount Pinatubo, 79
Mount Tambora, 79
Mozambique, 138, 268
MSG (Meteosat Second Generation), *91*
mudslides, 122, 138, *139*
multiregional model theory, 62
Murray-Darling Basin, 154, 155, *155*
Murray River, 155
musk oxen, 186, *186*
mussels, 184, *184*
mythology, and climate, 70

N

Nagoya, Japan, 197
NAO (North Atlantic Oscillation), 78–79, 81
Napa Valley, California, USA, 111
Nariokotome Boy, 60, 62
National Aeronautics and Space Administration
    (NASA), 90, 94, 118, 124
National Assessment of the Potential Consequences of
    Climate Variability and Change, 197, 199
National Botanical Institute (Capetown, South Africa),
    105–6
National Center for Atmospheric Research (Colorado,
    USA), 140
National climate policies, 242, 255
National Oceanic and Atmospheric Administration
    (NOAA), 90, 91, 161
Native Americans, 64, 66, 70, 74–75, 128, 151
natural coastal protectors, 166, 168 see also buffers
natural gas, 89
natural selection see evolution; human evolution
Nature, 226
Neanderthals, 60, 62–63
neap tides, 29
negative feedback, 26
neighbourhood electric vehicles (NEVs), 249
Nelesone, Panapasi, 165
neoglaciation, 67–68
Nepal, 59, 137, 224, 229
Neptune, 21
nesting, 174, 176–77
New Delhi, India, 105, 202
New Orleans, USA, 162, 199
New South Wales, Australia, 8, 265
newts, 177
New York City, New York, USA, 197, 198,
    199
New Zealand, 70, 244
Nicaragua, 229
Nicholls, Robert, 171
Nigeria, 142, 148, 214
Niger, River, 77, 149
Nile, River, 67–68, 149
Nile, River Delta, 158, 160
nitrogen, 20
nitrous oxide, 89
nival zone, 236
NOAA (National Oceanic and Atmospheric
    Administration), 90, 91, 161
Noah's flood, 66
Nordhaus, William, 258
North Africa, 40, 43, 59, 71, 73–74
North America
    extreme heat events, 101

formation of, 45, 48, 50, 58–59
    Little Ice Age and, 77, 78–79
    Little Optimum and, 74
    migration routes and, 63–64
    timing of ecological events in, 108
North American prairie pothole region, 232
North American tectonic plate, 41
North Atlantic Oscillation (NAO), 78–79, 81
northeasterly trade winds, 25, 25, 76
Northern Sea Route, 128, 128
North Pole, 54, 114, 117, 119, 123, 123,
    127, 227
North Sea, 108, 177, 213
Northwest Passage, 128
Norway, 122, 185, 186, 203
Norway spruce, 110
nothosaurs, 45
Novia Scotia, Canada, 188
Nubian ibex, 236
nuclear power, 252
Nukufetau, 164, 164–65, 165
numbats, 150, 150
Nunavet, Canada, 186, 186

O

obliquity (tilt of Earth), 18, 30, 30
Ocean City, Maryland, USA, 162, 171
ocean conveyor belt see thermohaline circulation
ocean currents, 25, 26, 26, 44, 59, 118, 118
oceans and seas
    climate change and, 26, 27, 44, 110
    first climate, 20
    harnessing energy of, 253–54
    life begins in, 32–33
    rise in levels see sea-level rise
    salinity of, 20, 44, 118, 169
    tides and, 29
ochre sea stars, 184, 184
octopuses, 108
Ogallala Aquifer, 144, 144
oil, 89, 111, 128
oil pipelines, 119, 122
oil spills, 129
Oklahoma, U.S.A., 80–81, 81
Ol Doinyo Lengai mountain, Tanzania, 54
Oligocene epoch, 51, 54
Oman, 169
Ontong Java, 48
open taiga biome, 227, 230
Ordovician period, 38, 40
organic carbon cycle, 30–31
origins of earth, 16–20
origins of species see evolution; human evolution

Orissa, India, 166, 168
ornithopods, 48
Osaka, Japan, 197
Oued Tensift River, 142
'Out of Africa' theory, 62
oxygen, 33, 33–34, 43
oysters, 190, 190
ozone layer, 33–34

P

Pachauri, Rajendra, 94
Pacific Northwest, U.S.A., 135, 135–36, 136,
    137, 140–41
Pacific Ocean, 59, 68, 210
Pacific salmon, 107, 110, 135–36, 136
Pacific tectonic plate, 58–59
pack ice, and polar bears, 126–27
Pakistan, 81, 125
Palau Islands, 18
Palenque, Mexico, 73
Paleocene epoch, 50
Paleomagnetism, 43, 43, 64, 91
Paleozoic era, 36–44, 46
Panama, 58–59
pandas, 231
Pangaea, 29, 43, 44, 45, 48
parasites, 186–91
Paris, France, 101, 101, 205, 205, 243
'park and ride' programmes, 243
Parmesan, Camille, 98, 107, 174, 258
Peary caribou, 227, 227
pelagic zone, 234
penguins, 114, 122, 122
permafrost, 119, 122–23, 227
Permian period, 28, 29, 42, 43–44
Peru, 68, 117, 134, 202, 203
pests, 186–91
Peterson, Townsend, 108
Phanerozoic eon, 36–51
    Cenozoic era, 50–51
    Mesozoic era, 45–49
    Paleozoic era, 36–44
phenology, 108, 174–77 see also migration
Phoenicia, 71
photolysis, 20
photosynthesis, 26, 33–34, 43
pied flycatchers, 176, 176–77
pikas, 182, 182–83, 183
Pineapple Express, 203
pine bark beetles, 188–91
Pivacharam, India, 261
plagues, 75, 77
plankton, 30, 177–78, 250

plate tectonics, 28, 28–29, 31, 48, 54, 160
Pleistocene ice ages, 52, 54, 56, 58–64
Pleistocene overkill hypothesis, 65
Pliocene epoch, 54, 58–59
Pluto, 21
Poland, 93, 166, 221
polar bears, 10–11, 92, 126, 126–27, 127, 129
polar ecosystems, 129
polar fronts, 25
polar winds, 25, 25
political chaos, and climate, 71, 74, 77–78
pollen accumulations, 64, 91
pollution, 93, 100, 101, 206, 243
pollution credits, 240
Pont du Gard Bridge, France, 72
populations, and storm impact, 171, 261–64
Portugal, 8, 8, 110
    heat wave (2003), 101, 105, 203, 214
positive feedback, 22–23, 26, 59, 118
Potts, Rick, 56
poverty, drought's role in, 141, 147
prairies, 232
Precambrian eon, 32–35, 42, 46
precession, 30
precipitation (rain), 132, 137 see also drought;
        floods; monsoons
    artificial (cloud seeding), 88, 88
    carbon cycle and, 30, 30–31, 31
    extreme weather events, 210, 213, 219, 221
    hydrologic cycle and, 26, 26
    signs of climate change, 92–93
    temperature and, 224, 226
predictions for Earth's climate, 261, 261
pressure gradients, 210
primates, 51, 54–58
Prince Rupert, Canada, 119
proboscideans, 51
proconsul, 56
Procopius, 71
Project Moses, 201
prominences, 23
Proterozoic ice ages, 32, 34–35, 38
Provincial Tree Improvement Programme
    (Canada), 266
Prudhoe Bay, Alaska, U.S.A., 180
public transport, 242–44
Punta Arenas, Chile, 114

Q

Quaternary period, 42, 59
Quelccaya ice cap, 117, 134, 134
Quezungal farming techniques, 267
Quito, Ecuador, 205, 205

R

radiation *see* solar radiation; ultraviolet radiation
radiometric dating, 46–47, 64
radiosonde, 90
Ragnarök, 70
ragweed, 100
rain *see* precipitation
rain barrels, 265, 267
rainforests, 185, 228–29, 231
rain gardens, 265
rainmakers, 70
rainwater harvesting, 265, 267
Rapid Climate Change Event (RCCE), 63, 80
reactive adaptation, 267
Recife, Brazil, *196,* 197
red foxes, 107, *107,* 108
red pandas, *231*
Red Sea, 28, 32, 62, 67, *168*
reducing emissions *see* emissions reduction
reindeer, 128
religion, and climate, 70
renewable energy, 240, 246, 251–55
reptiles, 43, 44, 45–49
Rhode Island, USA, 264
rice, 65, 66
Richtersveld National Park, Africa, *151*
Riding Mountain National Park, Canada, *265*
Rift Valley, Africa, 54, *54,* 56, 60, 62, 132
rising sea levels *see* sea-level rise
rising temperatures *see* global warming
Riyadh, Saudi Arabia, 206
Robinson, David, 117
Rocky Mountain Oilfield Testing Center, Wyoming,
USA, *250*
Rocky Mountains, 48, 67, 140
rodents, 51, 206–7
Rodinia, 35, 38
Roman Empire, *69,* 72–73
Rongbuk glacier, 120
Ross Sea, 122
runoff, *26,* 264–65
Russia, 81, 213

S

Saami culture, 128, 226
Saffir, Herbert, 217
Saffir-Simpson Hurricane Intensity Scale, 214, 217
Sahara Desert, 67, 149, 152, *224,* 233–34
    expansion of, 142, *142,* 147, *147,* 149, 152, *152*
    ice ages and, 59, 62
    sea level rise and, *158*
Sahel, 149, 152, 207, 213, 235

salinity of oceans, 20, 44, 118, 169
salmon, 107, 110, *135,* 135–36, *136,* 162
Salvador, Brazil, 207
sandstorms, *143,* 149, 152
Sanford, Eric, 184
San José, Costa Rica, 203
San Juan Island, 181, *181*
San Juan Mountains, *182*
Santa Barbara, California, U.S.A., 264
Santiago, Chile, 206, *206*
Sanz, Juan Jose, 176–77
São Paulo, Brazil, 206
Sargon of Akkad, 69
Saskatchewan, Canada, *224*
satellites, *90,* 90–91, *91,* 161, *161*
Saturn, 21
sauropods, 48
savannah biome, *107,* 235, 236
savannah hypothesis, 54, 56–57
Schneider, Stephen H., 94
Scotland, 78, 105, 110
seas *see* oceans and seas
sea floor sediments, *31,* 35, 64, 90–91
sea ice, thawing of, 111, 112–29
    albedo and sea levels, *117,* 117–18
    effects on industry and shipping, 111, 128–29
    impact on lives, 125, 128–29
    impact on native cultures, 111, 128
    impact on species, 10–11, 122, *126,* 126–27, *127*
    ocean currents and, 118
    permafrost and tundra, 119, 122–23
    signs of melting, 92, 114, 117, *124,* 127, *127,*
        158, 160
sea levels
    albedo and, *117,* 117–18
    migration routes and, 63–64
sea-level rise, 156–71, 221
    cities and, 196–201
    coastline and, 106, 160, 162, 163, 166, 168, 196
    coral reefs and, 103
    effects of, 162–71
    local variations in, 160
    measurement technology, 161
    mechanics of, 158, 160
    melting ice and snow and, 118, 119, 124
    rebounding landmasses and, 160, 162
    signs of, 92–93
seals, 129
sea stars, 184, *184*
Seattle, Washington, U.S.A., 203, 265
sea turtles, 109, *109*
sea urchins, 110, *186,* 188
sedimentary rocks, 46–47, 64
Senegal, 196–97, 204

Sesostris I, 67
Shanghai, China, 196, 206, *240*
Shark Bay, Australia, *32,* 42
sharks, 41, 50, *125*
Sharp's Island, Maryland, U.S.A., 163
shearwaters, 110, *110*
Shelley, Mary, 79
shellfish industry, 190, *190*
shipping, trans-Arctic, 111, *128,* 128–29
Shipwreck Coast, Australia, *134*
shoreline *see* coastline
short-tailed shearwaters, 110, *110*
Siberia, 40, 44, 67, 129
Silurian period, 40–41, *42*
silviculture, 110
Simpson, Bob, 217
Sinai Peninsula, 62
ski areas, 132
slow boil of chronic warming, 107–11
Small, Christopher, 171
smart cars, 245, *245,* 248–49
Snake River, *135,* 135–36, *136*
snow
    albedo of, 22–23, 40, *117,* 117–18
    as reservoir of water, 137, 140–41
snow, thawing of, 112–29
    albedo and sea levels, *117,* 117–18
    effects on industry and shipping, 111, 128–29
    impact on lives, 10–11, 125, 128–29
    impact on species, 10–11, 122, *126,* 126–27, *127*
    ocean currents and, 118, *118*
    permafrost and tundra, 119, 122–23
    signs of melting, 92, 114, 117, *124,* 127, *127,* 158,
        160
    ski areas, 132
Snowball Earth, 34
solar flares, 23
solar power, 246, *247, 248, 251, 252,* 253
solar radiation, 20, 22–23, 24, 29–30, *30,* 59
Solar system, 18, 20, 21
Solar Vector Magnetographs, *90*
Somalia, 146
Sonoma Valley, California, USA, 111
Sonoran Desert, 234
South Africa, 108, 151, *151,* 180, *233,* 235, *235*
South America
    deforestation of, 229, *229*
    formation of, 45, 48, 50, 51, 58–59
    Little Optimum and, 75–76
    migration routes and, 63–64
southeasterly trade winds, 25, *25*
South Equatorial Current, 76
Southern Oscillation, 68, 69, 91
South Polar Cap (Mars), *21*

South Pole, 43, 50, 51, 123, *123*
Spain, 105, 185, 243, 244
Sparks, Timothy, 176
species protection, 268
spike moss, *38*
squirrel monkeys, *228*
Sri Lanka, 264
starfish, 184
Steffen, Konrad, 124
steppes, 232
stilling wells, 161
Stonehenge, England, 66
storms (storm surges), 210–14, 220
    cost of, 214, 221
    flooding and, 163, 165, 171, 214, 221
    impact on populations, 171, 261–64
Stöwer, Willy, *80*
Strait of Gibraltar, 54
stromatolites, 32, *32,* 33, *47*
Stuttgart, Germany, 88–89
subtropical cells, 25, *25*
Succulent Karoo, Africa, 151, 233
Suez Canal, 67
Sumer, 65–66, 69
Sun
    formation of, 18
    role in climate, 20, 22–25
    solar system and, 21
    tides and, 29
Sundarban Islands, 169–71
sunspots, 23, *23,* 71, 78
surface air temperature change, 261, *261*
swamps, 45, 48, 166, 168 *see also* mangrove
    swamps
Sweden, 177
Swinoujscie, Poland, 93
Swiss Re, 106, 221
Switzerland, 77, *98, 101,* 132, 221, 254
    heat wave (2003), 101, 105, 203, 214
Syria, 194

T

taiga (boreal forests), 98, *107,* 227, 230, *230*
Tamil Nadu, India, 146, 266
Tanzania
    agricultural adaptation, 266, *266*
    cloud forests, 152
    glacier melt, 117, 121
    human evolution in, 54, *54,* 56
    rainwater harvesting, 267
    species protection, 268
Tarbosaurus bataar, *46*
Tasmania, 188, 231

Tassili N'Ajjer paintings, 67
tectonic plates, 28, 28–29, 31, 48, 54, 160
Tejo River, 110
Tell Leilan, Syria, 194
temperate deciduous forest biome, 230–31
temperate grassland biome, 232
temperate rain forest biome, 231
temperature rise see global warming
Teotihuacan, 71, 74
Tertiary period, 42, 50–51, 54
Texas, USA, 162, 216
TGV train, 243
Thames, River, 77
Thar Desert, 267
thermal expansion, 158, 160
thermohaline circulation, 25, 26, 26, 63, 118, 118
thermokarsts, 122
thermonuclear fusion, 18
theropods, 48
Thira, 71
Thomas, Chris D., 105, 226
Thuiller, Wilfried, 105–6
thunderstorms, 212, 212 see also storms
Tianjin, China, 196
Tibet, 59, 120–21, 137
tidal marshes, 166, 168
tidal power, 253–54, 254
tide gauges, 161, 161
tides, 29, 29, 33
Tigris, River, 69
Tikal, Mexico, 73
Timbuktu, 77
timing of ecological events, 108, 174–77
Titanic, 80, 80
Tiwanaku, 75–76
Tokyo, Japan, 197, 206, 243
TOPEX/Poseidon (T/P) satellite, 161
tornadoes, 212, 214, 218
tortoises, 45
toucans, 153
tourism, heat's effect on, 101, 105, 111
Toyota Prius, 249
trade winds, 25, 25, 68, 76
tradition, and climate, 70
traffic congestion, 243
trains, 242, 243, 243
trams, 243
trans-Arctic shipping, 111, 128, 128–29
transpiration, 26
tree frogs, 228
tree rings, 64, 90–91, 98
trees, 41–42, 266
    as carbon sinks, 89, 110, 246, 250
    in a changing climate, 185

death of, 110
Triassic period, 42, 45
Tribrachidium, 35
trilobites, 38, 38, 44
tropical rainforest biome, 185, 228–29, 236
tropical savannah biome, 235, 236
troposphere, 24–25
tropospheric ozone, 89
Tsho Rolpa, Nepal, 121
tsunami of 2004, 169, 197, 204, 261, 264
tundra, 119, 122–23, 226–27, 236
Turkana Boy, 60, 62
Turkey, 163
turnover pulse hypothesis, 56–57
turtles, 45, 109, 109
Tuvalu, 10, 163, 164, 164–65, 165
typhoons, 203, 210, 216–17
Tyrannosaurus rex, 48

U

Uaxactún, Mexico, 73
Uganda, 153
ultraviolet (UV) radiation, 22, 32, 33–34
ulva, 174
underground systems, 243
United Kingdom
    agriculture, 181
    breeding times, 177
    climate policies, 240, 242
    heat wave (2003), 101, 105, 203, 214
    ice ages and, 77–78
    rainwater harvesting, 267
    renewable energy, 254–55
United Nations Convention to Combat Desertification, 142
United Nations Educational, Scientific, and Cultural Organization (UNESCO), 120
United Nations Environment Programme (UNEP), 94, 95, 102, 132, 137, 142, 152
United Nations Intergovernmental Panel on Climate Change see Intergovernmental Panel on Climate Change
United Nations Population Division, 196, 196
United States. see also North America
    climate policy, 240, 242, 255
    droughts, 80–81, 81, 142, 143
    exploration of, 74, 76
    extreme heat events, 101
    green power, 252, 254
United States Geological Survey (USGS), 182, 199
United States National Hurricane Center, 217
upwelling, 25
Uranus, 21

urban areas see cities
urchins, 110, 186, 188
Uruguay, 166
Uzbekistan, 132

V

Van pool programmes, 245
Vanuatu, 251
variability selection hypothesis, 56–57, 60
velociraptors, 48
Venezuela, 138, 166, 203
Venice, Italy, 199–201, 200
Venice in Peril Fund Conference (2003), 201
Venus, 20, 21, 21
vertebrates, 38, 49
Vietnam, 140
Vietnam Red Cross Society, 264
Vikings, 74, 76
Vinland, 74
volcanoes (volcanism), 18, 20, 31, 38, 44, 48, 58–59, 71, 79
Vrba, Elizabeth, 56

W

Wadden Sea, 186
walking, 242, 244
Walpur, India, 210
Waltham, Tony, 199
Walther, Gian-Reto, 108
Washington, D.C., U.S.A., 205
water, 125, 130–55 see also drinking water; drought; floods
    consequences of changes in, 132, 134
    great drying, 147–49, 152–55
    hydropower and health, 203
    salmon and humans, 135–36
    species at risk, 150–51
    widespread problems with, 134–35
water hyacinths, 263, 263
water vapour, 20, 26, 26, 29–31
Watson, Robert T., 94
Watt, James, 86
wave energy, 253–54
weather
    artificial (cloud seeding), 88, 88
    extreme events see extreme weather events
    overview of, 24–25
    vs. climate, 11–12
weather forecasting, 70
Weil's disease, 207
West Antarctic ice sheet, 58, 114, 123–24, 224
westerlies, 25, 25, 78–79

Western Desert, Egypt, 149
West Nile virus, 100, 207, 207
wetland reclamation, 264–65
wetlands, 166, 168
Wetlands International, 265
whales, 51, 129
whitebark pines, 178, 179–80
white pine blister rust, 180
Whittier, Joan, 109
Whitty, Julia, 164–65
Whorf, Timothy, 29
Willoughby, Hugh, 217
wind power, 251, 251, 252, 252, 253
winds, 24–25, 25, 76, 210
    North Atlantic Oscillation, 78–79
wine and vineyards, 110–11, 181
Winthrop, Massachusetts, USA, 258
WMO (World Meteorological Organization), 92, 94, 101
Wolong Nature Reserve, China, 231
woodland biome, 235
Woodleigh crater, 42
work, energy-saving at, 246, 247, 250
World Meteorological Organization (WMO), 92, 94, 101
World War II, 81
World Wildlife Fund (WWF), 105–6, 107, 183, 226
Wright Glacier, 58
Wyoming, USA, 140, 190

X

Xia culture, 71

Y

Yanamarey glacier, 202
Yangtse River, 71, 169
Yanquing County, China, 143
Yap, 163
Yaxchilán, Mexico, 73
Yeager, Brooks, 183
Yellow River, 71
Yellowstone National Park, U.S.A., 145, 174, 180
Yohe, Gary, 98, 107, 174
Younger Dryas event, 26, 63, 63, 64
Yukon, 101

Z

zero-energy homes, 246, 247
zooplankton, 177–78
Zululand, South Africa, 235

ACKNOWLEDGMENTS & PICTURE CREDITS

Acknowledgments

The authors offer grateful thanks to: Miranda Smith of Rodale Books International; the Rodale Institute; Adrian Webster; Dr. Lara Hansen, Chief Scientist, World Wildlife Fund Climate Change Programme; the Ocean Research Academy; the Friday Harbor Marine Laboratory; Daniel Froehlich, ornithologist extraordinaire, Burke Museum at the University of Washington; Gail Greiner; Hannah Choi; Melinda Corey; Martha Corey-Ochoa; Wendy Glassmire, National Geographic Society; Harriet Mendlowitz and Anita Duncan, Photo Researchers, Inc.; Carolyn McGoldrick, AP Photos; Peter McQuarrie. Digital maps created by: Alex Reay, Advanced Illustration. Map consultant: Andrew Heritage, Heritage Editorial. Contributors: Chapter 6—Alison Fromme; Chapter 7—Jennifer Freeman; Chapter 8—Lisa Hayward; Chapter 10—Melinda Corey.

Picture Credits

t—top, **b**—bottom, **r**—right, **l**—left, **c**—center, **cf**—centerfold, **bg**—background

Photography Sources

AP—Associated Press **BAS**—British Antarctic Survey **ESA**—European Space Agency **NASA**—National Aeronautics and Space Administration **NGS**—National Geographic Society **NOAA**—National Oceanic and Atmospheric Administration **PR**—Photo Researchers, Inc. **SPL**—Science Photo Library

Half title Dr. Morley Read/SPL **Title page** Carsten Peter/NGS **4–5** Corbis **6** Robert Madden/NGS **9** Antonio Jose/Lusa **10** AP/Brian Cassey **11** AP/Aijaz Rahi **12–13** K Yamashita/PanStock/Panoramic Images/NGS

Chapter 1

14–15 Chris Butler/SPL **17**tl Corbis **17**tr Corbis **17**bl Dr. James P. McVey/NOAA Sea Grant Program **17**br AP/Billy Smith II **18–19** Jerry Lodriguss/PR **21**tl NASA/SPL **21**tr NASA **21**b NASA **22**tl NASA **22**bc James P. Blair/NGS **22**cf NASA Jet Propulsion Laboratory **23** NASA/SPL **24** Colin Cuthbert/SPL **25** Argosy Publishing **26** Advanced Illustration **27** Argosy Publishing **28** Advanced Illustration **29**l Andrew J. Martinez/PR **29**r Andrew J. Martinez/PR **30**t Wordstop Technologies **30**b Argosy Publishing **31** Argosy Publishing **32** Georgette Douwma/SPL **33** Wordstop Technologies. Data source: "Earth's Early Atmosphere" *Science*, February 12, 1993 **34** Michael Abbey/SPL **35**tr D.A. Peel/PR **35**br BAS/SPL

Chapter 2

37tl Sinclair Stammers/SPL **37**tr Vaughan Fleming/PR **37**bl Corbis **37**br Detlev Van Ravenswaay/SPL **38–39** Paul Nicklen/NGS **40** Argosy Publishing **42** Argosy Publishing **41** Simon Fraser/PR **43** Mark Garlick/SPL **44** NASA Goddard Space Flight Center **46**tl Sinclair Stammers/SPL **46**bl Mehau Kulyk/SPL **47**tl Sinclair Stammers/SPL **47**tr Jim Amos/SPL **47**c BAS/SPL **49** Erich Schrempp/SPL **50** Argosy Publishing **51** D. Van Ravenswaay/PR

Chapter 3

53tl Pascal Goetgheluck/PR **53**tr Royalty-Free/Punchstock **53**bl James Hanley/SPL **53**br Library of Congress Prints and Photographs Division **54–55** Carsten Peter/NGS **56** Pascal Goetgheluck/SPL **57**t Michael Nichols/NGS **57**b George Holton/PR **58**t NASA/Photo Researchers **58**b Gregory G. Dimijian, M.D./PR **59** Simon Fraser/SPL **61** Advanced Illustration **62** JM Labat/PR **63** Wordstop Technologies. Data source: *GISP2 Ice Core Temperature and Accumulation Data.* Richard B. Alley, 2000 **64**l Jan Hinsch/SPL **64**c Tom McHugh/PR **64**r Mark Newman/PR **66** Gilgamesh Relief/Schöyen Collection **67** Thomas J. Abercrombie/NGS **68** James Hanley/SPL **69** Argosy **70**l Prisma/Ancient Art and Architecture Collection Ltd. **70**r Werner Forman/Art Resource, NY **72**tl Petros Giannakouris/AP **72**cl Punchstock **72**cf Martin Reidl/SPL **73** Corbis **74** Bernhard Edmaier/SPL **75** Jean-Loup Charmet/PR **76** © North Wind/North Wind Picture Archives **77** Erich Lessing/Art Resource, NY **79** National Library of Ireland **80**t Wordstop Technologies. Data source: NASA, Goddard Institute for Space Studies, NY. **80**b Granger Collection-Titanic **81** Library of Congress Prints and Photographs Division

Chapter 4

82–83 AP **85**tl Geoff Woods **85**tr Jacques Jangoux/PR **85**bl John Reader/PR **85**br Image Source/Punchstock **86–87** Raymond Gehman/NGS **88**t Bill Curtsinger/NGS **88**b Novosti Press Agency/ SPL **89** K Yamashita/PanStock/Panoramic Images/NGS **90**tl ESA-Silicon World **90**cl David Parker/PR **90**cr NASA Marshall Space Flight Center (NASA-MSFC) **91**tl ESA/D. Ducros **91**br ESA **93**b Paul Nicklen/NGS **93**tl Courtesy of the National Snow and Ice Data Center **93**tr National Snow and Ice Data Center **95** © xela/Alamy Royalty Free

Chapter 5

97tl AP/Prakash Hatvalne **97**tr Digital Vision/Punchstock **97**bl AP/Jonathan Head **97**br Michael Lewis/NGS **98–99** AP/Keystone, Olivier Maire **100** Advanced Illustration. Data source: United Nations Environment Programme **101**l Wordstop Technologies. Data source: Department of Chronic Diseases and Traumatisms, France **101**r AP/Keystone, Fabrice Coffrini **102**tl Heather Perry/NGS **102**bl Tim Laman/NGS Image Collection **103**t NASA/SPL **103**b Wordstop Technologies. Data source: Hoegh-Guldberg, *Marine and Fresh Water Research,* Vol. 50, 1999 **104** AP/Richard Vogel **105** Annie Griffiths Belt/NGS **107**t World Wildlife Fund **107**b Paul Nicklen/NGS **108** Paolo Mazzei **109**cl Alexis Rosenfeld/PR **109**tr Brand X Pictures/Punchstock **109**br Carsten Peter/NGS **110** Mike Danzenbaker **111** B & C Alexander/PR

Chapter 6

113tl Maria Stenzel/NGS **113**tr Michael Van Woert, NOAA NESDIS, ORA **113**bl Norbert Rosing/NGS **113**br Gordon Wiltsie/NGS **114–115** Joe Scherschel/NGS **116** Argosy Publishing **118** Advanced Illustration. Data source: Arctic Climate Impact Assessment **119** AP/CP, *Prince Rupert Daily News* **120**tl Jodi Cobb/NGS; **120**bl AP/Gurinder Osan **120**cf AP/John McConnico **121**br George F. Mobley/NGS **121**cr NOAA Library Collection **122**t Bill Curtsinger/NGS **122**b Michael Van Woert, NOAA NESDIS, ORA **123**t Planetary Visions LTD/SPL **123**b Planetary Visions LTD/SPL **124** AP/ACIA **125**bl Nick Caloyianis/NGS **125**tr Simon Fraser/SPL **126**tl Lowell Georgia/NGS **126**bl George D. Lepp/PR **127**l Norbert Rosing/NGS **127**r Wordstop Technologies. Data source: Intergovernmental Panel on Climate Change **128** Advanced Illustration. Data source: Arctic Climate Impact Assessment

Chapter 7

131tl Jim Reed/Digital Vision/Punchstock **131**tr AP/Pavel Rahman; **131**bl Karen Kasmauski/NGS; **131**br Dr. Morley Read/SPL **133** AP/str **134**bl NOAA **134**cf Sam Abell/NGS **135** Advanced Illustration **136**tl Jim Richardson/NGS; **136**cf George F. Mobley/NGS **137** Wordstop Technologies. Data source: Scripps Institution of Oceanography/U.S. Geological Survey **138** AP/Jaime Puebla/Pool **139** Argosy Publishing **140** AP/The Rock Hill Herald, Tracy Smith **141** Steve McCurry/NGS **142**tl David Pluth/NGS **142**bl AP/Jalil Bounhar **143**cf Georg Gerster/PR **143**br AP/Xinhua, Wang Chengxuan **144**t Cary Wolinsky/NGS **144**b Wordstop Technologies. Data source: University of Akron, Ohio **145** AP/Harry N. Abrams, Inc., Yann Arthus-Bertrand **147** M-SAT Ltd/SPL **148** M-SAT Ltd/SPL **149** Georg Gerster/PR **150**tl Raymond K. Gehman/NGS **150**bl John Cancalosi/Nature Picture Library **150**c Advanced Illustration **151**tl Steve Winter/NGS **151**br Nigel J. Dennis/PR **152** Steve McCurry/NGS **153** Connie Bransilver/PR **154** E.R. Degginger/PR **155** AP/Rick Rycroft

Chapter 8

157tl Annie Griffiths Belt/NGS **157**tr Royalty-Free/Punchstock **157**bl James L. Stanfield/NGS **157**br AP/Eureka Times-Standard, Shaun Walker **159** ESA 2003 **160** Wordstop Technologies. Data source: United Nations Environment Programme **161**bl AP/NASA, Jet Propulsion Laboratories, Scott Michel **161**cr NOAA **162** James P. Blair/NGS **163**l Reza/NGS **163**r Steve Raymer/NGS **164**tl Peter McQuarrie **164**cl Peter McQuarrie **164**br Peter McQuarrie **165**bl Peter McQuarrie **165**tr AP Photo/Richard Vogel **166** Sam Abell/NGS **167** Argosy Publishing **168** Jeff Rotman/PR **169** Mark Boulton/PR **170** AP Photo/Pavel Rahman **171** Advanced Illustration. Data source: United Nations Environment Programme

Chapter 9

173tl Raymond Gehman/NGS **173**tr AP/San Diego Natural History Museum/Philip Roullard **173**bl Punchstock **173**br Simon Fraser/SPL **174–175** Joel Sartore/NGS **176** Wordstop Technologies. Data source: Both and Visser, *Nature,* May 17, 2001 **177** Roger Wilmshurst/PR **178**t Advanced Illustration. Data source: *Predicting Changes in Species Distributions under Global Warming Scenarios.* William W. Hargrove, Forrest M. Hoffman, Environmental Sciencies Division, Oak Ridge National Laboratory/U.S. Department of Energy **178**bl, br Advanced Illustration. Data source: *Poleward Shifts in Geographical Locations of Butterfly Species Associated with Global Warming.* Camille Parmesan, et al. **179**t Frank Zullo/PR **179**b Michael S. Quinton/NGS **180** AP/Cold Regions Research and Engineering Laboratory via *Anchorage Daily News* **181** Jennifer Hoffman, Ocean Research Academy, Everett, WA **182**t William Ervin/SPL **182**b Gregory Ochocki/PR **183** Michael S. Quinton/NGS **184**t AP/Rachel D'Oro **184**b Jim Richardson/NGS **185** Steve Winter/NGS **186** Norbert Rosing/NGS **187** David Hall/PR **188**l U.S. Department of Agriculture **188**r Alan Punton, Esq./SPL **189** Gregory Dimijian/SPL **190**tl Veronique Leplat/SPL **190**tr Susan Ford/Rutgers University **190**bl Nicole Duplaix/NGS **191** Kazuyoshi Nomachi/PR

Chapter 10

193tl Punchstock **193**tr George Chan/PR **193**bl AP/Vincent Thian **193**br AP/Dario Lopez-Mills **195** AP/Anat Givon **196** AP/Paulo Santos-Interfoto **197** AP/Firdia Lisnawati **200**t Bildarchiv Preussischer Kulturbesitz/Art Resource, NY **200**b Michael S. Yamashita/NGS **201** AP/Pavel Rahman **202** Data source: Center for Internatonal Earth Science Information Network, Columbia University, NY **204**t AP/Ron Edmonds, File **204**b AP/B.K. Bangash **205**tl AP/Franck Prevel **205**br Pablo Corral Vega/NGS **206** M.I.G./Baeza/PR **207**l CDC/SPL **207**r Darlayne A. Murawski/NGS

Chapter 11

209tl Michael Nichols/NGS **209**tr Corbis/Punchstock **209**bl James P. Blair/NGS **209**br Jim Reed/PR **210–211** James P. Blair/NGS **212** Argosy Publishing **213**tl Bruce Dale/NGS **213**br Jim Reed/SPL **215** Priit Vesilind/NGS **216**tl NOAA Photo Library; **216**cr Annie Griffiths Belt/NGS **216**bl Argosy Publishing **217**bg NASA/SPL **217**tr NOAA/SPL **218** Carsten Peter/NGS **219** Steve McCurry/NGS **221** Andrzej Gorzkowski/Alamy

Chapter 12

223tl Carol Baldwin/NOAA **223**tr Michael K. Nichols, NGS **223**bl Carsten Peter/NGS **223**br Bobby Model/NGS **224–225** John Eastcott and Yva Momatiuk/NGS **226**l Nigel J. Dennis/PR **226**r AP/Kent Gilbert **227** Ted Kerasote/PR **228**tl Michael Theberge/NOAA Ship *MacArthur* **228**c Gregory Dimijian/PR **228**br Paul Zahl/NGS **229**tl Advanced Illustration **229**bl Jacques Jangoux/PR **230** Richard Olsenius/NGS **231** Tom and Pat Leeson/PR **232** Georg Gerster/PR **233** Nigel J. Dennis/PR **234** Alexis Rosenfeld/PR **235**l Ray Ellis/PR **235**r Stephen J. Krasemann/PR **236** George Turner/PR **237** Advanced Illustration

Chapter 13

239tl Jason Edwards/NGS **239**tr Bill Bachmann/PR **239**bl Raul Touzon/NGS **239**br Punchstock **240–241** AP/Eugene Hoshiko **242** David R. Frazier/PR **243** AP/Claude Paris **244**l © SHOUT/Alamy **244–245**b K Yamashita/ PanStock/Panoramic Images/NGS **245**tr David R. Frazier/PR **247** Argosy Publishing. Data source: National Renewable Energy Laboratory/U.S. Department of Energy **248**tl Michael Nichols/NGS **248**br AP/Andy Wong **249**tl Martin Bond/PR **249**br Brad Lewis/PR **250**tl Gusto/PR **250**br AP/Casper Star-Tribune, Robert Hendricks **251**t Paul and Judy Wilcox **251**b NGS **252**l AP/East Oregonian/Don Cresswell **252**r AP/Esteban Felix **253** Argosy Publishing **254** Martin Bond/PR **255** Mark Segal/Panoramic Images/NGS

Chapter 14

257tl AP/The Maui News, Matthew Thayer **257**tr Punchstock **257**bl Georg Gerster/PR **257**br John Howard/PR **258–259** AP/Michael Dwyer **260** AP/Gautam Singh **261** Data source: Educational Global Climate Model (EdGCM), NASA, Goddard Institute for Space Studies, NY **262**tl Dick Durrance II/NGS **262**b AP/Rafiq Maqbool **263**tl AP/Pavel Rahman **263**tr Jacques Jangoux/PR **264** Scott Camazine/PR **265** Raymond Gehman/NGS **266** Emory Kristof/NGS **267** Susan McCartney/PR **268–269** Rich Reid/NGS

Appendices

270 (left to right) NASA; NASA; NOAA; Erich Schrempp/SPL; NOAA; NOAA; NOAA **271** (left to right) NOAA; NOAA; NOAA;NOAA; AP/Vincent Thian; NOAA; AP/Gautam Singh **274**bl Argosy Publishing. Data source: Arctic Climate Impact Assessment **274**tr, br Earth Observatory/NASA. **275**tl Intergovernmental Panel on Climate Change (IPCC). Data source: Annual mean change of precipitation and its range for scenario B2. IPCC Third Assessment Report, 2001 **275**tr Intergovernmental Panel on Climate Change (IPCC). Data source: Annual mean change of temperature and its range for scenario B2. IPCC Third Assessment Report, 2001